Practical Data Analysis
with JMP®

Second Edition

Robert H. Carver

support.sas.com/bookstore

The correct bibliographic citation for this manual is as follows: Carver, Robert. 2014. *Practical Data Analysis with JMP*®, *Second Edition*. Cary, NC: SAS Institute Inc.

Practical Data Analysis with JMP®, Second Edition

Copyright © 2014, SAS Institute Inc., Cary, NC, USA

ISBN 978-1-61290-823-6

SAS Institute Inc., SAS Campus Drive, Cary, North Carolina 27513-2414.

July 2014

SAS provides a complete selection of books and electronic products to help customers use SAS® software to its fullest potential. For more information about our offerings, visit **support.sas.com/bookstore** or call 1-800-727-3228.

Contents

About This Book

Purpose: Learning to Reason Statistically

We live in a world of uncertainty. Today more than ever before, we have vast resources of data available to shed light on crucial questions, but at the same time the sheer volume and complexity of the "data deluge" can distract and overwhelm us. The goal of applied statistical analysis is to work with data to calibrate, cope with, and sometimes influence uncertainty. Business decisions, public policies, scientific research, and news reporting are all shaped by statistical analysis and reasoning. Statistical thinking is an essential part of the boom in "big data analytics" in numerous professions. This book will help you to use and discriminate among some fundamental techniques of analysis, and it will also help you to engage in statistical thinking by analyzing real problems.

To be an effective analyst or consumer of other people's analyses, you must know how to use these techniques, when to use them, and how to communicate their implications. Knowing how to use these techniques involves mastery of computer software like JMP. Knowing when to use these techniques requires an understanding of the theory underlying the techniques, and practice with applications of the theory. Knowing how to effectively communicate with consumers of an analysis or with other analysts requires a clear understanding of the theory and techniques, as well as clarity of expression, directed toward a particular audience.

There was a time when a first course in statistics emphasized abstract theory, laborious computation, and small sets of artificial data—but not

practical data analysis or interpretation. Those days are thankfully past, and now we can address all three of the skill sets just cited.

Is This Book for You?

This book is intended to supplement an introductory college-level statistics course with real investigations of some important and engaging problems. Each chapter presents a set of self-paced exercises to help students learn the skills of quantitative reasoning by performing the kinds of analyses that typically form the core of a first course in applied statistics. Students can learn and practice the software skills outside of class. Instructors can devote class time to statistics and statistical reasoning, rather than to rudimentary software instruction. Both students and teachers can direct their energies to the practice of data analysis in ways that inform students' understanding of the world through investigations of problems that matter in various fields of study.

Though written with undergraduate and beginning graduate students in mind, some practitioners may find the book helpful on the job and are well-advised to read the book selectively to address current tasks or projects. Chapters 1 and 2 form a good starting point before reading later sections. Appendix B covers several data management topics that may be helpful for readers who undertake projects involving disparate data sources.

A Message for Instructors

I assume that most teachers view class time as a scarce resource. One of my goals in writing this book was to strive for clarity throughout, so that students can be expected to work through the book on their own and learn through their encounters with the examples and exercises. This book may be especially welcome for instructors using an inverted, or flipped, classroom approach.

Instructors might selectively use exercises as in-class demonstrations or group activities, interspersing instruction or discussion with computer work. More often, the chapters and scenarios can serve as homework exercises or assignments, either to prepare for other work, to acquire skills and understanding, or to demonstrate progress and mastery. Finally, some instructors may want to assign a chapter in connection with an independent analysis project. Several of the data tables contain additional variables that are not used within chapters. These variables may form the basis for original analyses or explorations.

The bibliography may also aid instructors seeking additional data sources or background material for exercises and assignments. Tips for classroom use of JMP are also available at the book's website, accessible via the author's page at http://support.sas.com/publishing/authors/carver.html.

A Message for Students

Remember that the primary goal of this book is to help you understand the concepts and techniques of statistical analysis. JMP provides an ideal software environment to do just that. Naturally, each chapter is "about" the software and at times you will find yourself focusing on the particular details of a JMP analysis platform and its options. If you become entangled in the specifics of a particular problem, step back and try to refocus on the main statistical ideas rather than software issues.

This book should augment, but not replace, your primary textbook or your classroom time. To get the maximum benefit from the book, work mindfully and carefully. Read through a chapter before you sit down at the computer. Each chapter should require roughly 30 minutes of computer time; work at your own pace and take your time.

The Application Scenarios at the end of each chapter are designed to reinforce and extend what you've learned in the chapter. The questions in this section are designed to challenge you. Sometimes, it is quite obvious how to proceed with your analysis; sometimes, you will need to think a bit before you issue your first command. The goal is to engage in statistical thinking, integrating what you have learned throughout your course. There is much more to data analysis than finding a numerical answer, and these questions provide an opportunity to do realistic analysis. Because the examples use real data, don't expect to find neat "pat" results; computations won't typically come out to nice round numbers.

JMP is a large program designed for many diverse user needs. Many of the features of the software are beyond the scope of an introductory course, and therefore this book does not discuss them. However, if you are curious or adventurous, you should explore the menus and Help system as well as the JMP website. You may find a quicker, more intuitive, or more interesting way to approach a problem.

Prerequisites

No prior statistical knowledge is presumed. A basic grounding in algebra and some familiarity with the Mac OS or Windows environment are sufficient to get you on your way. An open, curious mind is also helpful.

Scope and Structure of This Book

As a discipline statistics is large and growing; the same is true of JMP. One paperback book must limit its scope, and the content boundaries of this book are set intentionally along several dimensions. First, this book provides considerable training in the basic functions of JMP 11. JMP is a full-featured, highly interactive, visual, and comprehensive package. The book assumes that you have the software at your school or office. The software's capabilities extend far beyond an introductory course, and this book makes no attempt to "cover" the entire program. The book introduces students to its major platforms and essential features, and should leave students with sufficient background and confidence to continue exploring on their own. Fortunately, the Help system and accompanying manuals are quite extensive, as are the learning resources available online at http://www.jmp.com.

Second, the chapters largely follow a traditional sequence, making the book compatible with many current texts. As such, instructors and students will find it easy to use the book as a companion volume in an introductory course. Chapters are organized around core statistical concepts rather than software commands, menus, or features. Several chapters includes topics that some instructors may view as "advanced"—typically when the output from JMP makes it a natural extension of a more elementary topic. This is one way in which software can redefine the boundaries of introductory statistics.

This second edition also includes three new review chapters (Chapters 5, 9, and 17) that pause to recap concepts and techniques. One of the perennial challenges in learning statistics is that it is easy to lose sight of major themes as a course progresses through a series of seemingly disconnected techniques and topics. Some readers should find the review chapters to be helpful in this respect. The review chapters share a single large data set of World Development Indicators, published by the World Bank.

The scope and sequence of chapters has changed slightly since the first edition. In keeping with my choice to treat inference with categorical data before inference with quantitative data, Chi-Square tests now come before Two-Sample *t*-tests. There is more coverage of *data management* and analysis of data from *complex samples*, in recognition of the easy access to large public use data sets.

Third, nearly all of the data sets in the book are real and are drawn from those disciplines whose practitioners are the primary users of JMP software. Inasmuch as most undergraduate programs now require coursework in statistics, the examples span major areas in which statistical analysis is an important path to knowledge. Those areas include engineering, life sciences, business, and economics.

Fourth, each chapter invites students to practice the habits of thought that are essential to statistical reasoning. Long after readers forget the details of a particular procedure or the options available in a specific JMP analysis platform, this book may continue to resonate with valuable lessons about variability, uncertainty, and the logic of inference.

Each chapter concludes with a set of "Application Scenarios," which lay out a problem-solving or investigative context that is in turn supported by a data table. Each scenario includes a set of questions that implicitly require the application of the techniques and concepts presented in the chapter.

About the Examples

Real statistical investigations begin with pressing, important, or interesting questions, rather than with a set of techniques. Researchers do not begin a study by saying "Today is a good day to compute some standard deviations." Instead, they pose questions that can be pursued by analyzing data and follow a relatively straightforward protocol to refine the question, generate or gather suitable data, apply appropriate methods, and interpret their findings. The chapters in this book present questions that I hope you will find interesting, and then rely on data tables to search for answers. The questions and analyses become progressively more challenging through the book.

The Data Files

As previously noted, each of the data tables referenced within the book contains real data, much of it downloaded from public sites on the World Wide Web. In all there are nearly 50 different data tables, including eight that are new to this edition. Readers should download all of the JMP data tables via the author page at http://support.sas.com/publishing/authors/carver.html. Appendix A describes each file and its source. Many of the tables include columns (variables) in addition to those featured in exercises and examples. These variables may be useful for projects or other assignments.

JSL Scripts

In addition to the data tables, four scripts are included within the data file download. These are simulators and calculators written in the JMP Scripting Language (JSL) and are borrowed from the JMP Academic site's Interactive Tools for Learning and Teaching at http://www.jmp.com/academic/learning_modules.shtml.

Exercise Solutions

Solutions to the scenario questions are available via the author page at http://support.sas.com/publishing/authors/carver.html/; instructors who adopt the book will be able to access all solutions. Students and other readers can find solutions to the even-numbered problems at the same site.

Thanks and Acknowledgments

This first edition of this book began at the urging of Curt Hinrichs, the Academic Program manager for JMP. This led to conversations with Julie Platt, Editor-in-Chief at SAS Press, after which the project started to take shape. Throughout the writing, editing, and production process, many of their colleagues at both JMP and SAS Press have provided encouragement, advice, helpful criticism, and support. Thankfully, many of these same individuals have participated in both editions. The remarkable Stephenie Joyner has steered this project with a very steady hand and just the right amount of nudging. Our working relationship and friendship has deepened and strengthened over the past six years, and has made this work thoroughly enjoyable.

Shelley Sessoms, Stacey Hamilton, Shelly Goodin, and Mary Beth Steinbach at SAS Press have all brought their professional skills and good humor to bear through the various stages of launching the first edition. For this current edition, Cindy Puryear has led the way in shaping the marketing plan, and Mia Stephens and Curt Hinrichs continued to encourage and inspire me. The process of bringing the book to fruition was also enhanced by the efforts of Brad Kellam, Candy Farrell, Patrice Cherry, and Jennifer Dilley.

Many other marketing and technical professionals at JMP have shaped and informed the content of this book at critical points along the way. I am very grateful to John Sall, Xan Gregg, Jon Weisz, Jonathan Gatlin, Jeff Perkinson, Ian Cox, Chuck Pirrello, Brian Corcoran, Christopher Gotwalt, Gail Massari, Lori Harris, Mia Stephens, Kathleen Watts, Mary Loveless, Holly McGill, and Peng Liu for answering my questions, setting me straight, and listening to my thoughts.

I am especially thankful for the care and attention of several individuals who reviewed the first edition and made pointed and constructive suggestions: Mark Bailey, Michael Crotty, Tonya Mauldin, Paul Marovich, and Sue Walsh. For the second edition, Mark Bailey, Fang Chen, Volker Kraft, Tonya Mauldin, and Sue Walsh carried the reviewing burden, and did a superb job of it. Very special thanks to Brenna Leath whose work on the final copy-edits was extraordinary and gentle. Collectively, their critiques tightened and improved this book, and whatever deficiencies that may remain are entirely mine.

Naturally, the completion of a book requires time, space, and an amenable environment. I want to express public thanks to three institutions that provided facilities, time, and atmospherics suitable for steady work on this project. My home institution, Stonehill College, has been exceptionally supportive, particularly through the efforts of Provost Joe Favazza and my chairperson, Debra Salvucci, and Department Administrative Assistant Carolyn McGuinness. Colleagues Dick Gariepy and Michael Salé generously tested several chapters and problems in their classrooms, and Jan Harrison and Susan Wall of our IT Department have eased several technical aspects of this project as well.

Colleagues and students at the International Business School at Brandeis University have sharpened my pedagogy and inspired numerous examples found in the book. During a sabbatical leave from Stonehill, Babson College was good enough to offer a visiting position and a wonderful place to write the first edition. For that opportunity, thanks go to Provost Shahid Ansari, former chairperson Norean Radke Sharpe, then-chair Steve Ericksen, and colleagues John McKenzie and George Recck.

During the summer of 2013, the Stonehill Undergraduate Research Experience (SURE) program provided a grant to support this work with time, space, and finances. Carolyn Moodie (Class of 2015) was a superior and self-directed research collaborator, assisting in the critical phases of problem formulation, data identification, data cleaning and exploratory analysis. Carolyn also brought a keen eye to editorial tasks, and willingly gave her feedback on which topics student readers might find engaging. Thanks also to Bonnie Troupe for her skillful administration of the SURE program. During the spring and summer of 2013, Stonehill students Dan Doherty, Erin Hollander, and Tate Molaghan also pitched in with editorial and research assistance.

Special acknowledgement also goes to former Stonehill students from BUS207 (Intermediate Statistics) who "road tested" several chapters, and very considerable thanks to three students who assisted greatly in shaping prose and examples, as well as developing solutions to scenario problems: Frank Groccia, Dan Bouchard, and Matt Arey. My current students in BUS206 (Quantitative Analysis for Business) at Stonehill have also class-tested several chapters and exercises.

Several of the data tables came through the gracious permission of their original authors and compilers. I gratefully acknowledge the permission granted by good friend George Aronson for the Maine SW table; by Prof. Max A. Little for the Parkinsons vocal data; by Prof. Jesper Rydén for the Sonatas data table (from which the Haydn and Mozart tables were extracted); by Prof. John Holcomb for the North Carolina birth weight data; and by Prof. I-Cheng Yeh for the Concrete table and the two subsets from that data.

In recent years, my thoughts about what is important in statistics education have been radically reshaped by colleagues in the ISOSTAT listserve and the Consortium for the Advancement of Undergraduate Statistics Education and the US Conference on Teaching Statistics that CAUSE organizes every two years. The May 2013 CAUSE-sponsored workshop "Teaching the Statistical Investigation Process with Randomization-Based Inference" given by Beth Chance, Allan Rossman and Nathan Tintle influenced some of the changes in my presentation of inference. Over an even longer period, our local group of New England Isolated Statisticians and the great work of the ASA's Section on Statistics Education influence me daily in the classroom and at the keyboard.

Finally, it is a pleasure to thank my family. My sons Sam and Ben keep me modest and in contact with the mindsets of college-aged readers, and regularly provide inspiration and insight. My wife Donna—partner, friend, wordsmith extraordinaire—has my love and thanks for unflagging encouragement, support, and warmth. This book is dedicated to them.

Keep in Touch

We look forward to hearing from you. We invite questions, comments, and concerns. If you want to contact us about a specific book, please include the book title in your correspondence to saspress@sas.com.

To Contact the Author through SAS Press

By e-mail: saspress@sas.com

Via the Web: http://support.sas.com/author_feedback

SAS Books

For a complete list of books available through SAS, visit http://support.sas.com/bookstore.

Phone: 1-800-727-3228

Fax: 1-919-677-8166

E-mail: sasbook@sas.com

SAS Book Report

Receive up-to-date information about all new SAS publications via e-mail by subscribing to the SAS Book Report monthly eNewsletter. Visit http://support.sas.com/sbr.

Publish with SAS

SAS is recruiting authors! Are you interested in writing a book? Visit http://support.sas.com/saspress for more information.

About The Author

 Robert Carver is Professor of Business Administration at Stonehill College in Easton, Massachusetts, and Adjunct Professor at the International Business School at Brandeis University in Waltham, Massachusetts. At both institutions, he teaches courses on business analytics in addition to general management courses, and has won teaching awards at both schools. His primary research interest is statistics education. A JMP user since 2006, Carver holds an A.B. in political science from Amherst College in Amherst, Massachusetts and an M.P.P. and Ph.D. in public policy from the University of Michigan at Ann Arbor.

Learn more about this author by visiting his author page at http://support.sas.com/publishing /authors/carver.html. There you can download free book excerpts, access example code and data, read the latest reviews, get updates, and more.

Getting Started: Data Analysis with JMP

1

Overview

Statistical analysis and visualization of data have become an important foundation of decision making and critical thinking. Professionals in numerous walks of life—from medicine to government, from science to sports, from commerce to public health—all rely on the analysis of data to inform their work. In this first chapter, we take first steps into the important and rapidly-growing practice of data analysis.

Goals of Data Analysis: Description and Inference

The central goal of this book is to help you build your capacity as a statistical thinker through progressive experience with the techniques and approaches of data analysis, specifically by using the features of JMP. As such, before using JMP, we'll begin with some remarks about activities that require data analysis.

People gather and analyze data for many different reasons. Engineers test materials or new designs to determine their utility or safety. Coaches and owners of professional sports teams track their players' performance in different situations to structure rosters and negotiate salary offers. Chemists and medical researchers conduct clinical trials to investigate the safety and efficacy of new treatments. Demographers describe the characteristics of populations and market segments. Investment analysts study recent market data to fine tune investment portfolios. All of the individuals who are engaged in these activities have consequential, pressing needs for information, and they turn to the techniques of statistics to meet those needs.

There are two basic types of statistical analysis: *description* and *inference*. We do descriptive analysis in order to summarize or describe an ongoing process or the current state of a population—a group of individuals or items that is of interest to us. Sometimes we can collect data from every individual in a *population* (every professional athlete in a sport, or every firm in which we currently own stock), but more often we are dealing with a subset of a population—that is to say with a *sample* from the population. When we study on-going processes, we nearly always deal with samples.

If a company reviews the records of all of its client firms to summarize last month's sales to all customers, the summary will describe the population of customers. If the same company wants to use that summary information to make a forecast of sales for next month, the company needs to engage in inference. When we use available data to make a conclusion about something we cannot observe, or about something that hasn't happened yet, we are drawing an inference. As we will come to understand, inferential thinking requires risk-taking. Learning to measure and minimize the risks involved in inference is a large part of the study of statistics.

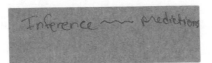
Inference ~~ predictions

Types of Data

The practice of statistical analysis requires *data*—when we "do" analysis, we're analyzing data. It's important to understand that analysis is just one phase in a statistical study. Later in this chapter we'll look at some data collected and reported by the World Population Division of the United Nations. Specifically, we will analyze the estimated life expectancy at birth for nations around the world in 2010. This set of data is a portion of a considerably larger collection spanning many years.

In this particular example we have four *variables* that are represented as four columns within a data table. A variable is an attribute that we can count, measure, or record. The variables in this example are country, region, year, and life expectancy. Typically, we'll record or capture multiple *observations* of each variable—whether we're taking repeated measurements of (say) stock prices, or recording facts from numerous respondents in a survey or individual countries around the globe. Each observation (often called a case or subject in survey data) occupies a row in a data table. In this example, the *observational units are countries*.

> Whenever we analyze a data set in JMP, we'll work with a *data table*. The *columns* of the table contain different variables, and the *rows* of the table contain observations of each variable. In your statistics course, you'll probably use the terms data set, variable, and observation (or case). In JMP, we more commonly speak of data tables, columns, and rows.

One of the organizing principles you'll notice in this software is the differentiation among *data types* and *modeling types*. The columns that you will work with in this book are all either *numeric* or *character* data types, much like data in a spreadsheet are numeric or labels.

In your statistics course, you may be learning about the distinctions among different kinds of quantitative and qualitative (or categorical) data. Before we analyze any date, we'll want to understand clearly whether a column is *quantitative* or *categorical*. JMP helps us keep these distinctions straight by using different *modeling types,* and JMP recognizes three such types:

- *Continuous* columns are inherently quantitative. That is to say, they are numeric so that you can meaningfully compute sums, averages, and so on. Continuous variables can assume an infinite number of values. Most measurements and financial figures are continuous data. Estimated average life expectancies (in years) are continuous.

- *Ordinal* columns reflect attributes that are sequential in nature or have some [ordering] (for example, small, medium, large). In our data table, we have [a] variable indicating the year, which in this case is 2010 for all [regions]. Ordinal columns can be either numeric or character data.

[handwritten note: Ordinal ~ sequential numbers]

- *Nominal* columns simply identify individuals or groups within the data. For example, if we are analyzing health data from different countries, we might want to label the nations and/or compare figures by continent. With our U.N. data, both [the] countries and their continental regions are nominal columns. [These vari]ables can also be numeric or character data. Names are nominal, as [are postal cod]es or telephone numbers.

[handwritten note: nominal ~ names/labels numeric or character]

As we'll soon see, understanding the differences among these modeling types is helpful in understanding how JMP treats our data and presents us with choices.

Starting JMP

Whether you are using a Windows-based computer or a Macintosh, JMP works in very similar ways. All of the illustrations in this book were generated in a Windows environment. Find JMP[1] among your programs and launch it. You'll see the opening screen shown in Figure 1.1. The software opens a **Tip of the Day** window each time you start the software (assuming no initial default settings have been changed). These are informative and helpful. You can elect to turn off the automatic messages by clearing the Show tips at startup check box in the lower-left part of the window. You'll be well advised to click the **Enter Beginner's Tutorial** button sooner rather than later to get a helpful introduction to the program (perhaps you should do so now or after reading this chapter). After you've read the tip of the day, click **Close**.

Figure 1.1: The JMP Opening Screen

The next window displayed by default is called the JMP Starter window, which is an annotated menu of major functions. It is worth your time to explore the JMP Starter window by navigating through its various choices to get a feel for the wide scope of capabilities that the software offers. As a new user, though, you may find the range of choices to be overwhelming.

In this book, we'll tend to close the JMP Starter window and use the menu bar at the top of the screen to make selections. Behind the JMP Starter window is the JMP Home Window. The home window is divided into four panes that can help you keep track of recently used files and currently open windows. One can customize this view, but this book shows the standard four-pane layout.

A Simple Data Table

In this book, we'll most often work with data that has already been entered and stored in a file, much like you would type and store a paper in a word-processing file or data in a spreadsheet file. In Chapter 2, you'll see how to create a data table on your own.

We'll start with the U.N. life expectancy data mentioned earlier.

1. Click **File ▶ Open**.

2. Navigate your way to the folder of data tables that accompany this book [2].

3. Select the file called **Life Expectancy 2010** and click **Open**.

The data table appears in Figure 1.2. Notice that there are four regions in this window including three vertically arranged *panels* on the left, and the *data grid* on the right.

Figure 1.2: The Life Expectancy 2010 Data Table

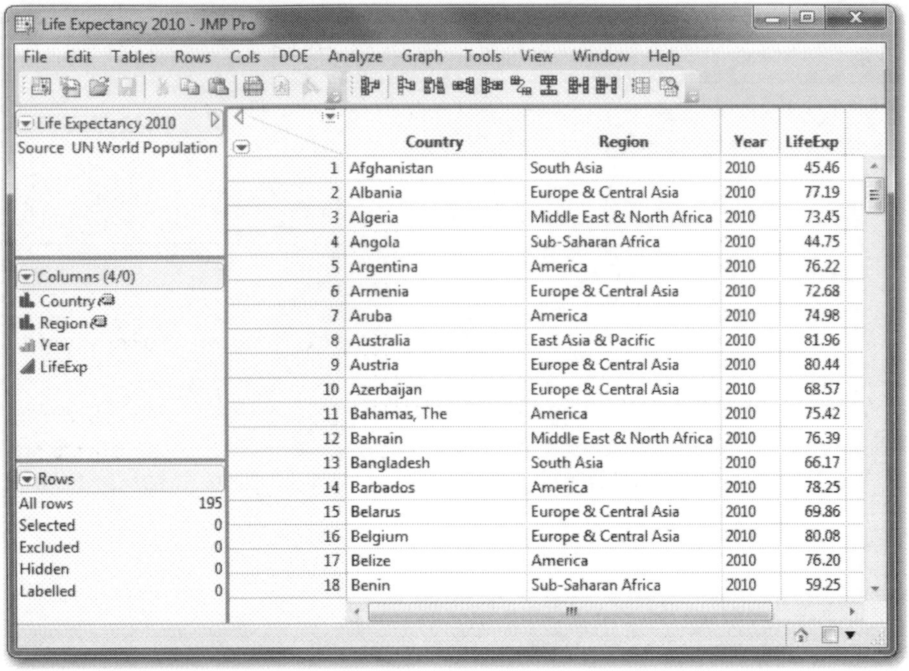

The three panels provide *metadata* (descriptive information about the data in the table) which is created at the time the data table was saved, and can be altered for various reasons. At this early stage, it may be helpful to understand the purpose of each panel.

Beginning at the top left, we find the *Table* panel, which displays the name of the data table file as well as optional information provided by the creator of the table. You'll see a small red triangle pointing downward next to the table name.

Red triangles indicate a context-sensitive menu, and they are an important element in JMP. We'll discuss them more in later chapters, but you should expect to make frequent use of these little red triangles.

Just below the red triangle, there is a note describing the data and identifying its source. You can open that note (called a *Table variable*) just by double-clicking on the word "Source." Figure 1.3 shows what you'll see when you double-click. A table variable contains metadata about the entire table.

Figure 1.3: Table Variable Dialog

Below the **Table** panel is the **Columns** panel, shown in Figure 1.4, which lists the column names, JMP *modeling types*, and other information about the columns. There are several important things to notice in this panel.

Figure 1.4: The Columns Panel

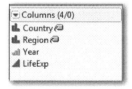

The notation **(4/0)** in the top box of the panel tells us that there are four columns in this data table, and that none of them are *selected* at the moment. In a JMP data table, we can select one or more columns or rows for special treatment, such as using the *label* property in the first two columns so that value labels will display within graphs. There is much more to learn about the idea of selection and *column properties*, and we'll return to it later in this chapter.

Next we see the names of the columns. To the left of the names are icons indicating the modeling type. In this example, first two red icons (these look like bar graphs) identify **Country** and **Region** as nominal data. The "price tag" icons indicate that these variables can act as labels to specifically identify observations that are displayed in a graph.

The green ascending bar icon next to **Year** indicates that year is to be analyzed as an ordinal variable. In this data table, all observations are from the same year, 2010, but in the original data set, we have observations at five-year intervals from 1950 through 2010. Hence, this is an ordinal variable.

Finally, the blue triangle next to **LifeExp** identifies the column as continuous data. Remember, it makes sense to perform calculations with continuous data.

At the bottom left, we find the **Rows** panel (Figure 1.5), which provides basic information about the number of rows (in this case 195, for 195 countries). Like the other two panels, this one provides quick reference information about the number of rows and their *states*. We'll come back to row states in a few pages.

Figure 1.5: Rows Panel

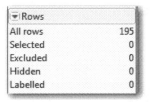

The top entry indicates that there are 195 observations in this data table. The next four entries refer to the four basic row states in a JMP data table. Initially, all rows share the same state, in that none has been selected, excluded, hidden or labeled. Row states enable us to control whether particular observations appear in graphs, are incorporated into calculations, or whether they are highlighted in various ways.

The *Data Grid* area of the data table is where the data reside. It looks like a familiar spreadsheet format, and it just contains the raw data for any analysis. Generally speaking, each column of a table contains either a raw data value (e.g., a number, date, or text) or the entire column contains a formula or the result of a computation. Unlike a spreadsheet, each cell in a JMP data table column must be consistent in this sense. You will not find some rows of a column representing one type of data and other rows representing a different type.

In the upper-left corner of the data grid, you'll see the region shown here. There is a triangular *disclosure button* (pointing to the left side here in Windows; on a Macintosh it is an arrowhead ▶). Disclosure buttons allow you to expand or contract the amount of information displayed on the screen. The disclosure button shown here lets you temporarily hide the three panels discussed above.

4. Try it out! Click the disclosure button to hide and then reveal the panels.

The red triangles offer you menu alternatives that won't mean much at this point, but which we'll discuss in the next section. The hotspot in the upper-right corner (above the diagonal line) relates to the columns of the grid, and the one in the lower-left corner to the rows. The very top row of the grid contains the column names, and the left-most column contains row numbers. The cells contain the data.

Graph Builder: An Interactive Tool to Explore Data

Our main interest within this data table is the life expectancy values around the world. As you peruse the list of values, you might notice that they vary. Variation is so common as to be unremarkable, but the very fact that they vary is what leads us to analyze them. We can imagine many reasons that life expectancy varies around the world; there are differences in nutrition, wealth, access to healthcare and clean water, education, political stability, and so forth. Are there systematic differences in different parts of the world?

We have a table displaying all 195 values, but it is hard to detect patterns by scanning up and down a long list. As a first step in analysis, we'll make some simple graph to summarize the table information visually. Software affords us many options to visualize a set of data, and can help us discover errors in the recording of the raw data, locate important patterns of variability, or identify possible connections between and among
~~h~~ **Builder** is an intuitive interactive platform for visualization.

Expectancy2010 data table window, click **Graph ▶ Graph Builder**.

The graph builder gives us a blank tableau on which we can create a JMP Report representing multiple columns in a single visual display. There are numerous options available, but in this first example, we'll look at just a few.

In analyzing this set of data, our primary interest lies in the variation of life expectancy. Following one of JMP's conventions, we'll think of this column as our Y variable.

2. To display **LifeExp** on the Y axis, click the LifeExp column in the panel of **Variables**, and drag it to the vertical **Y** drop zone in the Graph Builder window. When you do this, your screen should look like Figure 1.6.

Figure 1.6: Using the Graph Builder

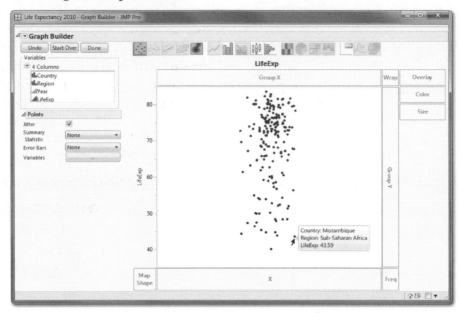

In this graph, each dot represents the value for one country. If you move your cursor to any dot and hover, the name of the country and other data appear. Notice that the reported life expectancies lie between roughly 40 years and 85 years, with a large number of countries enjoying life expectancies above 65 years.

By default, JMP "*jitters*" the points in this graph (see the checkbox next to **Jitter** just below the list of variables). This spreads the points apart to the left and right, so that identical or similar values do not overlap in the graph.

Now let's see how the values compared across different regions in the world.

3. One way to accomplish this is to drag **Region** to the **Color** drop zone.

This assigns a color code to each global region so that all of the counties in Europe, for example, are the same color. This immediately reveals that nearly all of the countries with brief life expectancy are in Sub-Saharan Africa. This fact was not at all obvious from the initial data table; that's what visualization can do for us.

4. Now move the cursor back to the list of columns and once again choose **Region**, and this time drag it to the **Group X** drop zone at the top of the tableau.

When you do this, you will now have six adjacent small graphs showing the values from each region. As you examine these graphs, you may notice that the values vary vertically within each region and that the patterns of variation are similar in some regions but dramatically different in others. The study of descriptive statistics largely revolves around common patterns of variation, comparisons of those patterns, and deviations from those patterns. Here again, it is very evident that the nations of Sub-Saharan Africa largely have the shortest life expectancies in the world. What other general patterns emerge?

Because the data are reported geographically, another useful way to examine the patterns is to overlay them on a map. Doing so magnifies a few key points.

5. In the **Graph Builder**, click the **Start Over** button in the upper left.

6. Drag the **Country** column to the lower left of the **Graph Builder** into the drop zone labeled **Map Shape**.

7. Now drag **LifeExp** over the map and release the mouse button. Alternatively you may drag **LifeExp** into the **Color** drop zone. Your map should now look like Figure 1.7. At this point, click the **Done** button.

Figure 1.7: Map of the World Colored by Life Expectancy

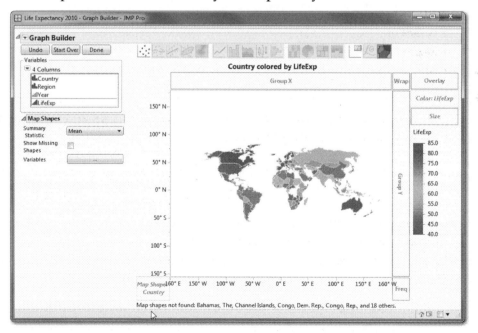

As the legend to the right indicates, the colors shaded dark red enjoy the longest life expectancies and dark blue countries have the shortest life expectancies. This map is an alternative method to see how life expectancy varies around the world.

Please note two limitations of this particular graph. You may have spotted a large white "hole" in the center of Africa. These are countries for which JMP found no data in our data table. Additionally, there is a notation at the bottom of the graph indicating that JMP did not recognize some of the country names, and hence did not display them on the map.

Using an Analysis Platform

Of course data analysis is not limited to graphing and mapping—there are numbers to be crunched, and JMP will do the heavy computational work. We have many pages ahead of us to learn how to request and to interpret many useful computations. With this set of data, we'll summarize life expectancy in different parts of the world. Don't worry about the details of these steps. The goal right now is just for you to see a typical JMP platform and its output.

1. Select **Analyze ▶ Fit Y by X**. This *analysis platform* lets us plot one variable (life expectancy) versus another (region).

> Why "fit" Y by X? Analysts often speak of *fitting* an abstract or theoretical *model* to a set of data. We can think of models as common or standard patterns of variation, and the process of model fitting begins with exploring how a Y column varies across categories or values of an X column.

2. In this dialog box (Figure 1.8), we'll cast **LifeExp** as the **Y** or **Response** variable[3] and **Region** as the **X** variable, or **Grouping** variable. Click **OK**.

Figure 1.8: Fit Y by X Dialog Box

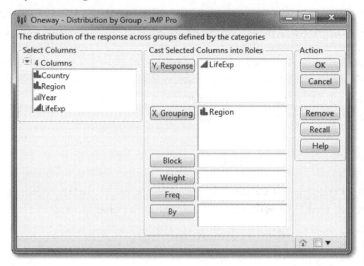

By design, the initial output of a JMP analysis platform includes one or more graphs. In this case, the initial report includes only a graph, as shown in Figure 1.9.

Figure 1.9: Initial Report of Life Expectancy by Region

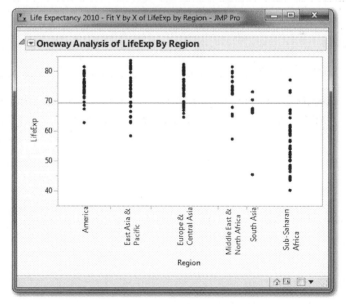

We saw a very similar graph earlier in the Graph Builder; in this graph, the points are not jittered and they are all black. There is one additional feature in this graph: the horizontal line at just below 70 years is the *mean* (average) of all 195 values.

It is clear that the overall grand mean of all countries does not really describe any of the continental regions. We might want to dig deeper and display the regional averages.

3.   Click the red triangular hot spot in the upper left next to **Oneway Analysis**, and choose **Display Options**, and check **Connect Means**.

Look again at the modified graph. The new blue line on your graph represents the mean life expectancy of the countries in each region. As a group, the nations of Europe and Central Asia appear to have the longest life expectancies, whereas countries in South Asia and Sub-Saharan Africa have far shorter life expectancies.

Finally, the graph now shows us the mean values of each group. Suppose we want to know the numerical values of the five averages.

4.   Click the hot spot once more, and this time choose **Means and Std Dev** (standard deviations).

This will generate a table of values beneath the graph, as shown in Figure 1.10. For the current discussion, we'll focus our attention only on the first three columns. Later in the book, we'll learn the meaning of the other columns. This table (below) reports the mean for each of the regions, and also reports the number of countries within each region.

Figure 1.10: Table of Means and Standard Deviations

Means and Std Deviations

Level	Number	Mean	Std Dev	Std Err Mean	Lower 95%	Upper 95%
America	39	74.8804	3.98288	0.6378	73.589	76.171
East Asia & Pacific	30	73.0844	6.57956	1.2013	70.628	75.541
Europe & Central Asia	49	75.6884	4.95994	0.7086	74.264	77.113
Middle East & North Africa	22	73.3942	5.47047	1.1663	70.969	75.820
South Asia	8	65.3113	8.38930	2.9661	58.298	72.325
Sub-Saharan Africa	47	55.0648	8.38674	1.2233	52.602	57.527

Row States

Our data table consists of 780 cells: four variables with 195 observations each, arrayed in four columns and 195 rows. One guiding principle in statistical analysis is that we generally want to make maximum use of our data. We don't casually discard or omit any portion of the data we've collected (often at substantial effort or expense). There are times, however, that we might want to focus attention on a portion of the data table or examine the impact of a small number of extraordinary observations.

By default, when we analyze one or more variables using JMP, every observation is included in the resulting graphs and computations. You can use **row states** to confine the analysis to particular observations or to highlight certain observations in graphs.

There are four basic row states in JMP. Rows can be one of the following:

- *Selected*: selected rows appear bolded or otherwise highlighted in a graph.
- *Excluded*: when you exclude rows, those observations are temporarily omitted from calculated statistics such as the mean. The rows remain in the data table, but as long as they are excluded they play no role in any computations.
- *Hidden*: when you hide rows, those observations do not appear in graphs, but are included in any calculations such as the mean.
- *Labeled*: The row numbers[4] of any labeled rows display next to data points in some graphs for easily identifying specific points. The user (you) can designate specific columns that contain useful labels.

Let's see how the row states change the output that we've already run by altering the row states of rows 3 and 4.

1. First, arrange the open windows so that you can clearly see both the **Fit Y by X report** window and the data table and click anywhere in the data table window to make it the active window.

2. Move your cursor into the column of row numbers the data table. Within this column your cursor will become a "fat cross" ⊕. Select rows 3 and 4 by clicking and dragging on the row numbers 3 and 4. You'll see the two rows highlighted within the data table.

Look at your graph. You should see nearly all of the points are gray, except for two black dots—one right near the mean value of the Middle East & North Africa and the other well below the average of Sub-Saharan Africa. That's the effect of selecting these rows. Notice also that the **Rows** panel in the Life Expectancy data window now shows that two rows have been selected.

3. Click on another row, and then drag your mouse slowly down the column of row numbers. Do you notice the rows highlighted in the table and the corresponding data points "lighting up" in the graph?

4. Press Esc or click in the triangular area above the row numbers in the data table to deselect all rows.

Next we will *exclude* two observations and show that the calculated statistics change when they are omitted from the computations. To see the effect, we first need to instruct JMP to automatically recalculate statistics when the data table changes.

5. Click the red triangle next to **Oneway Analysis** in the report window and choose **Script ▶ Automatic Recalc**.

6. Now let's exclude rows 3 and 4 from the calculations. To do this, first select them as you did before.

7. Select **Rows ▶ Exclude/Unexclude** (you can also find this option by clicking the red triangle above the row numbers; in Windows, you could also right-click). This will exclude the rows.

Now look at the analysis output. The number of observations in the Middle East drops from 22 to 21 and the mean value for that group has changed very slightly. Likewise, in Sub-Saharan Africa we have 46 rather than 47 observations, and mean life expectancy increased from 55.0648 years to 55.289 years. Toggle between the exclude and unexclude states of these two rows until you understand clearly what happens when you exclude observations.

8. Finally, let's *hide* the rows. First, be sure to unexclude rows 3 and 4 so that all 195 points appear in the graph and in the calculations. If you aren't sure if you have reversed earlier actions, choose **Rows ▶ Clear Row States** and then confirm in the **Rows** panel that the four row state categories show 0 rows.

9. Once again select rows 3 and 4 still selected and choose **Rows ▶ Hide/Unhide**. This will hide the rows (check out the very cool dark glasses icon).

Look closely at the graph and at the table of means. It is a subtle change, but the two black dots are gone, leaving only gray points. The numbers in the table of means are unaffected by hiding points. If you toggle the **Hide/Unhide** state you'll notice the black points come and go, but the number of observations in each region is stable.

10. Before continuing to the next section, clear all row states again (as in Step 8 above).

Exporting JMP Results to a Word-Processor Document

As a statistics student, you may often want or need to include some of your results within a paper or project that you're writing for class. As we wrap up this first lesson, here's a quick way to capture output and transfer it to your paper. To follow along, first open your word processing software, then write a sentence introducing the graph you've been working with. Next, return to the JMP oneway analysis report window.

> Windows users will notice that there are no menus in the report window. To reveal the menus, hover your cursor in the pale blue bar just above **Oneway Analysis of LifeExp by Region**.

Our analysis includes a graph and a table. To copy the graph only for your document, do this:

1. Select **Tools ▶ Selection**. Your cursor will now become an open cross. You could also click the open ("fat") cross button from the menu icon bar.

2. Move the open cross to the upper left of the graph and click. This should highlight the entire graph. If it doesn't, click and drag across the graph until it is entirely selected.

3. Select **Edit ▶ Copy**.

4. Now move to your word processor and paste your copied graph.

Figure 1.11: A Graph Pasted from JMP

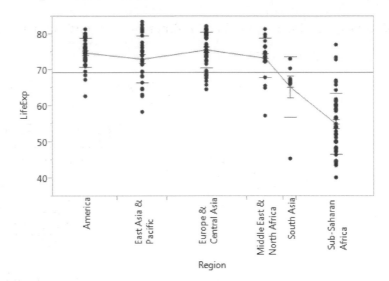

The graph should look like the one shown above in Figure 1.11. Note that the graph will look slightly different from its appearance within JMP, but this demonstration should illustrate how very easy it is to incorporate JMP results into a document.

Saving Your Work

As you work with any software, you should get in the habit of saving your work as you go. JMP supports several types of files, and enables you to save different portions of a session along the way. You've already seen that data tables are files; we've modified the **Life Expectancy 2010** data table and might want to save it.

Alternatively, you can save the *session script*, which essentially is a transcript of the session coded in the *JMP Scripting Language* (JSL) – all of the commands you issued, as well as their results. Later, when you restart JMP, you can open the script file, run it, and your screen will be exactly as you left it.

1. Select **File ▶ Save Session Script.** In the dialog box, choose a directory in which to save this JSL file, give the file a name, and click OK.

Leaving JMP

We've covered a lot of ground in this first session, and it's time to quit.

2. Select **File ▶ Exit JMP.**

Answer **No** to the question about saving changes to the data. Then you'll see this dialog box:

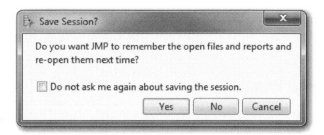

In this case, you can click **No**. In future work, if you want to take a break and resume later where you left off, you may want to click **Yes**. The next time you start the program, everything will look as it did when you quit.

Remember to run the Beginner's Tutorial from the **Tip of the Day** (or under the **Help** menu) before moving on to Chapter 2.

[1] This book's illustrations and examples are all based on JMP 11.0. In most instances, we show the default settings that come with JMP when it is newly installed.

[2] All of the data tables used in this book are available from http://support.sas.com/publishing/authors/carver.html. If you are enrolled in a college or university course, your instructor may have posted the files in a special directory. Check with your instructor.

[3] In Chapter 4, we will study *response variables* and *factors*. In this chapter, we are getting a first look at how analysis platforms operate.

[4] Columns can contain labels (for example, the name of respondent or country name), which are also displayed when a row is labeled.

Data Sources and Structures

Overview

This chapter is about data. More specifically, it is about how we make choices when we gather or generate data for analysis in a statistical investigation, and how we store that data in JMP files for further analysis. This chapter introduces some data storage concepts that are further developed in Appendix B. How and where we gather data are the foundation for what kind of conclusions and decisions we can draw. After reading this chapter, you should have a solid foundation to build upon.

Populations, Processes, and Samples

We analyze data because we want to understand something about variation within a *population*—a collection of people or objects or phenomena that we're interested in. Sometimes we're more interested in the variability of a *process*—an ongoing natural or artificial activity (like the occurrences of earthquakes or fluctuating stock prices). We identify one or more *variables* of interest and either count or measure them.

In most cases, it is impossible or impractical to gather data from every single individual within a population or every instance from an ongoing process. Marine biologists who study communication among dolphins cannot possibly measure every living dolphin. Manufacturers wanting to know how quickly a building material degrades in the outdoors cannot destroy all of their products through testing, or they will have nothing left to sell. Thanks to huge databases and powerful software, financial analysts interested in future performance of a particular stock actually can analyze every single past trade for that stock, but they cannot analyze trades that have not yet occurred.

Instead, we typically analyze a *sample* of individuals selected from a population or process. Ideally, we choose the individuals in such a way that we can have some confidence that they are a "second-best" substitute or stand-in for the entire population or ongoing process. In later chapters, we'll devote considerable attention to the adjustments we can make before generalizing sample-based conclusions to a larger population.

As we begin to learn about data analysis, it is important to be clear about the roles and relationship of a population and a sample. In many statistical studies, the situation is as follows (this example refers to a population, but the same applies to a process):

- We're interested in the variation of one or more attributes of a population. Depending on the scenario, we may wish to anticipate the variation, influence it or possibly just understand it better.

- Within the context of the study, the population consists of individual *observational units*, or *cases*. We cannot gather data from every single individual within the population.

- Hence, we choose some individuals from the population and observe them *to generalize about the entire population*.

We gather and analyze data from the sample of individuals instead of doing so for the whole population. The particular individuals within the sample are *not* the group we're ultimately interested in knowing about. We really want to learn about the variability within the population. When we use a sample instead of a population, we run some risk that the sample will misrepresent the population---and that risk is at the center of statistical reasoning.

Depending on just what we want to learn about a process or population, we also concern ourselves with the method by which we generate and gather data. Do we want to characterize or describe the extent of temperature variation that occurs in a particular part of the world? Or do we want to understand how patients with a disease respond to a specific dosage of a medication? Or do we want to predict which incentives are most likely to induce consumers to buy a particular product?

There are two kinds of generalizations that we might eventually want to be able to make, and in practice, statisticians rely on *randomization* in data generation as the logical path to generalization. If we want to use sample data to characterize patterns of variation in an entire population or process, it is valuable to select sample cases using a probability-based or random process. If we ultimately want to draw definitive conclusions about how variation in one variable is caused or influenced by another, then it is essential to randomly control or assign values of the suspected causal variable to individuals within the sample.

Therefore, the design strategy in sampling is terrifically important in the practical analysis of data. We also want to think about the timeframe for sampling. If we're interested in the current state of a population, we should select a *cross-section* of individuals at one time. On the other hand, if we want to see how a process unfolds over time, we should select a *time series* sample by observing the same individual repeatedly at specific intervals. Some samples (often referred to as *panel data*) consist of a list of individuals observed repeatedly at a regular interval.

Representativeness and Sampling

If we plan to draw general conclusions about a population or process from one sample, it's important that we can reasonably expect the sample to *represent* the population. Whenever we rely on sample information, we run the risk that the sample could misrepresent the population (in general, we call this *sampling error*). Statisticians have several standard methods for choosing a sample. No one method can guarantee that a particular sample accurately represents the population, but some methods carry smaller risks of sampling error than others. What's more, some methods have *predictable* risks of sampling error, while others do not. As you'll see later in the book, if we can predict the extent and nature of the risk, then we can generalize from a sample; if we cannot, we sacrifice our ability to generalize. JMP can accommodate different methods of representative sampling, both by helping us to select such samples and by taking the sampling method into account when analyzing data. At this point, we focus on understanding different approaches to sampling by examining data tables that originated from different designs. We will also take a first look at using JMP to select representative samples. In Chapters 8 and 21, we'll revisit the subject more deeply.

Simple Random Sampling

The logical starting point for a discussion of representative sampling is the *simple random sample* (SRS). Imagine a population consisting of N elements (for example, a lake with N = 1,437,652 fish), from which we want to take an SRS of n = 200 fish. With a little thought, we recognize that there are many different 20-fish samples that we might draw from the lake. If we use a sampling method that ensures that all 200-fish samples have the same chance of being chosen, then any sample we take with that method is an SRS. Essentially, we depend on the probabilities involved in random sampling to produce a representative sample.

Simple random sampling requires that we have a *sampling frame*, or a list of all members of a population. The sampling frame could be a list of students in a university, firms in an industry, or members of an organization. To illustrate, we'll start with a list of the countries in the world and see one way to select an SRS. For the sake of this example, suppose we want to draw a simple random sample of 20 countries for in-depth research.

> There are several ways to select and isolate a simple random sample drawn from a JMP data table. In this illustration, we'll first randomly select 20 rows, and then proceed to move them into a new data table. This is not the most efficient method, but it emphasizes the idea of random selection and it introduces two useful commands.

1. Open the data table called **World Nations**. This table lists all of the countries in the world as of 2005, as identified by the United Nations.

2. Select **Rows ▶ Row Selection ▶ Select Randomly…**. A small dialog opens (see Figure 2.1) asking either for a sampling rate of a sample size. If you enter a value between 0 and 1, JMP understands it as a rate. A number large than 1 is interpreted as a sample size, n. Enter 20 into the dialog.

Figure 2.1: Specifying a Simple Random Sample size of 20

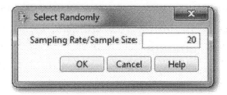

JMP randomly selects 20 rows of the 209 rows in this table. When you look at the Row panel in the Date Table window, you'll see that 20 rows have been selected. As you scroll down the list of counties, you'll see that 20 selected rows are highlighted. If you repeat Step 2, a different list of 20 rows will be highlighted because the selection process is random. With the list of 20 countries now chosen, let's create a new data table containing just the SRS of 20 countries.

3. Select **Tables ▶ Subset**. This versatile dialog box (see Figure 2.2) enables us to build a table using the just-selected rows, or to randomly sample directly.

4. As shown in the figure, choose **Selected Rows** and then change the **Output table name** to World Nations SRS. Then, click **OK**.

Figure 2.2: Creating a Subset of Selected Rows

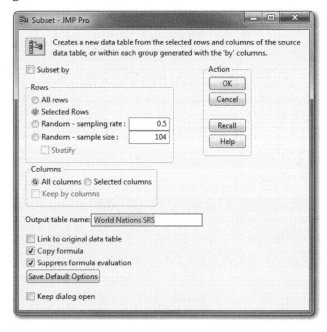

In the upper left corner of the data table window, there is a new hotspot and table variable ▾ Source . JMP inserts the JSL script that created the subset. Readers wishing to learn more about writing JMP scripts should click the red triangle, choose Edit, and see what a JSL script looks like.

Other Types of Random Sampling

As noted previously, simple random sampling requires that we can identify and access all *N* elements within a population. Sometimes this is not practical, and there are several alternative strategies available. It is well beyond the scope of this chapter to discuss these strategies at length, but Chapter 8 provides basic coverage of some of these approaches.

Non-Random Sampling

This book is about practical data analysis, and in practice many data tables contain data that were *not* generated by anything like a random sampling process. Most data collected within business and nonprofit organizations come from the normal operations of the organization rather than from a carefully constructed process of sampling. We can summarize and describe the data within a non-random sample, but should be very cautious about the temptation to generalize from such samples. Whether we are conducting the analysis or reading about it, we always want to ask whether a particular sample is likely to misrepresent the population or process from which it came. Voluntary response surveys, for example, are very likely to mislead us if only highly motivated individuals respond. On the other hand, if we watch the variation in stock prices during an uneventful period in the stock markets, we might reasonably expect that the sample could represent the process of stock market transactions.

Big Data

You may have heard or read about "Big Data"—high volume raw data generated by numerous electronic technologies like cell phones, supermarket scanners, radio-frequency identification (RFID) chips, or other automated devices. The world-spanning continuous stream of data carries huge potential for the future of data analytics and also presents many ethical, technical, and economic challenges. In general, data generated in this way are not random in the conventional sense and don't neatly fit into the traditional classifications of an introductory statistics course. Big data may include photographic images, video, or sound recordings that don't easily occupy columns and rows in a data table. Furthermore, streaming data is neither cross-sectional nor time series in the usual sense.

Cross-Sectional and Time Series Sampling

When the research concerns a population, the sampling approach is often *cross-sectional*, which is to say the researchers select individuals from the population at one period of time. Again, the individuals can be people, animals, firms, cells, plants, manufactured goods, or anything of interest to the researchers.

When the research concerns a process, the sampling approach is more likely to be *time series* or *longitudinal*, whereby a single individual is repeatedly measured at regular time intervals. A great deal of business and economic data is longitudinal. For example, companies and governmental agencies track and report monthly sales, quarterly earnings, or annual employment. The major distinction between time series data and streaming data is whether observations occur according to a pre-determined schedule or whether they are event-driven (*e.g.* when a customer places a cell phone call).

Panel studies combine cross-sectional and time series approaches. In a panel study, researchers repeatedly gather data about the same group of individuals. Some long-term public health studies follow panels of individuals for many years; some marketing researchers use consumer panels to monitor changes in taste and consumer preferences.

Study Design: Experimentation, Observation, and Surveying

If the goal of a study is to demonstrate a cause-and-effect relationship, then the ideal approach is a *designed experiment*. The hallmark features of an experiment are that the investigator controls and manipulates the values of one or more variables, randomly assigns *treatments* to observational units, and then observes changes in the *response* variable. For example, engineers in the concrete industry might want to know how varying the amount of different additives affects the strength of the concrete. A research team would plan an experiment in which they would systematically vary specific additives and conditions, then measure the strength of the resulting batch of concrete.

Similarly, consider a large retail company that has a "customer loyalty" program, offering discounts to its regular customers who present their bar-coded key tags at the checkout counter. Suppose the firm wants to nudge customers to return to their stores more frequently, and generates discount coupons that can be redeemed if the customer visits the store again within so many days. The marketing analysts in the company could design an experiment in which they vary the size of the discount and the expiration date of the offer, issue the different coupons to randomly-chosen customers, and then see when customers return.

Experimental Data—An Example[1]

In an experimental design, the causal variables are called *factors* and the outcome variable is called the *response* variable. As an illustration of a data table containing experimental data, open the data table called Concrete. Professor I-Cheng Yeh of Chung-Hua University in Taiwan measured the *compressive strength* of concrete prepared with varying formulations of seven different component materials. Compressive strength is the amount of force per unit of area, measured here in megapascals that the concrete can withstand before failing. Think of the concrete foundation walls of a building; they need to be able to support the mass of the building without collapsing. The purpose of the experiment was to develop an optimal mixture to maximize compressive strength.

1. Select **File ▸ Open**. Choose **Concrete** and click **OK**.

Figure 2.3: The Concrete Data Table

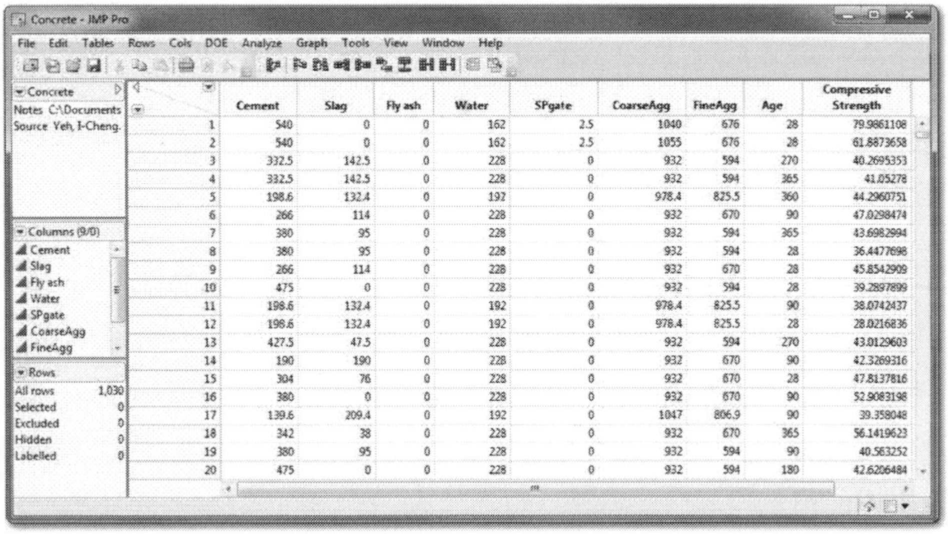

	Cement	Slag	Fly ash	Water	SPgate	CoarseAgg	FineAgg	Age	Compressive Strength
1	540	0	0	162	2.5	1040	676	28	79.9861108
2	540	0	0	162	2.5	1055	676	28	61.8873658
3	332.5	142.5	0	228	0	932	594	270	40.2695353
4	332.5	142.5	0	228	0	932	594	365	41.05278
5	198.6	132.4	0	192	0	978.4	825.5	360	44.2960751
6	266	114	0	228	0	932	670	90	47.0298474
7	380	95	0	228	0	932	594	365	43.6982994
8	380	95	0	228	0	932	594	28	36.4477698
9	266	114	0	228	0	932	670	28	45.8542909
10	475	0	0	228	0	932	594	28	39.2897899
11	198.6	132.4	0	192	0	978.4	825.5	90	38.0742437
12	198.6	132.4	0	192	0	978.4	825.5	28	28.0216836
13	427.5	47.5	0	228	0	932	594	270	43.0129603
14	190	190	0	228	0	932	670	90	42.3269316
15	304	76	0	228	0	932	670	28	47.8137816
16	380	0	0	228	0	932	670	90	52.9083198
17	139.6	209.4	0	192	0	1047	806.9	90	39.358048
18	342	38	0	228	0	932	670	365	56.1419623
19	380	95	0	228	0	932	594	90	40.563252
20	475	0	0	228	0	932	594	180	42.6206484

The first seven columns in the data table in Figure 2.3 represent factor variables. Professor Yeh selected specific quantities of the seven component materials, and then tested the compressive strength as the concrete aged. The eighth column, **Age**, shows the number of days elapsed since the concrete was formulated, and the ninth column is the response variable, **Compressive Strength**. In the course of his experiments, Professor Yeh repeatedly tested different formulations, measuring compressive strength after varying numbers of days. To see the structure of the data, let's look more closely at the two columns.

2. Choose **Analyze ▶ Distribution**.

3. As shown in Figure 2.4, select the **Cement** and **Age** columns and click **OK**.

Figure 2.4: The Distribution Dialog Box

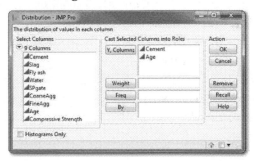

The Distribution platform generates two graphs and several statistics. We'll study these in detail in Chapter 3. For now you just need to know that the graphs, called *histograms*, display the variation within the two data columns. The lengths of the bars indicate the number of rows corresponding to each value. For example, there are many rows with concrete mixtures containing about 150 kg of cement, and very few with 100 kg.

4. Arrange the windows on your screen to look like Figure 2.5.

Figure 2.5: Selecting Rows by Selecting One Bar in a Graph

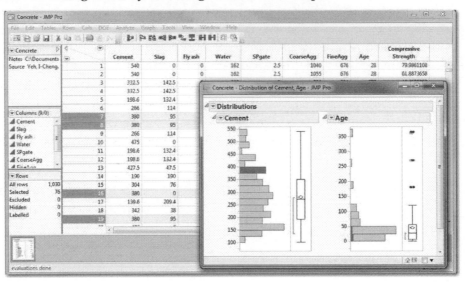

5. Move your cursor over the **Cement** histogram and click the bar corresponding to values just below 400 kg of cement.

When you click that bar the entire bar darkens, indicating that the rows corresponding to mixtures with that quantity (380 kg of cement, it turns out) are selected. Additionally, small portions of several bars in the **Age** histogram are also darkened, representing the same rows.

Finally, look at the data table. Within the data grid, several visible rows are highlighted. All of them share the same value of **Cement**. Within the **Rows** panel, we see that we've now altered the row state of 76 rows by selecting them.

Look more closely at rows 7 and 8, which are the first two of the selected rows. Both represent seven-part mixtures following the same recipe of the seven materials. Observation #7 was taken at 365 days and observation #8 at 28 days. These observations are *not* listed chronologically, but rather are a randomized sequence typical of an experimental data table.

Row 16 is the next selected row. This formulation shares several ingredients in the same proportion as rows 7 and 8, but the amounts of other ingredients differ. This is also typical of an experiment. The researcher selected and tested different formulations. Because the investigator, Professor Yeh, manipulated the factor values, we find this type of repetition within the data table. And because Professor Yeh was able to select the factor values deliberately and in a controlled way, he was able to draw conclusions about which mixtures will yield the best compressive strength.

Observational Data—An Example

Of course, experimentation is not always practical, ethical, or legal. Medical and pharmaceutical researchers must follow extensive regulations, can experiment with dosages and formulations of medications, and may randomly assign patients to treatment groups, but cannot randomly expose patients to diseases. Investors do not have the ability to manipulate stock prices (if they do and get caught, they go to prison).

Now open the data table called **Stock Index Weekly**. This data table contains time series data for six different major international stock markets. A stock *index* is a weighted average of the prices of a sample of publicly traded companies. In this table, we find the weekly index values for the following stock market indexes, as well as the average number of shares traded per week:

- Nikkei 225 Index – Tokyo
- FTSE100 Index – London
- S&P 500 Index – New York
- Hang Seng Index – Hong Kong
- IGBM Index – Madrid
- TA100 Index – Tel Aviv

The first column is the observation date; note that the observations are basically every seven days after the second week. All of the other columns simply record the index and volume values as they occurred. It was not possible to set or otherwise control any of these values.

Survey Data—An Example

Survey research is conducted in the social sciences, in public health, and in business. In a survey, researchers pose carefully constructed questions to respondents, and then the researchers or the respondents themselves record the answers. In survey data, we often find *coding schemes* where categorical values are assigned numeric codes or other abbreviations. Sometimes continuous variables, like income or age, are converted to ordinal columns.

As an example of a well-designed survey data table, we have a small portion of the responses provided in the 2006 administration of the National Health and Nutrition Examination Survey (NHANES). NHANES is an annual survey of people in the U.S. that looks at their diet, health, and lifestyle choices. The Centers for Disease Control (CDC) posts the raw survey data for public access.[2]

1. Open the **NHANES** data table.

As shown in Figure 2.6, this data table contains some features that we have not seen before. As usual, each column holds observations for a single variable, but these column names are not very informative. Typical of many large scale surveys, the NHANES Web site provides both the data and a *data dictionary* that defines each variable, including information about coding and units of measurement. For our use in this book, we've added some notations to each column.

Figure 2.6: NHANES Data Table

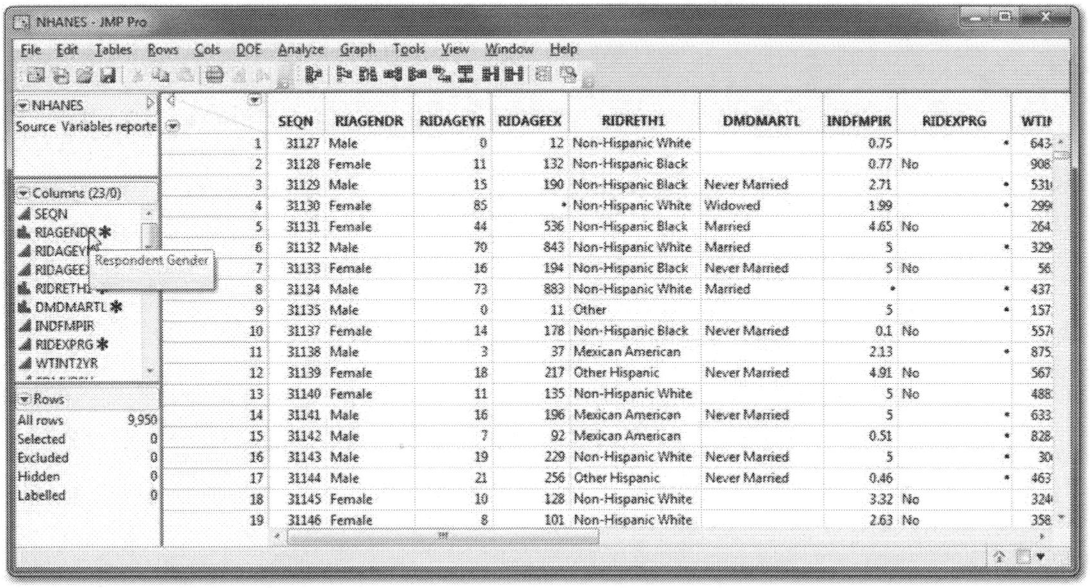

2. In the **Columns** panel, move your cursor to highlight the column named **RIAGENDR** and right-click. Select **Column info...** to open the dialog box shown in Figure 2.7.

Figure 2.7: Customizing Column Metadata

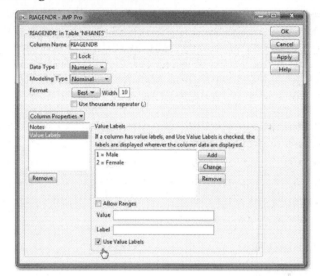

When this dialog box opens, we find that this column holds **numeric** data and its modeling type is **nominal**—which might seem surprising, since the rows in the data table all say **Male** or **Female**. In fact, the lower portion of the column info dialog box shows us what's going on here. This is an example of *coded data:* the actual values within the column are 1s and 2s, but when displayed in the data table a 1 appears as the word Male and a 2 as the word Female. This recoding is a *column property* within JMP.

> The asterisk next to **RIAGENDR** in the **Columns** panel indicates that this column has one or more special column properties defined.

3. At the bottom of the dialog box, clear the check box marked **Use Value Labels** and click **Apply**. Now, look in the Data Table and see that the column displays 1s and 2s rather than the value labels.

4. Click the word **Notes** under **Column Properties** (middle left of the dialog box). You'll see a brief description of the contents of this column. As you work with data tables and create your own columns, you should get into the habit of annotating columns with informative notes.

In this table, we also encounter *missing observations* for the first—but certainly not the last—time. Missing data is a common issue in survey data and it crops up whenever a respondent doesn't complete a question or an interviewer doesn't record a response. In a JMP data table a black dot (·) indicates missing numeric data; missing character data is just a blank. As a general rule, we want to think carefully about missing observations and *why* they occur in a particular set of data.

For instance, look at row 1 of this data table. We have missing observations for several variables, including the columns representing marital and pregnancy status, as well as pulse rate and blood pressure. Further inspection shows that respondent #31127 was a 12-month old baby boy.

Creating a Data Table

In this book, we'll almost always analyze the data tables that are available at support.sas.com/authors. Once you start to practice data analysis on your own, you'll often need to create your own data tables. Refer to Appendix B ("Data Management") for details about entering data from a keyboard, transferring data from the web, or reading in an Excel worksheet.

Raw Case Data and Summary Data

This book is accompanied by more than 50 data tables, most of which contain "case-wise" data, meaning that each row of the data table refers to one observational unit. For example, each row of the NHANES table represents a person. Each row of the Concrete table is one batch of concrete at a moment in time. Like most statistical software, JMP is intended to analyze raw data. However, sometimes data come to us in a partially-processed or summarized state. It is still possible to construct a data table and do some limited analysis of the data.

For example, consider public opinion surveys that have been reported in the news. In June, 2014, CNN released the results of an opinion poll of 1,003 adults in the United States[3]. One of the questions asked was, "Which of the following is the most important issue facing the country today?" The interviewers presented nine choices to respondents, randomly scrambling the order of the choices. Table 2.1 summarizes the responses as reported by CNN.

Table 2.1: Summary of a Survey About the Most Important Issues Facing the U.S. (Percentages may not sum to 100% due to rounding)

Response	May 29–June 1, 2014 Percentage	Jan. 14–15, 2013 Percentage
The economy	40	46
Health care	19	14
The federal budget deficit	15	23
The environment	8	2
Gun policy	6	6
Foreign policy	5	4
Immigration	5	3
Other	1	1
No opinion	1	-

We could easily transfer this into a data table with one major caveat that often confuses introductory students. It is crucial to understand how the structure of this table relates to its content. In JMP, we usually expect rows to represent observational units, columns to represent variables, and cells to contain data values. Each of those assumptions is incorrect with data structured this way.

To clarify our thinking, we should go back to asking how the raw data were generated. We know that respondents to this question—the observational units--were the 1,003 adults in the survey sample. The variable is the responses those people gave when asked the question, "Which of the following is the most important issue facing the country today?" Because their responses were descriptors of current issues, this is a nominal variable.

So, how does this clarify the contents of Table 2.1? The variable in this instance is the issue that a respondent believes to be most important. The left column in the table lists the nine different responses that were repeatedly selected by the 1,003 respondents. The percentage columns (corresponding to two samples taken approximately six months apart) appear to be continuous variables, sitting in two column looking very like numbers. However, these numbers are not measurements of observational units (people). They *summarize* the answers provided by the respondents, indicating the fraction of the sample giving each answer. Specifically the values are *relative frequencies* of each "level" of the nominal variable. In short, this data table represents a single nominal variable, summarizing the responses of two separate samples of individuals who are "invisible" when the information is presented this way. In later chapters, we'll learn how to use columns of frequencies. For now, our first consideration of data types and sources is complete.

Application

Now that you have completed all of the activities in this chapter, use the concepts and techniques that you've learned to respond to these questions.

1. Use your primary textbook or the daily newspaper to locate a table that summarizes some data. Read the article accompanying the table and identify the variable, data type, and observational units.

2. Return to the **Concrete** data table. Browse through the column notes, and explain what variables these columns represent: **Cement**, **SPgate,** and **FineAgg**.

3. In the **NHANES** data table, several nominal variables appear as continuous data within the **Columns** panel. Find the misclassified variables and correct them by changing their modeling types to nominal.

4. The **NHANES** data table was assembled by scientific researchers. Why don't we consider this data table to be experimental data?

5. Find the data table called **Military**. We will use this table in later chapters. This table contains rank, gender, and race information about more than a million U.S. military personnel. Use a technique presented in this chapter to create a data table containing a random sample of 500 individuals from this data table.

6. In Chapter 3, we will work with a data table called **FTSE100**. Open this table and examine the columns and metadata contained within the table. Write a short paragraph describing the contents of the table, explaining how the data were collected (experiment, observation, or survey), and define each variable and its modeling type.

7. In later chapters, we will work with a data table called **Earthquakes**. Open this table and examine the columns and metadata contained within the table. Write a short paragraph describing the contents of the table, explain how the data were collected (experiment, observation, or survey), and define each variable and its modeling type.

8. In later chapters, we will work with a data table called **Tobacco Use**. Open this table and examine the columns and metadata contained within the table. Write a short paragraph describing the contents of the table, explain how the data were

collected (experiment, observation, or survey), and define each variable and its modeling type.

9. Open the **Dolphins** data table, which we will work with in a later chapter. What are the variable(s), observational units, and data types represented in this table?

10. Open the data table **TimeUse**, which we will analyze more fully in later chapters. Write a few sentences to explain the contents of the columns named **marst, empstat, sleeping**, and **telff**.

11. Open the States data table, which contains statistics about the 50 U.S. states and the District of Columbia. Write a short paragraph describing the contents of the table, and in particular define the columns called **smoke**, **fed_spend**, and **nuclear**.

---

[1] In Chapter 21, we will learn how to design experiments. In this chapter, we'll concentrate on the nature of experimental data.

[2] Visit http://www.cdc.gov/nchs/surveys.htm to find the NHANES and other public-use survey data. Though the topic is beyond the scope of this book, readers engaged in survey research will want to learn how to conduct a database query and import the results into JMP. Interested readers should consult the section on "Importing Data" in Chapter 2 of the *JMP User Guide*.

[3] Visit http://i2.cdn.turner.com/cnn/2014/images/06/03/cnn.poll.obama.va.pdf for the full details of the poll. Interviews of 1,003 adults were conducted by telephone by ORC International between May 29 and June 1, 2014.

Describing a Single Variable

3

Overview

Once we have a framed some research questions and gathered relevant data, the next phase of an investigation is to examine the variability in the data. The goal of descriptive analysis is to summarize where things stand with each variable; in fact, the term *statistics* comes from the practice of characterizing the state of political affairs through the reporting of facts and figures. This chapter presents several standard tools that we can use to examine how a variable varies, to describe the pattern of variation that it exhibits, and to look as well for departures from the overall pattern.

The Concept of a Distribution

Data analysis generally focuses on one or more variables—attributes of the individual observations. When we speak of a variable's distribution, we are referring to a pattern of values. The distribution describes the different values the variable can assume, and how often it assumes each value.

In our first example, we'll continue to consider the variability of life expectancy around the world. The data we'll use come to us from the United Nations; in Chapter 1, we used a small portion of this data set.

Variable Types and Their Distributions

1. Select **File ▶ Open**, select the **Life Expectancy** data table, and click **Open**.

Before doing any analysis, make sure that you can answer these questions:

- What population does this data table represent?
- What is the source of the data?
- How many variables are in the table?
- What data type is each variable?
- What does each variable represent?
- How many observations are there?

Take special note of the way this particular data table has been organized. We have 13 annual observations for each country, and they are *stacked* one upon the other. Not surprisingly, JMP refers to this arrangement as stacked data.

As in Chapter 1, we'll raise some questions about how life expectancy at birth varies in different parts of the world. There are far too many observations for us get a general sense of the variation simply by scanning the table visually. We need some sensible ways to find the patterns among the large number of rows. We'll begin our analysis by looking at the *nominal* variable called **Region**.

Statisticians generally distinguish among four types of data:

Categorical Types	Quantitative Types
Nominal	Interval
Ordinal	Ratio

One reason that it is important to understand the differences among data types is that we analyze them in different ways. In JMP, we differentiate between nominal, ordinal, and continuous data. Nominal and ordinal variables are categorical, distinguishing one observation from another in some qualitative, non-measurable way. Interval and ratio data are both numeric. Interval variables are artificially constructed, like a temperature scale or stock index, with arbitrarily chosen zero points. Most measurement data is considered ratio data, because ratios of values are meaningful. For example, a film that lasts 120 minutes is twice as long as one lasting 60 minutes. In contrast, 120 degrees Celsius is *not* twice as hot as 60 degrees Celsius.

Distribution of a Categorical Variable

In its reporting, the United Nations identifies each country of the world with a continental region. There are six regions, each with a different number of countries. The variable Region is *nominal*—it literally names a country's general location on earth. Let's get familiar with the different regions and see how many countries are located in each. In other words, let's look at the distribution of Region.

1. Select **Analyze ▶ Distribution**. In the **Distribution** dialog box (Figure 3.1), select the variable Region as the **Y, Columns** variable. Click **OK**.

Figure 3.1: Distribution Dialog Box

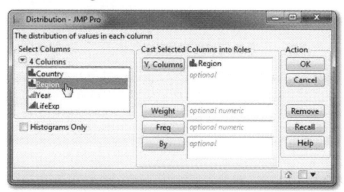

Any time you want to assign a column to a role in a JMP dialog, you have three options: you can highlight the column name in the **Select Columns** list and click the corresponding role button, you can double-click the column name, or you can click-drag the column name into the role box.

The result appears in Figure 3.2. JMP constructs a simple *bar chart* listing the six continental regions and showing a rectangular bar corresponding to the number of times the name of the region occurs in the data table. Though we can't immediately tell from the graph exactly how many countries are in each, it is clear that South Asia has the fewest countries and Europe & Central Asia has the most.

Figure 3.2: Distribution of Region

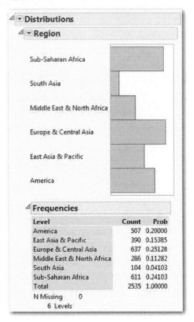

Below the graph is a *frequency distribution* (titled **Frequencies**), which provides a more specific summary. Here we find the name of each region, and the number of times each regional name appears in our table. For example, "America" occurs 507 times. As a proportion of the whole table, 20% of the rows (Prob. = 0.2000) represent countries in North, Central, or South America.

At this point, you might wisely pause and say, "Wait a second. There aren't 507 countries in the Americas!" And you would be right. Remember that we have stacked data, with 13 rows representing 13 years of data devoted to each country. Therefore, there are 507/13 = 39 countries in the region called America.

> Even though JMP handles the heavy computational or graphical tasks, always think about the data and its context and ask yourself if the results make sense to you.

Using the Data Filter to Temporarily Narrow the Focus

Because we know each country appears repeatedly in this data table, let's choose just one year's data to obtain a clearer picture of regional variation. *We can specify rows to display in a graph by using the* **Data Filter**. *This is a tool that allows us to select rows that satisfy specific conditions such as, only displaying data rows from the year 2010.*

> *This chapter illustrates the use of the Data Filter to temporarily select rows in a data table for all active analyses. This is known as the global Data Filter. Alternatively, when you click the red triangles in most analysis reports, you will find a Script option with a local Data Filter that applies only to the current report. The local Data Filter is illustrated in later chapters, but curious readers should explore it at any time.*

1. Select **Rows ▶ Data Filter**. In the list of **Columns**, select **Year** and click the **Add** button.

2. The dialog box takes on a new appearance (Figure 3.3). It now displays a list of years contained in the table. Near the top of the dialog, check **Show** and **Include** so that only the rows we select for **2010** will appear in all graphs and be included in any computations. Other rows will be hidden and excluded.

Figure 3.3: Choosing 2010 in the Data Filter

3. Scroll down the list of **Year** levels and highlight **2010**. As noted in the dialog, this selects 195 rows and temporarily suppresses the others.

4. Minimize the **Data Filter**. If you look in the data table of Life Expectancy, you will see that most rows now have two icons (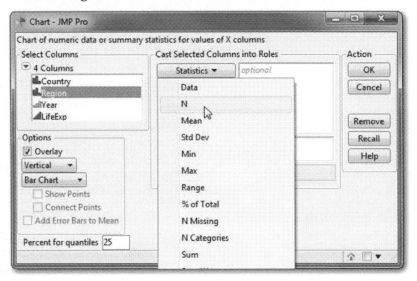) indicating that they are excluded and hidden. The rows from 2010 are highlighted and will remain so until we clear the **Data Filter** or take another action that selects other rows.

Using the Chart Command to Graph Categorical Data

The Distribution dialog generates both a graph and some numerical summaries. We can also create other graphs for a categorical variable as follows:

1. Select **Graph ▶ Chart.** The **Chart** dialog box enables you to create bar, pie, line, and other chart styles. The default is a vertically oriented bar chart.

2. Select the **Region** column, click the **Statistics** button, and choose **N** (see Figure 3.4). In this way, we're requesting a bar chart showing the number (n) of observations for each region.

Figure 3.4: Chart Dialog Box

The resulting bar chart appears in Figure 3.5. It presents exactly the same information as the graph that we created using the **Distribution** command. There are two differences between the graphs in Figure 3.2 and Figure 3.5. First, one is oriented vertically and the other horizontally. Also, in Figure 3.5, the bars have clear space between them, indicating that the horizontal axis represents distinct and separate categories.

We can alter and customize the graph in several ways, most of which you can explore by clicking on the red triangle next to **Chart**.

Figure 3.5: A Typical Bar Chart

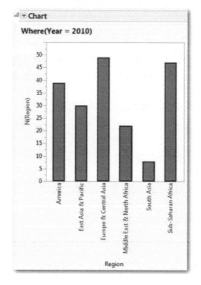

It is always good practice to help a reader by giving a graph an informative descriptive title. The default title "Chart," though accurate, isn't very helpful. In JMP, it is easy to alter the titles of graphs and other results.

3. Move your cursor to the title **Chart** just above the graph and double-click. You can now customize the title of this chart to make it more informative. Type Observations per Region, replacing "Chart" as the title.

With most categorical data, JMP automatically reports values in alphabetical sequence[1] (America, East Asia & Pacific, and so on). We can revise the order of values to suit our purposes as well. Suppose that we want to list the regions approximately from West to East, North to South. In that case, we might prefer a sequence as follows:

America
Europe & Central Asia
Middle East & North Africa
Sub-Saharan Africa
South Asia
East Asia & Pacific

To change the default sequence of categorical values (whether nominal or ordinal), we return to the data table.

4. Select **Region** from the data grid or the columns panel, right- click, and select **Column Info**.

5. Click **Column Properties**, and select **Value Ordering**.

6. Select a value name and use the **Move Up** and **Move Down** buttons to revise the value order to match what we've chosen. Then, click **OK**.

Now look at the bar chart. You'll see that it is unchanged—so far.

7. To apply the revised ordering, click the hotspot next to the chart title, and select **Script ▶ Redo Analysis**. This creates a revised bar chart in a new window, and this one has the desired column sequence.

The Chart command offers several graphing formats. Perhaps a pie chart might do a better job of showing how countries are distributed around the world.

8. Click the hotspot once again, and choose **Pie Chart**.

The effect should speak for itself. Experiment with the other charting options. With categorical data, your choices are limited. Still, it's worth a few minutes to become familiar with them. When you are through exploring, restore the graphic to a bar chart and leave it open. We will return to this graph in a few pages.

Using the Graph Builder to Explore Categorical Data

In Chapter 1 we met the Graph Builder, and we will use it throughout this book. It is most useful when working with multiple variables, but even with a single nominal variable it provides a quick way to generate multiple views of the same data. Because interactivity is such an important feature of the tool, this section of the chapter provides few step-by-step directions; you should interact with the tool and think about the extent to which different graphing formats and options communicate the information content of the variable called Region.

1. Select **Graph ▶ Graph Builder.** The Region column identifies groups of countries. Drag it to the **X** drop zone.

Within the Graph Builder, you can freely reposition a column from one drop zone to another. Hover the cursor over the column name until the cursor changes to the hand shape 🖑, then click-drag it to any other drop zone. What's more, there is also an Undo button.

You can also use the same variable in more than one drop zone. For example, you might also color the bars by region.

With Region on the X axis, you'll see six clumps of black points above the six region names. This is not very informative.

At the top of the Graph Builder is a selector bar of icons (see Figure 3.6) representing different graph types. The graphing options available depend on the kind of data we've placed on the graph. Hence, some icons are dimmed, but with Region on the X axis, we can opt for any of the highlighted option.

Figure 3.6: Graphing Options for a Nominal Column

2. Spend some time using different graphing formats. Which ones do you think do the best job of clearly and fully summarizing the number of countries within each region?

Distribution of a Quantitative Variable

The standard graphing choices expand considerably when we have quantitative data—particularly for continuous variables or discrete variables with many possible values. For one thing, it's not at all obvious where we would "slice the pie," or just where to set boundaries between the bars in a bar chart.

As such, most graphing tools for quantitative data establish arbitrary boundaries, which we can adjust appropriately to the context of the data. As a way of visualizing the distribution of a continuous variable, the most commonly used graph is a *histogram*. A histogram is basically a bar chart with values of the variable on one axis and frequency on the other. Let's illustrate.

In our data set, we have estimated life expectancy at birth for each country for 13 different years. We just used the Data Filter to isolate the data for 2010 so let's continue to explore the state of the world in 2010.

Using the Distribution Platform for Continuous Data

As before, we'll first use the **Distribution** platform to do most of the work here.

1. Select **Analyze ▶ Distribution.** Cast **LifeExp** into the role of **Y,Columns**.

2. When the distribution window opens, click the hotspot next to **Distributions**, and select **Stack**. This will re-orient the output horizontally making it a bit easier to interpret.

The histogram (Figure 3.7) is one representation of the distribution of life expectancy around the world in 2010, and it gives us one view of how much life expectancy varies. Above the histogram is a *box plot* (also known as a box-and-whiskers plot), which will be explained later in this chapter.

Figure 3.7: A Typical Histogram

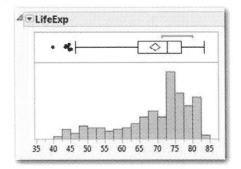

As in the bar charts we've studied earlier, there are two dimensions in the graph. Here, the horizontal axis displays values of the variable and the vertical axis displays the *frequency* of each small interval of values. For example, we can see that only a few countries have projected life expectancies of 40 to 45 years, but many have life expectancies between 70 and 75 years.

When we look at a histogram, we want to develop the habit of looking for four things: the *shape*, the *center* (or central tendency), the *dispersion* of the distribution and *unusual observations*. The histogram can very often clearly represent these three aspects of the distribution.

Shape: Shape refers to the symmetry of the histogram and to the presence of peaks in the graph. A graph is symmetric if you could find a vertical line in the center defining two sides that are mirror images of one another. In Figure 3.7, we see an *asymmetrical* graph. There are few observations in the *tails* on the left, and most observations clump together on the right side. We say this is a *left-skewed* (or negatively skewed) distribution.

Many distributions have one or more peaks—data values that occur more often than the other values. Here we have a distinct peak around 70 to 75 years. Some distributions have multiple peaks, and some have no distinctive peaks at all. In short, we might describe the shape of this distribution as "single-peaked and left-skewed."

Center (*or central tendency*): Where do the values congregate on the number line? In other words, what values does this variable typically assume? As you may already know, there are several definitions of center as reflected in the mean, median, and mode statistics. Visually, we might think of the center of a histogram as the half-way point of the horizontal axis (the median, which is approximately 73 in this case), as the highest-frequency region (the modal class, which is 72.5–75), perhaps as a kind of visual balancing point (the mean, which is approximately 65–70), or in some other way. Any of these interpretations have legitimacy, and all respond to the question in slightly different ways.

Dispersion (*or spread*)**:** While the concept of center focuses on the typical, the concept of spread focuses on departures from the typical. The question here is "how much does this variable vary?" and again there are several reasonable ways to respond. We might think in terms of the lowest and highest observed values (from about 40 to 85), in terms of a vicinity of the center (for example, "life expectancy tends to vary in most countries between about 65 and 85"), or in some other relative sense.

Unusual Observations: We can summarize the variability of a distribution by citing its shape, center, and dispersion, but in some distributions there may be a small number of observations that deviate substantially from the pattern. Look at the boxplot above the histogram. There are about six dots at the far left end; these represent a small number of nations with extraordinarily brief life expectancies. We refer to such values as *outliers*.

3.   Hover the cursor over the left-most point in the boxplot. You will see a pop-up note that this is Swaziland, with a life expectancy of only 40.12 years in 2010.

Often, it's easiest to think of these four characteristics by comparing two distributions. For example, Figure 3.8 shows two histograms using the life expectancy data from 1950 and 2010. We might wonder how human life expectancy has changed since 1950, and in these two histograms, we can note differences in shape, center, dispersion and unusual observations.

Figure 3.8: Comparing Two Distributions

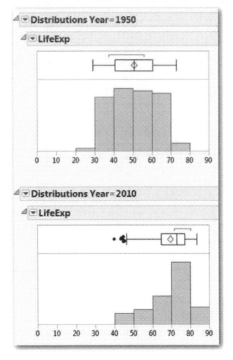

To create the results shown in Figure 3.8, do the following:

4. Return to the original **Life Expectancy** data table.

5. Re-open the Data Filter dialog (either choose **Windows** and find the filter or **Rows ▶ Data Filter**).

6. To add 1950, hold down the **CTRL** key and highlight **1950**.

7. From the menu bar, choose **Analyze ▶ Distribution**.

8. Select **LifExp** as **Y**, just as you did earlier.

9. Choose **Year** as the **By** column, and click **OK**.

This creates the two distributions with vertically oriented histograms. When you look at them, notice that the axis of the first one runs from 25 to 75 years, and the axis on the second graph runs from 40 to 85 years.

To facilitate the comparison, it is helpful to orient the histograms horizontally in a stacked arrangement and to set the axes to a uniform scale, an option that is available in the hotspot menu next to **Distributions.** This makes it easy to compare their shapes, centers, and spreads at a glance.

10. In the **Distribution** report, while pressing the CTRL key, click the uppermost red triangle and select **Uniform Scaling**.

> If you click the hotspot without pressing the CTRL key, the uniform scaling option would apply only to the upper histogram. Pressing the CTRL key has the effect of applying the choice to all graphs in the window.

11. Click the hotspot once again and choose **Stack**.

The histograms on your screen should now look like Figure 3.8. How does the shape of the 1950 distribution compare to that of the 2010 distribution? What might have caused these changes in the shape of the distribution?

We see that people tend to live longer now than they did in 1950. The location (or central tendency) of the 2010 distribution is largely to the right of the 1950 distribution. Additionally these two distributions also have quite different spreads (degrees of dispersion). We can see that the values were far more spread out in 1950 than they are in 2010 and that there were no outliers in 1950. What does that reveal about life expectancy around the world during the past 60 years?

Taking Advantage of Linked Graphs and Tables to Explore Data

When we construct graphs, JMP automatically links all open tables and graphs. If we select rows either in the data table or in a graph, JMP selects and highlights those rows in all open windows.

1. Within the 2010 life expectancy histogram, place the cursor over the right-most bar and click. While pressing the Shift key, also click the adjacent bar. Now you should have selected the two bars representing life expectancies of over 70 years. How many rows are now selected? Look in the **Rows** panel of the **Data Table** window.

2. Now find the first window with the Distribution of **Region**. Notice that some bars are partially highlighted. When you selected the two bars in the histogram, you were indirectly selecting a group of countries. These countries are grouped within the bar chart as shown, revealing the parts of the world where people tend to live longest.

Customizing Bars and Axes in a Histogram

When we use the **Distribution** platform to analyze a continuous variable, JMP determines how to divide the variable axis and how to create "bins" for grouping observations. These automatic choices can affect the appearance of the distribution and there are several ways to customize the appearance of a histogram.

We can alter the number of bars in the histogram, creating new boundaries between groups of observations and shifting observations from one bar to the next.

1. In the **Distribution** report window, hover over the pale blue bar just above **Distributions Year = 1950** to display the menu bar. From the menu bar, select the hand tool 🖑 (also called the *Grabber*).

2. Position the hand anywhere over the bars in the 2010 histogram, and click-drag the tool straight up and down. In doing so, you'll change the number and width of the bars, sometimes dramatically changing the shape of the graph.

Think about this: the apparent shape of the distribution depends on the number of bars we create. By default, the software chooses an initial number of bars, or *bins*, to categorize the continuous variable. However, that initial choice should not be the final word. As we adjust the number of bins, we should watch closely to see how the shape changes, looking for a rendering that accurately and honestly displays the overall pattern of variation.

One was to resolve the issue by using a *shadowgram*. A shadowgram visually averages a large number of bin widths into a diffuse image with no distinct bars at all. Here's how:

3. Click the red triangle next to **LifeExp** in the 2010 histogram.

4. Choose **Histogram Options ▶ Shadowgram**. Figure 3.9 shows the result.

Figure 3.9: A Shadowgram for a Continuous Variable

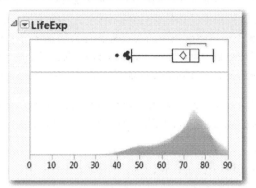

> You should notice that there are several **Histogram Options**. While you are here, explore them—see what there is to see.

We can change the scale of the horizontal axis interactively. Initially, JMP set the left and right endpoints and the limits changed when we chose uniform scaling. Suppose we want the axis to begin at 30 and end at 100.

5. Move the cursor to the left end of the horizontal axis, and notice that the hand now points to the left (this is true whether or not you've previously chosen the hand tool). Click and drag the cursor slowly left and right, and see that you're scrunching or stretching the axis values. Stop when the minimum value is 30.

6. Move the cursor to the right end of the axis, and similarly set the maximum at 100 years just by dragging the cursor.

Finally, we can "pan" along the axis. Think of the border around the graph as a camera's viewfinder through which we see just a portion of the entire infinite axis.

7. Without holding the mouse button, move the cursor toward the middle of the axis until the hand points upward. Now click and drag to the left or right, and you'll pan along the axis.

Exploring Further with the Graph Builder

Our original data table contains values for 13 years, and we've now compared the variation in life expectancy for two years. The Graph Builder can allow us to make a quick visual comparison over 13 years.

1. First, we want to clear our earlier filtering so that we can now access all years. Re-open the **Data Filter**, click **Clear,** and close the **Data Filter**.

2. Select **Graph ▶ Graph Builder**.

3. Drag **LifeExp** to the **X** drop zone.

4. Find the menu bar at the top of the **Graph Builder** window, and locate the **Histogram** button ![histogram icon] near the center. Click it.

5. Drag **Year** to the **Wrap** drop zone, and click the **Done** button. Your graph should look like Figure 3.10.

Figure 3.10: Longer Lives in Most of the World, 1950 to 2010

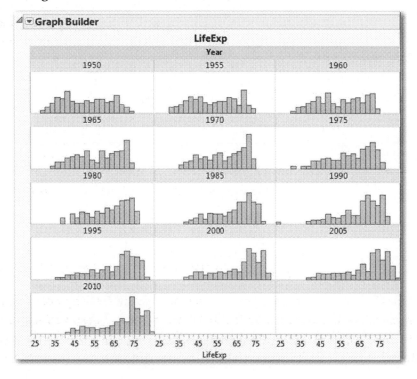

What do you see as you inspect these small multiple histograms? Can you see life expectancies gradually getting longer in most countries? There were two peaks in 1950: many countries with short lives, and many with longer lives. The lower peak slowly flattened out as the entire distribution has crept rightward. Also notice the unusual observation in 1990. Tragically, this is the "fingerprint" of the Rwandan genocide.

Summary Statistics for a Single Variable

Graphs are an ideal way to summarize a large data set and to communicate a great deal of information about a distribution. We can also describe variation in a quantitative variable with *summary statistics* (also called summary measures or descriptive statistics). Just as a distribution has shape, center, and dispersion, we have summary statistics that capture information about the shape, center, or dispersion of a variable.

Let's look back at the distribution report for our sample of 2010 life expectancies in 195 countries of the world. Just to the right of the histogram, we find a table of *quantiles* followed by a list of *Summary Statistics*.

> You now have several open reports and your screen may be cluttered. You can find any open report in the **Window** menu, in the **Window List** tree of the **JMP Home Window,** or (for Windows users) locate a thumbnail image of the report along the bottom of your **Data Table**.

Figure 3.11: Quantiles and Summary Statistics

Quantiles			Summary Statistics	
100.0%	maximum	83.529	Mean	69.47079
99.5%		83.529	Std Dev	10.417608
97.5%		81.9905	Std Err Mean	0.7460204
90.0%		80.15	Upper 95% Mean	70.942142
75.0%	quartile	76.966	Lower 95% Mean	67.999438
50.0%	median	72.902	N	195
25.0%	quartile	64.622		
10.0%		51.7618		
2.5%		44.7239		
0.5%		40.119		
0.0%	minimum	40.119		

Quantile is a generic term; you may be more familiar with *percentiles*. When we sort observations in a data set, divide them into groups of observations, and locate the boundaries between the groups, we establish quantiles. When there are 100 such groups, the boundaries are called percentiles. If there are four such groups, we refer to quartiles.

For example, we find that the 90th percentile is 80.15 years. This means that 90% of the observations have life expectancies shorter than 80.15 years. JMP also labels five quantiles known as the *five-number summary*. They identify the minimum, maximum, 25th percentile (1st quartile or Q1), 50th percentile (median), and 75th percentile (3rd quartile or Q3). Of the 195 countries listed in the data table, one-fourth have life expectancies shorter than 64.622 years, and one-fourth have life expectancies longer than 76.966 years.

Summary Statistics refer to the common descriptive statistics shown in Figure 3.11. At this stage in your study of statistics, three of these statistics are useful, and the other three should wait until Chapter 8.

- The *mean* is the simple arithmetic average of the observations, usually denoted by the symbol $\bar{X}$ and computed as follows:

$$\bar{X} = \frac{\sum_i X_i}{n}$$

  Along with the *median*, it is commonly used as a measure of central tendency; in a symmetric distribution, the mean and median are quite close in value. When a distribution is strongly left-skewed like this one, the mean will tend to be smaller than the median. In a right-skewed distribution, the opposite will be true.

- The *standard deviation* (**Std Dev**) is a measure of dispersion, and you might think of it as a typical distance of a value from the mean of the distribution. It is usually represented by the symbol s, and is computed as follows:

$$s = \sqrt{\frac{\sum_i (X_i - \bar{X})^2}{(n-1)}}$$

  We'll have more to say about the standard deviation in later chapters, but for now, please note that it must be greater than or equal to zero, and that highly dispersed variables have larger standard deviations than consistent variables.

- n refers to the number of observations in the sample.

Outlier Box Plots

Now that we've discussed the five-number summary, we can interpret the box plot just above the histogram back in Figure 3.7. The key to interpreting an outlier box plot is to recognize that it is a diagram of the five-number summary.

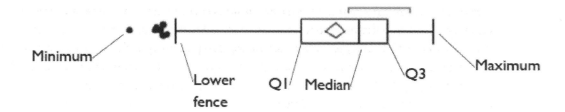

In a box plot there is a rectangle with an intersecting line. Two edges of the rectangle are located at the first (Q1) and third (Q3) quartile values, and the line is located at the median. In other words, the rectangular box spans the *interquartile range* (IQR). Extending from the ends of the box are two lines called *whiskers*. In a distribution that is free of outliers, the whiskers reach to the minimum and maximum values. Otherwise, the plot limits the reach of the whiskers by the *upper* and *lower fences*, which are located 1.5 IQRs from each quartile. In our distribution, we have a cluster of seven low-value outliers.

JMP also adds two other features to the box plot. One is a diamond that represents the location of the mean. If you imagine a vertical line through the vertices of the diamond, you've located the mean. The other two vertices are positioned at the upper and lower *confidence limits* of the mean. We'll discuss those in Chapter 11.

The second additional feature is a red bracket above the box. This is the *shortest half bracket*, representing the smallest part of the number line comprising 50% of the cases. We can divide the observations in half in different ways. The median gives the upper and lower halves; the IQR box gives the middle half. This bracket gives the shortest half.

A box plot very efficiently conveys information about the center, symmetry, dispersion and outliers for a single distribution. When we compare box plots across several groups or samples, the results can be quite revealing. In the next chapter, we'll look at such box plots and other ways of summarizing two variables at a time.

Application

Now that you have completed all of the activities in this chapter, use the techniques you've learned to respond to these questions.

1. *Scenario:* We'll continue our analysis of the variation in life expectancy at birth in 2010. Reset the Data Filter to show and include 2010.

    a. When we first constructed the **Life Exp** histogram, we described it as single-peaked and left-skewed. Use the hand tool to increase and reduce the number of bars. Adjust the number of bars so that a second peak appears. Describe what you did, and where the peaks are located.

b. Rescale the axes of the same histogram and see if you can emphasize the two peaks even more (in other words, have them separated distinctly). Describe what you did to make these peaks more distinct and noticeable.

c. Based on what you've seen in these exercises, why is it a good idea to think critically about an analyst's choice of scale in a reported graph?

d. Using the lasso tool, highlight the outliers in the box plot for **LifeExp**. Which continent or continents are home to the seven countries with the shortest life expectancies in the world? What might account for this?

2. *Scenario:* Now let's look at the distribution of life expectancy 25 years before 2010. Use the **Data Filter** to choose the observations from 1985.

a. Use the Distribution platform to summarize **Region** and **LifeExp** for this subset. In a few sentences, describe the distribution of **LifeExp** in 1985.

b. Compare the five-number summaries for life expectancy in 1985 and in 2010. Comment on what you find.

c. Compare the standard deviations for life expectancy in 1985 and 2010. Comment on what you find.

d. You'll recall that in 2010, the mean life expectancy was shorter than the median, consistent with the left-skewed shape. How do the mean and median compare in the 1985 data?

3. *Scenario:* The data file called **Sleeping Animals** contains data about the size, sleep habits, lifespan, and other attributes of different mammalian species.

a. Construct box plots for **Lifespan** and **TotalSleep**. For each plot, explain what the landmarks on each plot tell you about the distribution of each variable. Comment on noteworthy features of the plot.

b. Which distribution is more symmetric? Explain specifically how the graphs and descriptive statistics helped you come to a conclusion.

c. According to the data table, "Man" has a maximum life span of 100 years. Approximately what percent of mammals in the data set live less than 100 years?

d. Sleep hours are divided into "dreaming" and "non-dreaming" sleep. How do the distributions of these types of sleep compare?

e. Select the species that tend to get the most total sleep. Comment on how those species compare to the other species in terms of their predation, exposure, and overall danger indexes.

f. Now use the **Distribution** platform to analyze the body weights of these mammals. What's different about this distribution in comparison to the other continuous variables that you've analyzed thus far?

g. Select those mammals that sleep in the most exposed locations. How do their body weights tend to compare to the other mammals? What might explain this comparison?

4. *Scenario:* When financial analysts want a benchmark for the performance of individual equities (stocks), they often rely on a "broad market index" such as the S&P 500 in the U.S. There are many such indexes in stock markets around the world. One major index on the London Stock Exchange is the FTSE 100, and this set of questions refers to data about the monthly values of the FTSE 100 from January 1, 2003 through December 1, 2007. In other words, our data table called **FTSE100** reflects monthly market activity for a five-year period.

a. The variable called **Volume** is the total number of shares traded per month (in millions of shares). Describe the distribution of this variable.

b. The variable called **Change%** is the monthly change, expressed as a percentage, in the closing value of the index. When **Change%** is positive, the index increased that month; when the variable is negative, the index decreased that month. Describe the distribution of this variable.

c. Use the **Quantiles** to determine approximately how often the FTSE declines. (Hint: What percentile is 0?)

d. Use the **Chart** command to make a **Line Graph** (you'll need to find your own way to make a line graph rather than a bar chart) that shows closing prices over time. Then, use the Distribution platform to create a histogram of closing prices. Each graph summarizes the **Close** variable, but each graph presents a different view of the data. Comment on the comparison of the two graphs.

e. Now make a line graph of the monthly percentage changes over time. How would you describe the pattern in this graph?

5. *Scenario:* Anyone traveling by air understands that there is always some chance of a flight delay. In the United States, the Department of Transportation monitors the arrival and departure time of every flight. The data table **Airline Delays** contains a sample of nearly 15,000 flights for two airlines destined for four busy airports.

a. The variable called **Dest** is the airport code for the flight destination. Describe the distribution of this variable.

b. The variable called **Delay** is the actual arrival delay, measured in minutes. A positive value indicates that the flight was late and a negative value indicates that the flight arrived early. Describe the distribution of this variable.

c. Notice that the distribution of **Delay** is skewed. Based on your experience as a traveler, why should we have anticipated that this variable would have a skewed distribution?

d. Use the **Quantiles** to determine approximately how often flights in this sample were delayed. (Hint: What percentile is 0?)

6. *Scenario:* For many years, it has been understood that tobacco use leads to health problems related to the heart and lungs. The **Tobacco Use** data table contains recent data about the prevalence of tobacco use and of certain diseases around the world.

a. Use an appropriate technique from this chapter to summarize and describe the variation in tobacco usage (**TobaccoUse**) around the world.

b. Use an appropriate technique from this chapter to summarize and describe the variation in cancer mortality (**CancerMort**) around the world.

c. Use an appropriate technique from this chapter to summarize and describe the variation in cardiovascular mortality (**CVMort**) around the world.

d. You've now examined three distributions. Comment on the similarities and differences in the shapes of these three distributions.

e. Summarize the distribution of the region variable and comment on what you find.

f. We have two columns containing the percentage of males and females around the world who use tobacco. Create a summary for each of these variables and explain how tobacco use among men compares to that among women.

7. *Scenario:* The **States** data table contains measures and attributes for the 50 U.S. states and the District of Columbia.

a. Pop2010 is the population of the state as recorded by the 2010 United States census. Summarize the data in this column, commenting on the center, shape, and spread of the distribution. Note any outliers.

b. Construct boxplots for homeownership and poverty. For each plot, explain what the landmarks tells you about the distribution of each variable and comment on noteworthy features of the plot.

c. Income is the average income per capita, and med_income is the median family income in the state. Use an appropriate technique from this chapter to summarize the data in these two columns and comment on what you see.

d. Tr_deaths is the rate of traffic deaths per 100,000 persons in the state. Summarize the responses and comment.

e. Smoke is the percentage of smokers within the population of a state. Use an appropriate technique to summarize the distribution of this variable. Identify the outlying states and suggest a reason for the fact that these particular states are outliers.

[1] There are exceptions to this general principle, as with months of the year or days of the week, for example.

Describing Two Variables at a Time

Overview

Some of the most interesting questions in statistical inquiries involve *covariation*: how does one variable change when another variable changes? After working through the examples in this chapter, you will know some basic approaches to *bivariate* analysis, that is, the analysis of two variables at a time.

Two-by-Two: Bivariate Data

Chapter 3 covered techniques for summarizing the variation of a single variable: Univariate distributions. In many statistical investigations, we're interested in how two variables vary together and, in particular, how one variable varies in response to the other. For example, nutritionists might ask how consumption of carbohydrates affects weight loss; marketers might ask whether a demographic group responds positively to an advertising strategy. In these cases, it's not sufficient to look at one univariate distribution or even to look at the variation in each of two key variables separately. We need methods to describe the *covariation of bivariate* data, which is to say we need methods to summarize the ways in which two variables vary together.

The organization of this chapter is simple. We have been classifying data as categorical or continuous. If we focus on two variables in a study and conceive of one variable as a *response* to the other *factor*, there are four possible combinations to consider, shown in Table 4.1. The next three sections discuss the three of the four possibilities: We might have two categorical variables, two continuous variables, a continuous response with a categorical factor, or a categorical response to a continuous factor.

Table 4.1: Chapter Organization—Bivariate Factor-Response Combinations

	Continuous Factor	**Categorical Factor**
Continuous Response	Third section to follow	Second section to follow
Categorical Response	See Chapter 19	Next Section

In this chapter, we'll introduce several common methods for three ways to pair bivariate data. The first examples relate to a serious issue in civil (non-military) air travel: the periodic collisions between wildlife and commercial airplanes. According to the U.S. Federal Aviation Administration (FAA) so-called wildlife-aircraft strikes have cost hundreds of lives in the past century and account for significant financial losses as well in damage to aircraft. These collisions present environmental, public safety, and business issues for many interested parties. The FAA maintains a database to monitor the incidence of wildlife-aircraft strikes. For the years 2000 to 2009, the database contains nearly 60,000 reports of strikes in North America. The state reporting the largest number of events was California.

For this chapter, we will use a subset of the database, looking only at bird strikes associated with three California airports: Los Angeles International, Sacramento (the state capital), and San Francisco International. All of the available data is in the data table called **FAA Bird Strikes CA**. This data table contains 21 columns providing attributes for each of 2111 bird strikes at or near the three airports.

1. Open the **FAA Bird Strikes CA** data table now, and scroll through the columns. Our analysis will make use of several columns, each of which we'll explain as we work through the examples.

> This data table has a characteristic we have not seen previously: *missing data.*
> Although the FAA attempts to record a full set of variables for each strike, sometimes
> the data is not known to the individual reporting the incident. In a JMP data table,
> missing categorical data appears as a blank cell, and missing continuous data is a dot.
> In a bivariate analysis we think in term of a pair of observations for each bird strike.
> JMP will only analyze those incidents that have a complete pair of observations for
> whichever two columns we select.

Describing Covariation: Two Categorical Variables

At what point in a flight do bird strikes most often occur? Was it the same at all three
airports? In our data table, we have two variables that identify the airport (the **Airport
code** and **City**) and another ordinal variable identifying the **Phase of the Flight** at
which the strike happened. We can use JMP to investigate the covariation of these
categorical variables using a few different approaches.

1. Select **Analyze ▶ Distribution.** You should be quite familiar with this dialog by
    now. Select the **City** and **Phase of Flight** columns, cast them into the **Y** role, and
    click **OK**.

JMP produces two univariate graphs with accompanying frequency tables for each of the
two columns. In the default state, they reveal no information about patterns of
*co*variation, but there are a few important things to notice.

Wildlife strikes are rare during the taxi phase (in either direction between the terminal
and runway). Similarly, though the FAA database does contain strike reports while
flights are en route, these three airports did not report any such strikes. As a
consequence, the en route phase of flight does not even appear in the right-hand panel of
this graph.

In contrast, about 46% of the strikes occurred during the approach phase of the flight.
Clicking the Approach bar highlights the bar in the right-hand graph and also highlights
all approach-related observations in the left-hand graph. If the relevant dynamics are
similar at the three airports, then we'd expect somewhat less than half of each of the three
bars in the left chart to be darkened. However this is not the case for Los Angeles. We'll
explore this finding shortly.

2. Click the bar representing Approach (to the airport just prior to landing), we see
    something interesting, as shown in Figure 4.1.

Before going further, look at the frequency tables under each graph. Each table tallies the number of observations in each column category as well as the number of *missing observations*, and the number of distinct values (*levels*) within the column. The data table identifies the airport city for every bird strike, but only has data about the flight phase for 1,841 incidents. The other 270 are missing—they are unknown to history.

Figure 4.1: Two Linked Univariate Distributions

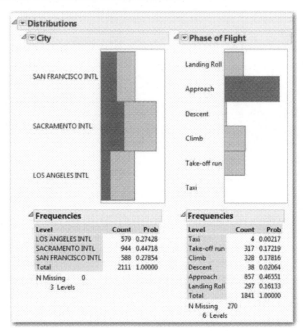

The issue of missing data is quite common in observational and survey data, though introductory courses often bypass it as a topic. In a bivariate analysis, we'll need two values for each observation. If one is present but the other is missing, then that observation will be omitted from the analysis.

It behooves the analyst to think about *why* a particular value is missing, and whether missing observations share a common cause. The very fact that observations are missing could be informative in its own right. For the current discussion, it is only important to realize that univariate analyses may include observations that are excluded in a related bivariate analysis.

3. We can generate a different view of the same data by using the **By** feature of the **Distribution** platform. Click the red triangle next to the word **Distributions** in

the current window and select **Script ▶ Relaunch Analysis**. This reopens the dialog box we used earlier. The two columns still appear in the **Y, Column** box.

4. Drag the column name City from its current position in the **Y, columns** box into the **By** box and click **OK**.

The ability to re-launch and modify an analysis is a very handy feature of JMP, encouraging exploration and discovery—and allowing for easy correction of minor errors. Experiment with this feature as you continue to learn about JMP.

5. Before discussing the results, let's make one modification to enhance readability. Click the uppermost red triangle icon in the new window, and select **Stack**. This orients the graphs horizontally and stacks all of the results vertically. Your results now look like Figure 4.2. On your screen, you need to scroll to see the results for San Francisco; in the figure below, we show only L.A. and Sacramento.

Figure 4.2: Distribution of Phase of Flight BY City

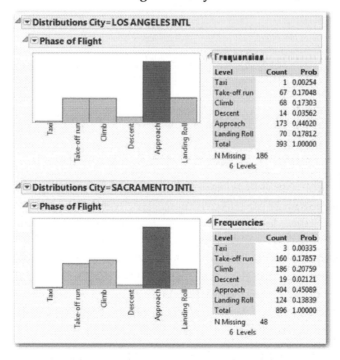

Figure 4.2 shows the results for two of the three airports. In general, the relative frequency of strikes is similar at both airports, though strikes during the Climb phase were more common in Sacramento than in Los Angeles, and just the opposite is true for strikes during the Landing Roll (the phase immediately after touching down, while the aircraft slows along the runway).

Based on Figure 4.1 we had the impression that strikes during the approach were less common in Los Angeles than in the other two airports. We now see that about 44% of the strikes in L.A. occurred during the Approach phase, compared to 45% for Sacramento and 51% for San Francisco. Hence, L.A. and Sacramento are really quite similar.

Why the difference between the two graphs? Missing observations. In L.A., we have Phase of Flight information about 393 strikes, but not for another 186 strikes. In other words, we have complete pairs of data for slightly more than two-thirds of all strikes (there were 393 + 186 = 579 strikes in L.A.). In contrast, the Phase of Flight was recorded for 95% of the strikes in Sacramento and 94% in San Francisco. Figure 4.1 essentially treated "unknown" as another phase of flight.

Another common way to display covariation in categorical variables is a *crosstabulation* (also known as a two-way table, a crosstab, a joint-frequency table, or a contingency table). JMP provides two different platforms that create crosstabs, and we'll look at one of them here.

6. Select **Analyze ▶ Fit Y by X**. Select **Phase of Flight** as **Y, Response** and **City** as **X, Factor** as shown in Figure 4.3.

Figure 4.3: Fit Y by X Contextual Platform

The **Fit Y by X** platform is contextual in the sense that JMP selects an appropriate bivariate analysis depending on the modeling types of the columns that you cast as Y and X. The four possibilities are shown in the lower left of the dialog box.

Figures 4.4 and 4.5 show the results of this analysis. We find a *mosaic plot*—essentially a visual crosstab—and a contingency table. In the mosaic plot, the vertical axis represents phase of flight, and the horizontal represents airports. The width and height of the rectangles in the mosaic are determined by the number of events in each category, and colors indicate the different phases of flight.

Figure 4.4: A Mosaic Plot of Wildlife Strikes at Three Airports

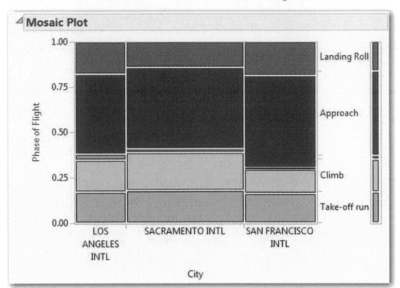

In this graph, it is quite clear that Sacramento had the greatest number of reported strikes that included phase of flight information, and L.A. had the fewest. Though the data table includes six distinct phases of flight, we see labels for only the four most commonly appearing in the data. Finally, the height of boxes is mostly consistent across cities, indicating that strikes tend to occur during the same phases of flight at the three airports. The exceptions to the general pattern are that strikes during the climb phase are relatively common in Sacramento, whereas strikes during the landing roll are less prevalent there. And, as noted earlier, San Francisco has a relatively high number of strikes during the approach phase.

The plot provides a clear visual impression, but if we want to dig into the specific numerical differences across regions, we turn to the Contingency Table (Figure 4.5). Again, in your course, you may be learning that tables like these are called crosstabs or crosstabulations, or perhaps joint-frequency tables or two-way tables. All of these terms are synonymous.

Figure 4.5: A Crosstabulation of the Bird Strike Data

Contingency Table

City	Phase of Flight						
Count / Total % / Col % / Row %	Taxi	Take-off run	Climb	Descent	Approach	Landing Roll	
LOS ANGELES INTL	1	67	68	14	173	70	393
	0.05	3.64	3.69	0.76	9.40	3.80	21.35
	25.00	21.14	20.73	36.84	20.19	23.57	
	0.25	17.05	17.30	3.56	44.02	17.81	
SACRAMENTO INTL	3	160	186	19	404	124	896
	0.16	8.69	10.10	1.03	21.94	6.74	48.67
	75.00	50.47	56.71	50.00	47.14	41.75	
	0.33	17.86	20.76	2.12	45.09	13.84	
SAN FRANCISCO INTL	0	90	74	5	280	103	552
	0.00	4.89	4.02	0.27	15.21	5.59	29.98
	0.00	28.39	22.56	13.16	32.67	34.68	
	0.00	16.30	13.41	0.91	50.72	18.66	
	4	317	328	38	857	297	1841
	0.22	17.22	17.82	2.06	46.55	16.13	

Across the top of this contingency table, we find the six values of the **Phase of Flight** column, and the rows correspond to the three airports. Each cell of the table contains four numbers representing the number of countries classified within the cell, as well as percentages computed for the full table, for the column, and for the row. For example, look at the lower-right cell of the table. The numbers and their meanings are as follows:

103	103 reported incidents were strikes during the landing roll at San Francisco (SFO).
5.59	5.59% of all 1841 reported incidents fall into this cell (landing roll at SFO).
34.68	34.68% of the 297 landing roll events occurred at SFO.
18.66	18.66% of the 552 events at SFO occurred during landing roll.

Describing Covariation: One Continuous, One Categorical Variable

We have just been treating **Phase of Flight** as a response variable, inquiring whether the prevalence of strikes during a particular phase varies depending on which airport was the site of the incident. With only minor differences, we found that strikes occur most often during the approach phase of the flight. We might wonder what it is about the approach phase that might account for the relatively high proportion of bird strikes—perhaps the altitude? The speed? With the data available to us here, we can begin to answer such a question. To illustrate, let's examine the distribution of flight speeds during the different flight phases. Bear in mind that the observational units in our data table are incidents involving wildlife strikes. We are not observing all flights, and in particular we aren't observing incident-free flights (which constitute the vast majority of air travel). Hence, we can only describe aspects of those flights that did strike birds.

1. Before performing a **Fit Y by X** analysis, open the **Graph Builder**. Drag **Speed** to the **Y** drop zone, and **Phase of Flight** to the **X** drop zone.

What do you see in your graph? As we might expect, the columns of jittered points tend to rise and fall with the familiar phases of flight. You might also notice that each of the speed distributions is asymmetric; there tend to be concentrations of many points at a relatively high or low speed. To visualize this more distinctly, we could make boxplots or histograms of Speed by Phase of Flight. Instead, let's learn a new type of graph:

2. In the menu bar at the top of the Graph Builder, click the **Contour** icon ( ). This will create a *violin plot*, so called because some of the resulting shapes (see Figure 4.6) look a bit like violins.

Figure 4.6: Violin Plot of Aircraft Speeds by Phase of Flight

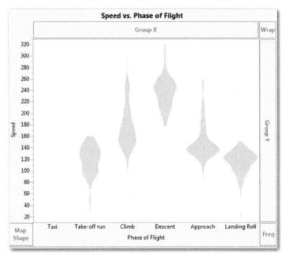

A violin plot shows the range of values for a variable; for example, those strikes which occurred during takeoff had recorded speeds between 30 and 160 mph. The narrow portion of the violin indicates very few takeoff strikes below 70 mph. The bulges in the violin indicate higher frequencies. It appears that strikes during takeoff are relatively common in the vicinity of 120 to 140 mph. Looking across all phases, one might say that except during descent (when strikes are uncommon), strikes are relatively frequent at speeds between approximately 120 and 160 mph regardless of flight phase.

For a more refined set of numerical summaries, do this:

3. Select **Analyze ▶ Fit Y by X**. Select **Speed** as the **Y** column and **Phase of Flight** as **X** and click **OK**.

4. From the earlier graphs, we know that the speed distributions are skewed. As such, quantiles provide a better summary than means. Hence, in the resulting **Oneway Analysis** report, click the red triangle and select **Quantiles**.

5. Below the graph should look like Figure 4.7.

Figure 4.7: Quantiles for Speed by Phase of Flight

Quantiles							
Level	Minimum	10%	25%	Median	75%	90%	Maximum
Take-off run	30	90	100	120	140	150	160
Climb	90	140	150	170	200	240	300
Descent	180	205	212.5	247.5	250	265	320
Approach	85	125	134.25	140	160	200	300
Landing Roll	20	90	100	120	130	137.1	150

The table reports seven quantiles for each flight phase. We readily see strong similarities between speeds at the time of strikes for take-off run and landing roll, phases when the aircraft is on the ground. Quantiles during climb and approach are also comparable, but speeds during descent are uniformly higher than other phases.

Describing Covariation: Two Continuous Variables

To illustrate the standard bivariate methods for continuous data, we'll now shift to a different set of data. In earlier chapters, we looked at variation in life expectancy around the world. We'll now look at data related to variation in birth rates and the risks of childbirth in different nations as of 2005. We'll rely on a data table with five continuous columns; two of them measure the relative frequency of births in each country, and the other three measure risks to mothers and babies around birth. Initially we'll look at two of the variables: the columns labeled **BirthRate** and **MortMaternal**. A country's annual birth rate is defined as the number of live births per 1,000 people in the country. The maternal mortality figure is the average number of mothers who die as a result of childbirth, per 100,000 births.

As we did in the previous sections, let's start by simply looking at the univariate summaries of these two variables.

1. Open the **Birthrate 2005** data table.

2. Select **Analyze ▶ Distribution**. Cast **BirthRate** and **MortMaternal** in the **Y** role and click **OK**.

3. Within your **BirthRate** histogram, select different bars or groups of bars, and notice which bars are selected in the maternal mortality histogram. The results should look much like Figure 4.8.

Figure 4.8: Linked Histograms of Two Continuous Distributions

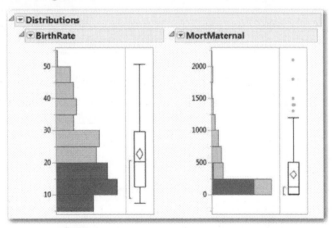

The general tendency is that countries with low birth rates also have low maternal mortality rates, but as one rate increases so does the other. We can see this tendency more directly in a *scatterplot*, an X-Y graph showing ordered pairs of values.

4. In the data table, press the Esc key to clear the de-select rows you selected by clicking on histogram bars. Then choose **Analyze ▶ Fit Y by X**. Cast **MortMaternal** as **Y** and **BirthRate** as **X** and click **OK**.

Your results will look like those shown in Figure 4.9.

Figure 4.9: A Scatterplot

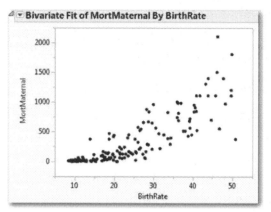

> By now you also have had enough experience with **Graph Builder** to know that you
> can easily create a similar graph with that tool. You should feel free to explore data
> with **Graph Builder**.

This graph provides more information about the ways in which these two variables tend
to vary together. First, it is very clear that these two variables do indeed vary together in
a general pattern that curves upward from left to right; maternal mortality increases at an
accelerating rate as the birth rate increases, though there are some countries that depart
from the pattern. A large number of countries are concentrated in the lower left, with low
birth rates and relatively low maternal mortality.

As we'll learn in Chapters 15 through 19, there are powerful techniques available for
building models of patterns like the one visible in Figure 4.9. At this early point in your
studies, the curvature of the pattern presents some unwelcome complications. Figure 4.10
shows another pair of variables whose relationship is more linear. We'll investigate this
relationship and meet three common numerical summaries of such bivariate covariation.

5.   Click the red triangle next to the word **Bivariate** in the current window and
     select **Script ▶ Relaunch Analysis**.

6.   Remove **MortMaternal** from the **Y, Response** role and replace it with **Fertil**.
     This will produce a scatterplot (seen in modified fashion below in Figure 4.10).

Earlier we noted that birth rate counts the number of live births per 1,000 people in a
country. Another measure of the frequency of births is the *fertility rate*, which is the mean
number of children that would be born to a woman during her lifetime in each country.

When we look at this relationship in a scatterplot, we see that the points fall in a distinctive upward sloping pattern that is generally straight. We can also calculate three different numerical summaries to characterize the relationship. Each of the statistical measures compares the pattern of points to a perfectly straight line. The first summary is the equation of a straight line that follows the general trend of the points (see Chapter 15 for a full explanation). The second summary is a measure of the extent to which variation in X is associated with variation in Y, and the third summary measures the strength of the linear association between X and Y.

7. In the scatterplot of **Fertility** vs. **Birthrate**, click the red triangle next to **Bivariate Fit** and select **Fit Line**.

8. Then click the red triangle again and select **Histogram Borders**.

Figure 4.10: Scatterplot with Histogram Borders and Line of Best Fit

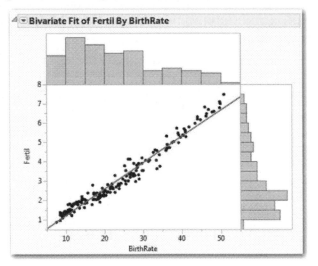

Now your results look like Figure 4.10. The consequence of these two customizations is that along the outside borders of the bivariate scatterplot, we see the univariate distributions of each of our two columns. Additionally, we see a red *fitted line* that approximates the upward pattern of the points.

Below the graph, we find the equation of that line:

Fertil = -0.111182 + 0.1360134*BirthRate

The slope of this line describes how these two variables co-vary. If we imagine two groups of countries whose birth rates differ by one birth per 1,000 people, the group with the higher birth rate would average about 0.136 more births per woman.

9. Below the linear fit equation you'll find the **Summary of Fit** table (not shown here). Locate the first statistic called **Rsquare**.

Rsquare (r^2) is a goodness-of-fit measure; for now, think of it as the proportion of variation in Y that is associated with X. If fertility were perfectly and completely determined as a linear function of birth rate, then r^2 would equal 1.00. In this instance, r^2 equals 0.967; approximately 97% of the cross-country variability in fertility rates is accounted for by differences in birth rates.

A third commonly used summary for bivariate continuous data is called *correlation*, which measures the strength of the linear association between the two variables. The coefficient of correlation, symbolized by the letter r, is the square root of r^2 (if the slope of the best-fit line is negative, then r is $-\sqrt{r^2}$). As such, r always lies within the interval [–1, +1]. Values near the ends of the interval indicate strong correlations, and values near zero are weak correlations.

10. Select **Analyze ▶ Multivariate Methods ▶ Multivariate**. Cast both columns (**BirthRate** and **Fertil**) as **Y** and click **OK**. The upper part of your results will look like Figure 4.11.

Figure 4.11: A Correlation Matrix

We find the correlation between birth rate and fertility rate to equal 0.9833, which is a very strong correlation. JMP uses a color scheme when reporting correlations. Strong positive values are in bright blue, and strong negative correlations are in bright red. Weaker values take on paler hues of blue and red, passing through gray near zero.

More Informative Scatter Plots

This chapter has introduced ways to depict and summarize bivariate data, but sometimes we want or need to incorporate more than two variables into an analysis. We already know that with **Graph Builder** we can color points to represent another variable, and we can wrap or overlay additional variables in a graph. We can use a **By** column in a **Fit Y by X** analysis to add a third factor into an investigation.

Another very useful visualization tool called a *bubble plot* provides a way to visualize as many as seven columns in one graph. This section provides a brief first look at bubble plots in JMP. We'll use the tool further in Chapter 5.

Think of a bubble plot as a "scatterplot plus…". In addition to the X-Y axes, the size and color of points can represent variables. Other variables can be used to interactively label points, and in cases where we have repeated measurements over time, the entire graph can be animated through time. Try this first exercise with the 2005 birth rate data:

1. **Graph ▶ Bubble Plot**. Cast **Fertil** as **Y** and **Birthrate** as **X**, just as before.

2. Next, cast **Country** as **ID, Region** as **Coloring,** and **Sizes** as **MortMaternal**. Click **OK**.

3. In the lower left of the **Bubble Plot** window, find the slider control next to **Bubble Size** and slide it to the left until your graph looks like Figure 4.12.

Figure 4.12: Enhancing a Scatterplot by Using a Bubble Plot

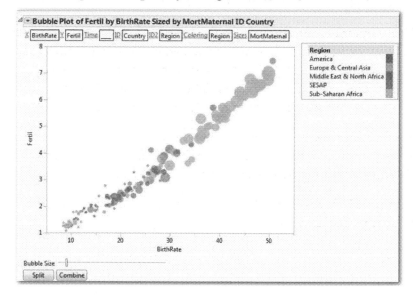

Compare this graph to Figure 4.10. The Y and X variables are the same, but this picture also conveys further descriptive generalizations; the highest birth and fertility rates are in sub-Saharan Africa where maternal mortality rates are also the highest in the world. As you move your cursor around the graph, you'll see popups identifying the points. You can also notice that countries with similar birth and fertility rates might have quite different maternal mortality rates (bubble sizes). Bubble plots pack a great deal of information into a compact, data-rich display.

Application

Now that you have completed all of the activities in this chapter, use the concepts and techniques you've learned to respond to these questions.

1. *Scenario:* We'll continue to examine the World Development Indicators data in **BirthRate 2005**. We'll broaden our analysis to work with other variables:

   - **Provider:** Source of maternity leave benefits (public, private or a combination of both).

   - **Fertil:** Average number of births per woman during child-bearing years.

   - **MortUnder5:** Deaths, children under 5 years per 1,000 live births.

- **MortInfant:** Deaths, infants per 1,000 live births.

  a. Create a mosaic plot and contingency table for the **Provider** and **Region** columns. Report on what you find.

  b. Use appropriate methods to investigate how fertility rates vary across regions of the world. Report on what you find.

  c. Create a scatterplot for **MortUnder5** and **MortInfant**. Report the equation of the fitted line and the rsquare value, and explain what you have found.

  d. Is there any noteworthy pattern in the covariation of **Provider** and **MatLeave90+**? Explain what techniques you used, and what you found.

2. *Scenario:* How do prices of used cars differ, if at all, in different areas of the United States? How do the prices of used cars vary according to the mileage of the cars? Our data table **Used Cars** contains observational data about the listed prices of three popular compact car models in three different metropolitan areas in the U.S. The cities are Phoenix, AZ, Portland, OR, and Raleigh-Durham-Chapel Hill, NC. The car models are the Chrysler PT Cruiser Touring Edition, the Honda Civic EX, and the Toyota Corolla LE. All of the cars are two years old.

   a. Create a scatterplot of price versus mileage. Report the equation of the fitted line, the rsquare value, and the correlation coefficient, and explain what you have found.

   b. Use the **Graph Builder** to see whether the relationship between price and mileage differs across different car models.

   c. Describe the distribution of prices across the three cities in this sample.

   d. Within this sample, are the different car models equally favored in the three different metropolitan areas? Discuss your analysis and explain what you have found.

3. *Scenario:* High blood pressure continues to be a leading health problem in the U.S. We have a data table (**NHANES**) containing survey data from a large number of people in the U.S. in 2005. For this analysis, we'll focus on only the following variables:

   - **RIAGENDR:** respondent's gender

   - **RIDAGEYR:** respondent's age in years

   - **RIDRETH1:** Respondent's racial or ethnic background

   - **BMXWT:** Respondent's weight in kilograms

- **BPXPLS:** Respondent's resting pulse rate
- **BPXSY1:** Respondent's systolic blood pressure ("top" number in BP)
- **BPXD1:** Respondent's diastolic blood pressure ("bottom" number in BP)

    a. Create a scatterplot of systolic blood pressure versus age. Within this sample, what tends to happen to blood pressure as people age?

    b. Compute and report the correlation between systolic and diastolic blood pressure. What does this correlation tell you?

    c. Use either a bubble plot to incorporate gender (Color) and pulse rate (Bubble Size) into the graph. Comment on what you see.

    d. Compare the distribution of systolic blood pressure in males and females. Report on what you find.

    e. Compare the distribution of systolic blood pressure by racial/ethnic background. Comment on any noteworthy differences you find.

    f. Create a scatterplot of systolic blood pressure and pulse rate. One might suspect that higher pulse rate is associated with higher blood pressure. Does the analysis bear out this suspicion?

4. *Scenario:* Despite well-documented health risks, tobacco is used widely throughout the world. The **Tobacco** data table provides information about the several variables for 133 different nations in 2005, including these:

- **TobaccoUse:** Prevalence of tobacco use (%) among adults 18 and older (both sexes)
- **Female:** Prevalence of tobacco use among females, 18 and older
- **Male:** Prevalence of tobacco use among males, 18 and older
- **CVMort:** Age-standardized mortality rate for cardiovascular diseases (per 100,000 population in 2002)
- **CancerMort:** Age-standardized mortality rate for cancers (per 100,000 population in 2002)

    a. Compare the prevalence of tobacco use across the regions of the world, and comment on what you see.

    b. Create a scatterplot of cardiovascular mortality versus prevalence of tobacco use (both sexes). Within this sample, describe the relationship, if any, between these two variables.

      c.   Create a scatterplot of cancer mortality versus prevalence of tobacco use (both sexes). Within this sample, describe the relationship, if any, between these two variables.

      d.   Compute and report the correlation between male and female tobacco use. What does this correlation tell you?

      e.   Create a bubble plot to modify your scatterplot from item c, above, to augment the display to incorporate region (color) and cardiovascular mortality (bubble size). Comment on what you find in the graph.

5.   *Scenario:* Since 2003 the U.S. Bureau of Labor Statistics has been conducting the biennial American Time Use Survey. Census workers use telephone interviews and a complex sampling design to measure the amount of time people devote to various ordinary activities. We have some of the survey data in the data table called **TimeUse**. Our data table contains observations from approximately 16,000 respondents in 2003 and 2007. Use the Data Filter to select and include just the 2007 responses.

      a.   Create a crosstabulation of employment status by gender, and report on what you find.

      b.   Create a crosstabulation of marital status by gender, and report on what you find.

      c.   Compare the distribution of time spent sleeping across the employment categories. Report on what you find.

      d.   Now change the data filter to include all rows. Compare the distribution of time spent on personal e-mail in 2003 and in 2007. Comment on your findings.

6.   *Scenario:* The data table **Sleeping Animals** contains information about the sizes, life spans, and sleeping habits of various mammal species. The last few columns are ordinal variables classifying the animals according to their comparative risks of being victimized by predators and the degree to which they sleep in the open rather than in enclosed spaces.

      a.   Create a crosstabulation of predation index by exposure index, and report on what you find.

      b.   Compare the distribution of hours of sleep across values of the danger index. Report on what you find.

      c.   Create a scatterplot of total sleep time and life span for these animals. What does the graph tell you?

    d.   Compute the correlation between total sleep time and life span for these animals. What does the correlation tell you?

7. *Scenario:* Let's return to the data table **FAA Bird Strikes CA**. The FAA includes categorical variables pertaining to the number of birds struck, the size of the birds struck, and the general weather conditions.

    a.   Create a crosstabulation of number of birds struck by sky conditions (**Sky**), and report on what you find.

    b.   Create a crosstabulation of number of birds struck by the precipitation conditions (**Precipitation**), and report on what you find.

    c.   Investigate the relationship between the number of birds struck and the speed of the aircraft. How do you describe that relationship?

    d.   Investigate the relationship between the number of birds struck and the height of the aircraft. How do you describe that relationship?

8. *Scenario:* The data table called **USA Counties** contains demographic, economic, commercial, educational and other data about each of the 3,143 counties in the United States as of 2010.

    a.   Create a scatterplot of median household income (Y) versus percent of the population with a bachelor's degree. Comment on what you see.

    b.   Compute and report the line of best fit for these data. Use that line to estimate the median household income in counties with 25% of the population holding bachelor's degrees.

    c.   Create and report on a scatterplot, between the percentage of households where a foreign language is spoken in the home (**foreign_spoken_at_home**) and the percentage of households with a foreign-born member (**foreign_born**). How do you explain the distinctive pattern in the graph?

    d.   Compute and explain the correlation coefficient for the two variables in item c above.

    e.   Estimate the line of best fit using the population as determined by the 2010 US Census as Y and the 2000 population count as X. Think about the slope of this line. What does it tell us about what happened to the average of US counties' populations between 2000 and 2010?

    f.   The point representing Cook County, Illinois is distinctive in that it lies below the red estimated line (2000 population was 5,194,675). According to this fitted line, what was unusual about Cook County in comparison to other counties of the United States?

Review of Descriptive Statistics

Overview

The prior four chapters introduced several foundational concepts of data analysis and have also led you through a series of illustrative analyses. In this chapter, we pause to pull together what we have learned about descriptive analysis. This is the first in a series of short review chapters, each of which shares the common goals of recapitulating recent material and calling upon you, the reader, to apply the principles, concepts and techniques that you've recently studied.

In the review chapters, you'll find fewer step-by-step instructions. Instead, there are guiding questions to remind you of the analytical process that you'll want to follow. Refer to earlier chapters if you have forgotten how to perform a task. The examples in this and later review chapters are based on the World Development Indicators, collected and published by the World Bank.

The World Development Indicators[1]

The World Bank was established in 1944 to assist with the redevelopment of countries after World War II. It has evolved into a group of five institutions concerned with interconnected missions of economic development. One important goal of its work is the alleviation of poverty world-wide, and as part of that mandate, the World Bank annually publishes the World Development Indicators (WDI) gathered from 214 nations. Earlier, we looked at some of the WDI data about birth rates and life expectancy.

In the public sector as well as in business, policy makers rely on accurate, current data to gauge progress and to evaluate the impact of policy decisions. The WDI data informs policy-making by many agencies globally, and the World Bank's annual data collection and reporting play an important role in the U.N.'s Millennium Development Goals.

Millennium Development Goals[2]

At the start of the millennium, the United Nations sponsored a Millennium Summit which led to the adoption of the Millennium Declaration by 189 member states. The declaration laid out an ambitious set of goals "including commitments to poverty eradication, development, and protecting the environment. Many of these commitments were drawn from the agreements and resolutions of world conferences and summits organized by the U.N. during the preceding decade"(2014).

A critical part of this project was the adoption of eight specific goals, "supported by 18 quantified and time-bound targets and 48 indicators, which became known as the Millennium Development Goals (MDGs)" (2014). Some of these indicators are identical to the World Bank's WDI's.

The goals, listed below, "guide the efforts of virtually all organizations working in development and have been commonly accepted as a framework for measuring development progress" (2014):

- Eradicate extreme poverty and hunger
- Achieve universal primary education

- Promote gender equality and empower women

- Reduce child mortality

- Improve maternal health

- Combat HIV/AIDS, malaria, and other diseases

- Ensure environmental sustainability

- Develop a Global Partnership for Development

We return to our introductory example from Chapter 1 about variation in life expectancies world-wide, and further investigate variables that might give us insight into why people in some regions tend to live long than in others. We will analyze some of the WDIs from the year 2010 to explore and to understand their variability, as well as plausible associations between variables. In doing so, we'll engage in an extended exercise in statistical reasoning and review several of the descriptive techniques that were presented in the first four chapters.

Questions for Analysis

In chapter 1, we speculated about possible factors that might contribute to variation in life expectancies. At that time, we wondered about the impacts of education, health care, basic sanitation, nutrition, political stability, and wealth.   Relying on several of the WDIs in our data, we will focus on four *constructs* as measured by the variables listed in Table 5.1. Notice that several of these variables are indirect measures of the constructs; statisticians sometimes refer to such variables as *proxies* to indicate that they "stand in" for a difficult-to-measure concept or attribute. Many of the WDIs are reported for all or nearly all of the nations, while others are reported more sparsely. For this chapter, we'll use indicators that are reported for the large majority of the 214 countries. We'll also examine life expectancies in different regions of the world.

Table 5.1 Ten Variables[3] Used in this Chapter

Construct	Variable
Wealth	Gross Domestic Product per capita Percent of population in rural areas[4] World Bank Income Group
Health Care	Health expenditures per capita Percent of children ages 12-23 months receiving DPT immuniza-tion

Construct	Variable
Education	Percent of eligible children enrolled in primary education
	Percent of eligible children enrolled in secondary education
	Percent of eligible children enrolled in tertiary (post-high school) education
Basic Sanitation	Percent of population with access to an improved water source
	Percent of population with access to improved sanitation facilities

Our research questions in this chapter are:

- How did life expectancy vary around the world in 2010?

- How did each of the variables listed in the table above vary in 2010?

- How do we best describe the co-variation between each factor and life expectancy?

Applying an Analytic Framework

In any statistical analysis, it is essential to be clear-minded about the research questions and about the nature of the data we'll be analyzing. We've mentioned these before, and this is a good time to review them in the context of a larger research exercise.

Data Source and Structure

We know these indicators are published by the World Bank for the purpose of monitoring economic development, identifying challenges, and assessing policy interventions. The figures are determined by analysts at the World Bank, for the most part using estimates gathered and provided by governmental agencies from each country. Each of the annual indicators is best described as **observational** data, as opposed to experimental or survey.

In this data table we have 40 variables, or data series, observed for 214 countries for each year from 1990 through 2011. Hence, we have 22 years of repeated observations for 214 countries, giving us 4,708 rows. Some cells of the table are empty, reflecting difference in national statistical infrastructure or, in some cases, simply reflecting periods before the World Bank began monitoring a particular development indicator.

Recall the difference between *cross-sectional* data (many observations selected at one time from a population or process), and *time series* or *longitudinal* data (regularly repeated observations of a single observational unit). In this table, we actually have a combination: we have 22 sets of repeated cross-sectional samples.

Observational Units

Each row in the data table represents a country in a particular year. We will be dealing with aggregated totals and rates rather than with individual people.

Variable Definitions and Data Types

Before we dive into analysis, it is critical to understand what each variable measures or represents, as well as the type of each variable. We need to understand the measurements in order to interpret them; we need to know the data types because that guides the choice of summary or descriptive techniques that are applicable.

Table 5.2 Data Types and Definitions for all Variables

Col. name	Extended Description	Data Type[5]
life_exp	Mean life expectancy of a baby born in observation year. Estimated number of years of life assuming no change in patterns of mortality from the year of birth onward.	C
Region	Each country is assigned to one of 7 regions.	N
gdp_pc	Gross Domestic Product per capita, expressed in constant (inflation-adjusted) 2000 U.S. dollars. Equals annual GDP divided by midyear estimate of population size.	C
pop_rural	Percent of population in rural areas. Determined by statistical standards of each nation.	C
Income Group	The Bank defines five groups of countries according to per capita income in each nation. Among the highest-earning countries, they distinguish between countries that are members of the Organisation for Economic Co-operation and Development (OECD) and those that are not. OECD members tend to be "developed" nations.	O
hlth_exp_pcap	Health expenditures per capita. This is the sum of all public and private expenditures expressed in 2006 purchasing power parity (PPP) rates, divided by national population for the year.	C
imm_dpt	Percent of children ages 12-23 months receiving DPT immunization. DPT immunization is a combination of vaccines to combat three serious infectious diseases, identified by their initials: diphtheria, pertussis, and tetanus. DPT immunization has proven effective in eradicating or reducing incidence of these life-threatening conditions in children.	C
prim_enroll	Percent of eligible children enrolled in primary education.[6]	C
sec_enroll	Percent of eligible children enrolled in secondary education.	C
ter_enroll	Percent of eligible children enrolled in tertiary (post-high school) education.	C
h20_safe	Percent of population with reasonable access to an adequate amount of water from an improved source.	C
sani_acc	Percent of population with at least adequate access to sanitation facilities for disposal of human waste.	C

Preparation for Analysis

This chapter relies entirely on data found in the JMP data table called **WDI**. Open the data table now. It contains 40 variables and 4,708 observations (214 countries times 22 years of data for each).

> In the chapter, we're going to work with about a dozen columns. Instructors may want to assign additional exercises incorporating other columns, and inquisitive readers may want to apply the approach of this chapter to further explore additional columns.

We can easily restrict our attention to the columns listed in Table 5 by just selecting the relevant columns as we work through menus. However, we also want to graph and compute statistics using only those rows containing 2010 data. For this, we'll apply the **Data Filter** and subset the table so that we are only working with the 2010 data[7].

1. From the **WDI** data table window, click **Row ▶ Data Filter**. Filter by **Year**, and **Select, Show,** and **Include** the 214 rows in which **Year** equals 2010.   Now minimize the **Data Filter**.

2. **Tables ▶ Subset**. The default setting is to use the **Selected Rows**, which is what we want. Click **OK**.

Univariate Descriptions

Our purpose in this analysis is to more fully understand the variability in life expectancy around the world by looking at how it does or does not co-vary with other indicators. As a first step, we'll look at the distributions of each variable while taking advantage of the way that JMP interconnects all open graphs. Most of the variables listed in Table 5.1 are continuous, so we can anticipate examining a series of histograms looking for center, shape, and dispersion of each variable. The first illustration also includes one ordinal variable, Income Group.

1. Select **Analyze ▶ Distribution**. Choose the following columns: **life_exp, Income Group, gdp_pc,** and **pop_rural**. Click **OK**. This will produce the four side-by-side analyses.

2. Hold down the CTRL key and click the red triangle next to **life_exp**. Choose **Histogram Options ▶ Shadowgram** to more clearly suggest the shape of each distribution.

3. Click the red triangle next to the column for the one ordinal variable. Choose

Histogram Options ▶ Separate Bars to indicate that each category is distinct from the others.

At this point, the graphs in the Distributions report should look like Figure 5.1.

Figure 5.1: Life Expectancy and Wealth Variables

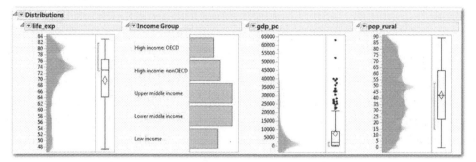

Among the three continuous variables, we immediately notice three quite different shapes: life expectancy is strongly skewed negatively (*i.e.* the long tail reaches to lower values) with multiple peaks. GDP per capita is also strongly skewed, but it is skewed positively, with numerous outliers. Finally, the percentage of the population in rural areas is generally symmetric with multiple small peaks, but an overall mound shape.

The distribution of **Income Groups** is approximately uniform, with a similar number of countries within each category. This is largely by design; "income group" is not a naturally occurring attribute, but rather a designation assigned by the World Bank based on gross national income per capita.

4. To facilitate further comparison, let's rearrange the Distributions so that the analysis of the four variables is stacked vertically and the graphs are oriented horizontally. Click the red triangle next to **Distributions**, and choose **Stack**.

Look back at the continuous variables. You should be able to compose a few brief sentences to characterize the center (mean, median) and dispersion (standard deviation, range, interquartile range) of each variable.

In the next section of this chapter we'll more systematically investigate how each of these indicators varies with life expectancy. Within this set of univariate distributions, we can get some inkling of the patterns of variation.

5. Before looking at the remaining variables, move your cursor into the **Income Group** bar graph, and click the bar corresponding to **Low income** countries. The

shadowgram for life expectancy will look like Figure 5.2.

Figure 5.2: Distribution of Life Expectancy with Low Income Nations Shaded

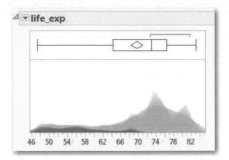

We see that the poorest countries make up a large portion of the lower tail of the distribution. Whereas the entire world has a large number of countries with life expectancies heavily concentrated roughly between 70 and 80 years, the lower income countries have a relatively flat distribution with three modal peaks, and they nearly all have life expectancies between 48 and 68 years.

6. Click through the other bars of the **Income Group** graph and observe the changes in the remaining graphs.

7. Now move on to the other sets of variables, repeating this analysis by generating groups of descriptive analysis for **life_exp** plus the variables measuring each of the constructs listed in Tables 5.1 and 5.2.

    Specifically, generate the distributions for these variable lists:

    *   **life_exp, hlth_exp_pcap, imm_dpt**
    *   **life_exp, prim_enroll, sec_enroll, ter_enroll**
    *   **life_exp, h20_safe, sani_acc**

Once again, examine the distributions for noteworthy features and for insights into possible relationships with life expectancy. In particular, consider these questions:

*   Which variables have symmetric distributions, and what might account for the symmetry?

*   Which variables have skewed distributions, and what might account for the skewness?

*   Which variable have distributions with multiple modes or peaks?

*   Which variables have an unusual number of outlying observations?

- Which variables appear to be most closely associated with variation in life expectancy?

Explore Relationships with Graph Builder

We are now ready to investigate the nature and strength of relationships between life expectancy and the ten other variables. In addition, because we've selected variables to represent four constructs, we'll find it useful to examine how pairs of these variables co-vary.

In Chapter 1, we became aware that life expectancy varies differently in different regions of the world. In this section, we will use region to color-code countries while investigating how other variables relate to life expectancy.

1. **Graph ▶ Graph Builder**. Drag **life_exp** to the **Y** drop zone.

2. Drag **Region** to the **Color** drop zone.

3. Drag **Income Group** to the **Group X** drop zone.

4. Finally, right-click and **Add** a **Box Plot** to the graph. The result will look like Figure 5.3.

Figure 5.3: Variability in Life Expectancy by Income Group

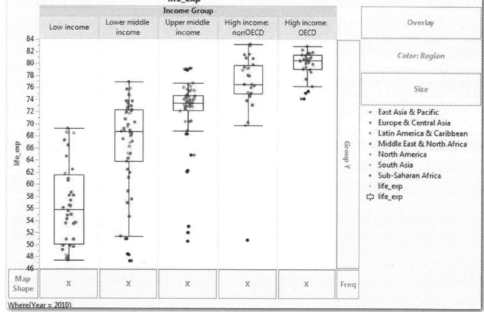

What do you notice in this graph? The most important features include:

- Life expectancy, indicated by the medians and the location of the boxes, generally increases with income group.

- The two low-income groups also show more dispersion; low income countries differ more than higher-income countries.

- The coloration of points seems to indicate an association between income level and region: for example, Sub-Saharan Africa dominates the lower-income groups and is mostly absent from the High Income groups.

The remaining nine variables are all continuous, and can all be analyzed in similar fashion. We'll illustrate the approach using GDP per capita, and leave the remaining analyses to the reader.

5. To clear the **Graph Builder** report, click the **Start Over** button in the upper left. Alternatively, you can replace the current variables by dragging a new column directly over the current on in the drop zones, or open a new **Graph Builder** from the menu.

6. Again drag **life_exp** to the vertical **Y** drop zone.

7. Now drag **gdp_pc** to the **X** drop zone on the horizontal axis.

By default, JMP places a *smoother* on the scatterplot to trace a pattern formed by the points. You can leave the smoother in place or remove it by clicking on the smoother icon, which is second from the left on the icon bar. In this graph (Figure 5.4) we see that there is fairly strong curvilinear relationship; countries with higher per capita GDP tend to have longer life expectancies, but the relationship is better described as a curve than as a line.

Figure 5.4: Relationship between Life Expectancy and GDP per Capita

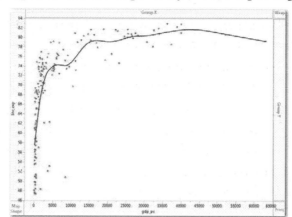

8. Drag **pop_rural** to the **X** drop zone and drop it directly over **gdp_pc** to replace the horizontal variable. How does life expectancy vary with the relative size of the rural population? Does this make sense to you?

9. Continue to swap out **X** variables proceeding through our list of candidates, as listed below. For each variable, make note of the form, direction, and strength of the relationship with life expectancy, and think about which graphs do or do not surprise you.

- **hlth_exp_prcap**
- **imm_dpt**
- **prim_enroll**
- **sec_enroll**
- **ter_enroll**
- **h20_safe**
- **sani_acc**

> JMP's Column Switcher provides a quick way to swap out variables. Click the red triangle next to Graph Builder, and choose Script ▶ Column Switcher to check it out.

By now you have rather quickly looked at ten relationships, nine of which were between pairs of continuous variables. These nine visual inspections should lead to the conclusions summarized in Table 5.3. We might generate graphical and numerical summaries to describe the covariation of life expectancy with any of these variables, but based on the preliminary explorations, it may be most useful to focus for now on the clear association between life expectancy and the ordinal variable **Income Group** and on the strongest linear associations: access to improved sanitation, secondary school enrollments, and the percentage of people living in rural areas.

Table 5.3: Summary of Graph Builder Explorations

Column	Form	Direction	Strength
gdp_pc	curvilinear	positive	strong
pop_rural	linear	negative	moderate
hlth_exp_pcap	curvilinear	positive	strong
imm_dpt	positive	linear (?)	moderate-weak
prim_enroll	none	vertical	weak
sec_enroll	linear	positive	strong
ter_enroll	curvilinear	positive	moderate-strong
h20_safe	linear	positive	moderate-strong
sani_acc	linear	positive	strong

Further Analysis with the Multivariate Platform

Recall from Chapter 4 that the Multivariate platform computes correlations and generates multiple small scatterplots for pairs of continuous columns. As we narrow our investigation, we can summarize the relationships using correlations.

1. **Analyze ▶ Multivariate Methods ▶ Multivariate**. Cast **life_exp**, **pop_rural**, **sec_enroll** and **sani_acc** into the **Y** role.

Your results should look like Figure 5.5.

Figure 5.5: Multivariate Summary of Three Relationships

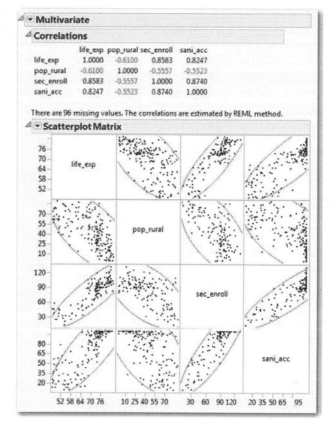

Secondary enrollments have the strongest correlation with life expectancy ($r = 0.8583$), followed by access to improved sanitation ($r = 0.8247$) and percentage of population living in rural areas ($r = -0.6100$). The red ellipses in the scatterplot matrix help to emphasize the comparative strength of the relationships. The matrix arrangement also makes it clear that there is also a strong linear relationship between secondary enrollments and access to sanitation. Thinking ahead, this set of variables will not enable us to logically make the case that education or sanitation is associated with longer life expectancy.

Further Analysis with Fit Y by X

To generate both graphical and numerical descriptive summaries for bivariate relationships, the simplest approach is to use the **Fit Y by X** platform. It takes into account the data types of whichever variables we select for the X and Y roles, and performs appropriate analysis.

1. Use **Analyze ▶ Fit Y by X** to cast life expectancy as **Y**, and income group, rural population percentage, secondary enrollment percentage and percent with access to improved sanitation as **X**. Note that you can select one Y and four X columns to generate four analyses at once.

Look at the four graphs in the report window. The left-most graph has an ordinal horizontal axis, with points (unjittered) vertically arrayed within each category. The other three graphs are scatterplots. For the first graph, we want to compute quantiles, means, and standard deviations by income group; for the others, a simple linear regression equation is the most suitable descriptive summary.

2. Click the red triangle next to **Oneway Analysis of life_exp by Income Group.** Select **Quantiles**. Click the hotspot again and select **Means and Std Dev**. This will produce the tables shown in Figure 5.6.

Figure 5.6: Summary Statistics of Life Expectancy by National Income Group

Quantiles

Level	Minimum	10%	25%	Median	75%	90%	Maximum
Low income	47.4022	47.95891	50.1318	55.86659	61.5708	68.43602	69.3
Lower middle income	47.36507	52.07633	63.83727	68.76483	72.30639	74.5348	76.90095
Upper middle income	50.65366	62.28668	72.26065	73.45854	74.68599	76.55141	79.19261
High income: nonOECD	50.8408	70.19062	74.98734	76.57322	79.7191	81.68512	83.15938
High income: OECD	74.20732	75.59268		79.1	80.56585	81.45122	81.86537	82.84268

Means and Std Deviations

Level	Number	Mean	Std Dev	Std Err Mean	Lower 95%	Upper 95%
Low income	36	56.6940	6.87477	1.1458	54.368	59.020
Lower middle income	51	66.7316	7.67118	1.0742	64.574	68.889
Upper middle income	49	71.9116	6.18207	0.8832	70.136	73.687
High income: nonOECD	28	76.3624	6.06286	1.1458	74.011	78.713
High income: OECD	31	79.9205	2.15687	0.3874	79.129	80.712

Notice that the measures of location (mean and quantiles) consistently show an association between rising life expectancy and rising income levels; high income non-OECD nations fare better than high-income OECD nations.

Now we want to find regression lines for the continuous relationships. You could use the hotspots to add a linear fit to each graph separately, but it's more efficient to do the following:

3. Hold the CTRL key on your keyboard (command key on a Mac) and click the red triangle next to **Bivariate Fit of of life_exp by pop_rural**. Select **Fit Line**.

> Pressing the CTRL key while clicking a red triangle applies the choice to all similar analyses open in the report window.

Within the results, you should locate three estimated linear fits with their associate R^2 values, as follows (I have rounded the estimated slopes and intercepts):

- life_exp = 80.86 – 0.26 (pop_rural) $R^2 = 0.39$
- life_exp = 47.24 + 0.30 (sec_enrolll) $R^2 = 0.73$
- life_exp = 48.93 + 0.28 (sani_acc) $R^2 = 0.70$

Summing Up: Interpretation and Conclusions

With observational data, we cannot draw definitive cause-and-effect conclusions. At this point in your study of statistics, we must limit ourselves to descriptive statements about patterns observed within the available data for 2010. There are many ways one might verbally summarize what we have seen in these analyses. Below are several statements that characterize the descriptive findings in this chapter:

- Across the World Bank income groups, there was a strong tendency for life expectancies in higher-income countries to exceed those in lower-income countries. In the well-off countries, life expectancies were consistently close to 80 years on average. Lower-income countries exhibited far more variability, but at the lowest end of the income scales, life expectancies were in the vicinity of 55 years.

- The strongest observed association ($r = 0.86$) in 2010 was between life expectancy and secondary school enrollments. Countries with one additional percentage point in secondary enrollment had a mean life expectancy 0.2995 years (3.6 months) longer than comparable countries with lower enrollment.

- We also saw a strong ($r = 0.83$) correlation between life expectancy and access to improved sanitation. Increased access to sanitation (one additional percent of a population) was associated with an increase in life expectancy of 0.277 years, or a little more than three months.

- Finally, there was a weak association between life expectancy and the percent of a country's population living in rural areas. Countries with comparatively large rural populations had shorter life expectancies. Each additional percent of rural population living was associated with a reduction of 0.259 years, or just over three months.

Visualizing Multiple Relationships

To conclude this chapter we briefly adopt a more ambitious descriptive task: visualizing six WDI variables in *one* screen graphic. In the previous section, we noted that life expectancy varied predictably with three different continuous variables in the year 2010. This in turn raises other questions: Were these patterns unique to 2010, or did they appear in other years? To what degree are, for example, secondary enrollments related to the relative size of the rural population? How does access to sanitation compare across income groups?

1. Return to the original WDI table that we opened at the beginning of the chapter. You may recall that we have been working only with rows of data from 2010, and this table contains observations from 1990 through 2011.

2. We can construct a bubble plot with the following variable assignments (see Figure 5.7 below): **Y** is **life_exp**, **X** is **sec_enroll**, , **ID** is **Country Name**, **Time** is **year**, bubble **Sizes** are determined by **pop_rural**, and **Coloring** is **Income Group**. Click **OK**.

Figure 5.7: Role Assignments for a Six-Variable Bubble Plot

3. You will now see a bubble plot with **life_exp** on the vertical axis and **sec_enroll** on the horizontal. The default bubble sizes are a little too large for optimal viewing of what comes next, so make them slightly smaller by grabbing the slider control next to **Bubble Size** in the lower left portion of the window, and slide it slightly leftward. The bubbles will shrink.

4. Just below the Bubble Size slider control you'll see three animation controls; press the "Play" button ▶ in the middle, and watch the patterns unfold over the years from 1990 through 2011.

This graph contains all of the variables we've investigated over the past several pages, but there are two major differences between this bubble plot and the previous visual displays. First, all variables are represented simultaneously by taking advantage of several visual *primitives*: position, size, and color. Second, the entire picture is animated over time, powerfully representing changes as they occurred. Hence this graphing tool opens the door for complex multivariate descriptive summaries of large volumes of data.

[1] World Bank History, http://go.worldbank.org/65Y36GNQB0, accessed June 21, 2013.

[2] World Bank Millennium Development Goals, http://data.worldbank.org/about/millennium-development-goals, accessed June 17, 2013.

[3] See Table 5.2 for fuller definitions of each variable.

[4] I have classified this measure—percent of population living in rural areas—as an indicator of wealth because poverty is often, but not always, associated with rural populations.

[5] C = Continuous, O = Ordinal, N = Nominal

[6] The three enrollment indicators are calculated by taking the total enrollment at each grade level divided by the number of children who are the appropriate age for that level. Accordingly, some countries have enrollment percentages in excess of 100% because of overage or underage children in school. Often, this occurs when students repeat grades.

[7] We could rely on the Data Filter alone in this chapter, but with so many tasks to follow, there is a chance that the reader might inadvertently clear the filter. With a subset of the table, all should go smoothly.

Elementary Probability and Discrete Distributions

6

Overview

The practical analysis of data occurs within the context of models and among the most valuable models come from the world of probability. Some of the patterns we observed in conducting descriptive analysis reflect or mimic patterns that are characteristic of specific probability models or theories. This chapter offers a brief introduction to a few probability models that will help build bridges to the statistical topics of later chapters.

The Role of Probability in Data Analysis

In the prior chapters, we've discussed ways to summarize collections of data visually and numerically. At this point, our attention turns from the analysis of empirical information—data collected through observation—to the consideration of some useful theoretical constructs and models. *Probability theory* is a large and useful body of material and is the subject of entire courses in its own right. In an elementary applied statistics course and in this book, probability is especially germane in two respects. First, sometimes the goal of data analysis is to approximate the likelihoods of events or outcomes that are important to a decision-maker. For example, people who plan communications or transportation networks want to know the likelihood of traffic at different network locations during the day.

Second, probability theory provides the bridge between purely descriptive analysis and *statistical inference*, or the practice of drawing general conclusions from a single sample. Whenever we rely on one sample to represent a population, we need to be aware that our sample is one of many possible samples we might have taken. Each sample imperfectly represents the parent population, and samples vary in their representativeness. Because it's impossible to know for certain just how representative any sample is, sampling inherently risks drawing erroneous conclusions about a population. To manage the risk, we'll (eventually) use probability theory.

In this chapter, we'll confine our attention to a few probability topics that pertain to these two uses. If you are taking an introductory statistics course, your principal text might include additional topics. The next section of this chapter selectively reviews some of the elementary vocabulary and rules of probability theory. It does not provide an exhaustive treatment, but recapitulates some of the terms and ideas covered by your major text.

Elements of Probability Theory

In the next section of this chapter, we'll look at some data collected by Professor Marianne Rasmussen of the University of Southern Denmark who observed the behavior of dolphins off the coast of Iceland. Let's use this activity—observing dolphins in their typical comings and goings—as an illustrative framework for defining some terms and concepts.

Professor Rasmussen's basic activity was to watch the behavior of dolphins at different times of day and to record their activities. She identified three basic activities: traveling quickly, feeding, or socializing. Dolphins tend to engage in these activities in groups, so she also tallied how many dolphins engaged in each activity. She also noted the time of day: morning, noon, afternoon, or evening.

Consider the activity of observing the behavior of one dolphin. In the language of probability, an *event* is an observable result of an activity. So, we might define the event *A* as that of observing a dolphin who is traveling quickly, or an event *B* as observing a feeding dolphin (by the same token, we could call these events *T* or *F*). In Dr. Rasmussen's study, we have three possible events of interest.

Similarly, we can define events as the outcomes of noting the time of day, and we have four possible events. For each of the two research activities in this study, we have pre-defined lists of possible events (traveling, feeding, or socializing and morning, noon, afternoon, and evening). For the purposes of this study, any time she spots a dolphin it will be doing one of the three defined activities, and she does her watching four times a day.

Probability of an Event

We'll assign a numerical value to the *probability of an event* as being proportional to its relative frequency of occurrence in the long run. For any event *A* we can write *Pr(A)* to represent this value. There are just two rules in assigning probabilities to events:

- The probability of an event is a value between 0 and 1, that is, $0 \pounds Pr(A) \pounds 1$. An event that is impossible has a probability of 0, and an event that is certain to happen has a probability of 1.

- For any activity, if *S* is the complete set of all possible outcomes, $Pr(S) = 1$. For the dolphins, this means that if Dr. Rasmussen sees a dolphin, it will surely be traveling, feeding, or socializing.

Relying on these two rules, we can always define an event $\overline{A}$ as the *complement* of event A. The complement of an event refers to all possible outcomes excluding the event, and $Pr(\overline{A}) = 1 - Pr(A)$. To illustrate, if event *A* is seeing a feeding dolphin, then event $\overline{A}$ is seeing a dolphin who is not feeding, or rather, one who is traveling or socializing.

Rules for Two Events

In many situations, we're interested in two events at a time, and ask questions about the interaction or co-occurrence of two events. This section briefly reviews several important definitions and possible relationships between two events.

- **Mutually exclusive events:** If two events cannot possibly occur together, we call them mutually exclusive or *disjoint*. In the dolphin study, socializing and traveling are mutually exclusive.

- **Joint probability:** Joint probability refers to the simultaneous occurrence of two events. Often it is expressed as the probability of events *A and B* occurring together, or the probability of the *intersection* of events *A* and *B*.

- **Union of two events:** The union of two events refers to the situation in which one or the other event occurs, including their joint occurrence. Often it is expressed as *A or B*.

- **Conditional Probability:** Sometimes the occurrence or non-occurrence of event *B* affects the likelihood of event *A*. Conditional probability is the probability of one event under the condition that another event has occurred. For example, in the dolphin study, let's designate event *A* as "observing a dolphin who is feeding." In general, there is some probability at any time of observing a feeding dolphin; call it *Pr(A)*. We might ask whether that probability is the same at different times of day. For instance, we might want to know the probability of observing a feeding dolphin if we are looking at noon. Customarily, we'd speak of the conditional probability of feeding *given that* it is noon, and we'd write this as *Pr(A|B)* or *Pr(Feeding|Noon)*.

  NOTE: Conditional probability is different from joint probability. When we ask about the probability that we observe a dolphin feeding *and* it is noon, the implication is that we don't know either the behavior or the time. When we ask about the probability of observing a feeding dolphin *given that* it is noon our situation is different. We know that it's noon, but we don't yet know what the dolphin is doing.

- **Independence:** Two events are *independent* if the occurrence of one provides no information about the occurrence of the other. More formally, if $P(A \mid B) = P(A)$, we say that events A and B are independent. So, if dolphins are more likely to eat at one time of day than another, we'd find that the event feeding is not independent of the event noon.

Assigning Probability Values

We've learned that a probability value is a number between 0 and 1, and it's time to discuss how we determine which value should attach to which events. There are three basic methods for assigning probabilities—they are known as the Classical, Relative

Frequency, and Subjective methods. The approach that we use depends on the circumstances of the particular activity we're conducting.

- The *Classical* method applies if we know and can list the possible outcomes of an activity, and we can assume that all basic outcomes share the same likelihood. If there are n equally likely outcomes, then the probability of each outcome is $1/n$. Among other things, the classical method is applicable to games of chance (like card games) and to activities like simple random sampling.

  Computations using classical probabilities often refer to events that can happen in multiple ways. So, for example, when rolling two dice, we can think of the elementary outcomes for each die as being the six distinct faces. In dice games, we sum the spots on the two dice. In the context of the game, the events of interest—rolling a seven, or doubles, or eleven—are not equally likely, but the fundamental outcomes for each die do share the same probability of 1/6.

- The *Relative Frequency* method applies to situations in which we know and can list the possible outcomes, but cannot assume that the outcomes are equally likely. This method relies on empirical observation of a large number of repetitions of the situation, and then computing relative frequencies. Observing dolphin behavior is one such situation. Other examples include rates of disease incidence, batting averages, and college graduation rates.

- The *Subjective* method applies to situations in which we lack information about possible outcomes or their relative frequency or both. In such instances, expert opinion forms the basis for probability assignments. Problems requiring expert opinion are beyond the scope of an introductory course, and will receive no further attention here.

For elementary classical problems involving one or two events, a calculator is the tool of choice. For relative frequency problems, we are well served to use software like JMP to construct univariate frequency tables or bivariate contingency tables. We'll use the dolphin example to illustrate.

Contingency Tables and Probability

1. Select **File ▶ Open**. Select the **Dolphins** data table.

   This data table (Figure 6.1) is different from those we've seen before. The rows here actually represent the behavior of 189 dolphins, but consolidate observations of the same time and behavior into a single row. The column Groups shows the *joint frequency* of each pair of time and behavior events. For instance, throughout the entire observation period, Dr. Rasmussen observed six dolphins traveling in the morning and 28 feeding in the morning.

Figure 6.1: Dolphins Data Table

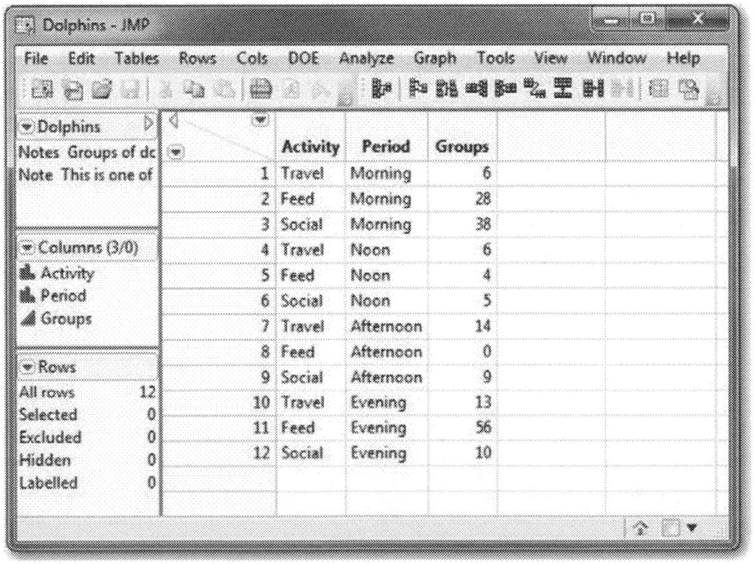

When we analyze these columns, we'll just need to specify that the rows represent frequencies rather than individual observations.

2. Select **Analyze ▶ Fit Y by X**. You have worked with this dialog (not shown) many times by now. With this particular data table, we'll make use of the **Freq** role for the first time.

3. Cast Activity as **Y** and Period as **X** and then cast Groups as **Freq**.

We could have reversed the X and Y roles, but here we want to think about the logic of the situation. It's reasonable to think that animals naturally incline toward different behaviors at different times of day. In terms of dependence and independence, it makes more sense to ask if a dolphin's behavior depends on the time of day than it does to ask if the time of day depends on the behavior.

Because we're interested in assigning probability values, we'll focus first on the contingency table in the output report. Look closely at the table, which is reproduced with annotations in Figure 6.2.

Figure 6.2: Contingency Table for Dolphins Data

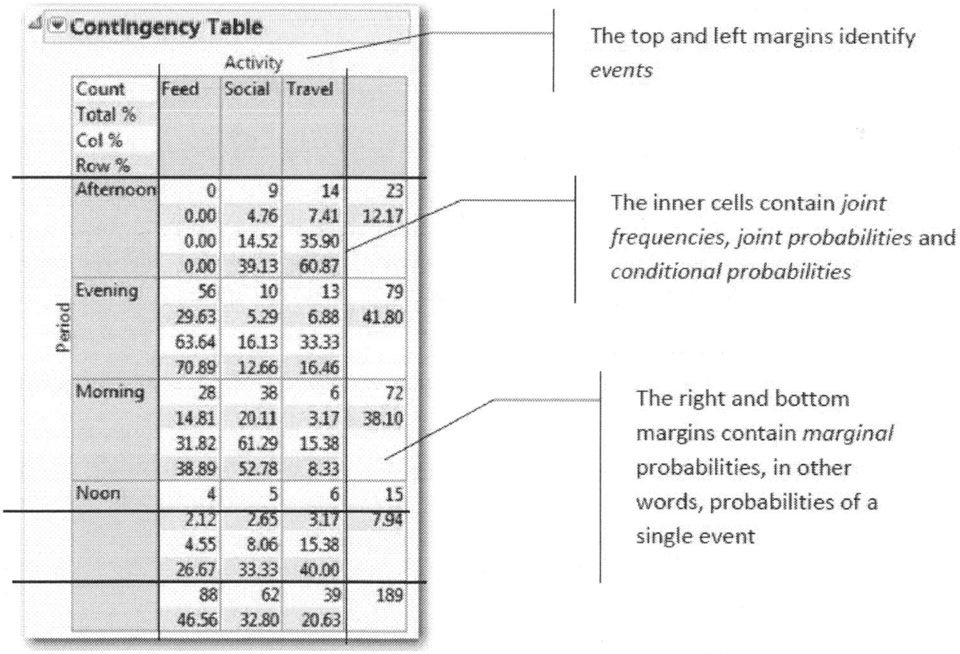
Contingency Table
Activity
Period
The top and left margins identify *events*
The inner cells contain *joint frequencies, joint probabilities* and *conditional probabilities*
The right and bottom margins contain *marginal* probabilities, in other words, probabilities of a single event

Period	Count / Total % / Col % / Row %	Feed	Social	Travel	
Afternoon	Count	0	9	14	23
	Total %	0.00	4.76	7.41	12.17
	Col %	0.00	14.52	35.90	
	Row %	0.00	39.13	60.87	
Evening	Count	56	10	13	79
	Total %	29.63	5.29	6.88	41.80
	Col %	63.64	16.13	33.33	
	Row %	70.89	12.66	16.46	
Morning	Count	28	38	6	72
	Total %	14.81	20.11	3.17	38.10
	Col %	31.82	61.29	15.38	
	Row %	38.89	52.78	8.33	
Noon	Count	4	5	6	15
	Total %	2.12	2.65	3.17	7.94
	Col %	4.55	8.06	15.38	
	Row %	26.67	33.33	40.00	
		88	62	39	189
		46.56	32.80	20.63	

We'll think of the table as having three main regions, as indicated in Figure 6.2. This particular study defined seven different events, which are all named in the left and upper margins of the table. To find the probability (relative frequency) of any one event, we look in the opposite margin. For instance, 88 of 189 dolphins were observed to be feeding. The relative frequency was 46.56%, so we find that *Pr(Feed)* = 0.4656[1].

Within the interior cells, the upper number is the joint frequency showing the number of times that Dr. Rasmussen observed a particular behavior at a particular time. The next number is the *joint probability*. In her observations, she did not see any dolphins feeding in the afternoon, but she did see dolphins feeding in the evening. The joint probability *Pr(Feed and Evening)* = 0.2963.

The upper-left cell of the table reports that the next two numbers within each cell are the column percentage and the row percentage. In terms of assigning probabilities, these are *conditional* values. A column percentage is the relative frequency of the row event within the column only. In other words, restricting ourselves to the condition represented by the column, it is the relative frequency of each row. Similarly, the row percentages give the probability of the column events conditioned by each row.

So, *Pr(Noon | Feed)* = .0455, but *Pr(Feed | Noon)* = 0.2667.

With this contingency table, we can easily check for independence of any two events by comparing the conditional probability *Pr(A | B)* to the marginal probability *Pr(A)*. If they are equal, the events are independent; if they are unequal, the events are not independent. So, for instance, we've just noted that *Pr(Feed | Noon)* = 0.2667. Our earlier glance at the bottom margin of the table shows that *Pr(Feed)* = 0.4656. Since these two probabilities are unequal, we'd conclude that feeding behavior is *not* independent of the noon condition.

Looking at the mosaic plot (not reproduced here) you might notice that feeding is much more common at evening than at any other time of day, and we might be tempted to conclude that dolphins tend to eat in the evening. Because we're looking at sample data rather than at definitive information about all dolphins at all times, this is an oversimplification, but it is a useful guide at this point in the course. In Chapter 12, we'll see a more sophisticated way to interpret these results.

Discrete Random Variables: From Events to Numbers

Thus far in this chapter, we have framed the research questions in terms of categorical variables. In some instances, the research variables are numerical rather than categorical: during a flu outbreak, how many children in the school system will become ill? In a random sample of customer accounts, how many billing errors will we find? In such cases, we're focusing on a *discrete* variable—one that has countable integer values. If we're talking about a flu outbreak, we may find 0, 1, 2, 3... sick children and each of these values may occur with a different probability.

If the variable of interest takes on discrete numerical values rather than simple categories, we'll refer to the *probability distribution* of the variable. In practice, we can often use one of several *theoretical discrete distributions* to build a model of such situations. It is well beyond the scope of this book to provide a full treatment of discrete random variables. Refer to your principal text for a full discussion of probability distributions. The following two sections assume that you are familiar with the concept of a random variable, you know which theoretical distribution you want to apply to a problem, and you want to use JMP either to calculate probability values for a random variable or to generate simulated random data conforming to a known discrete distribution. The goal here is to introduce how JMP handles discrete distributions, but not to provide an exhaustive treatment of the subject.

Three Common Discrete Distributions

There are quite a large number of theoretical discrete distributions. This section and the next demonstrate how JMP implements only three of them, but others work similarly. In this section, we'll see how to build a data table that computes probability values for each distribution; in the next section we'll simulate random observations that conform to each distribution.

For these illustrations, we'll consider three random variables, each of which can assume values starting at 0 up to a maximum value of 20. To begin, we'll set up a new data table whose first column is a list of integers from 0 to 20.

1. Select **File ▶ New ▶ Data Table** (or Ctrl-N).

2. Double-click the word **Untitled** in the table panel and type in a new name for this data table: Discrete Distributions.

3. In the **Columns** panel, highlight **Column 1**. Right click and choose **Column Info**. Type the **Column name** Value k and click the **Missing/Empty ▼** button and choose **Sequence Data** as shown below in Figure 6.3.

Figure 6.3: Creating a Column of Consecutive Integers

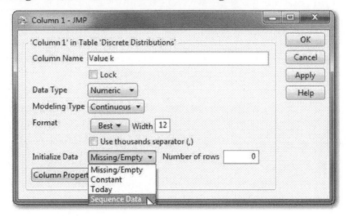

4. This expands the dialog box adding some new options. Replace the default value of **From** with 0, and replace the **To** value with 20.

Integer Distribution

The integer distribution (sometimes called the *uniform discrete distribution*) applies to situations in which there are *n* numerical outcomes of consecutive integers, each of which is equally likely. A roulette wheel is a good example of a process generating values following an integer distribution.

In our example, we'll consider an integer random variable that assumes values of 0 to 20. With 21 possible values and the characteristic of equal probabilities, we don't need software to tell us that for each value the probability equals 1/21. Note that if we know the number of possible values (*n*) we can do the computation; we refer to *n* as the *parameter* of the integer distribution. We'll start here though to demonstrate how JMP handles such calculations for simple and not-so-simple formulas.

1. Select **Cols ▶ New Column**. Rename **Column 2** as Integer; click the **Column Properties**, select **Formula**. This will open the **Formula Editor** dialog.

2. As shown below in Figure 6.4, within the red rectangle that initially contains the words no formula, type 1/21 and then click anywhere in the formula editor dialog box that is outside the blue outline where you just typed.

Figure 6.4: Using the Formula Editor to Create a Simple Formula

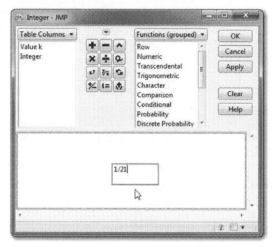

3. The formula editor changes appearance to show your typing as a numerator over a denominator. Click **OK** twice. The second column now shows the probability of each possible value of the random variable in column 1.

Binomial

Binomial distributions describe random variables that represent the number of times we observe a binary outcome in repeated attempts. More specifically, a binomial variable is the number of observed outcomes ("successes") in n trials where the following is true:

- Each trial is conducted in the same fashion.

- There are just two complementary outcomes possible on each trial, traditionally called *success* and *failure*. A success is the outcome of interest to the investigator.

- The probability, p, of the outcome of interest is the same each trial.

- Each trial is independent of the one before.

Imagine a survey researcher who intends to email 20 volunteers, asking each to complete a short online survey. She wants some idea of how many volunteers might proceed to take the survey. Imagine further that the probability that each volunteer takes the survey is 0.30 (researchers should be so lucky), and that all volunteers are strangers who do not consult with each other. The number of survey-takers is a binomial variable.

The computation of binomial probabilities is far more complex than for integer probabilities, but JMP does all of the work for us once we specify the two parameters of the binomial distribution, *n* and *p*. Following our hypothetical example, let's consider a binomial random variable with 20 trials and probability 0.3 of success.

1. Set up a third column (**Cols ▶ New Column**) named **Binomial**. Again, create a formula for this column just as you did in the previous step.

2. Instead of immediately typing in the formula, we'll use a built-in function. In the **Functions** list of the formula editor, scroll down to find **Discrete Probability** and select **Binomial Probability** (see Figure 6.5).

Figure 6.5: Locating the Discrete Probability Functions in the Formula Editor

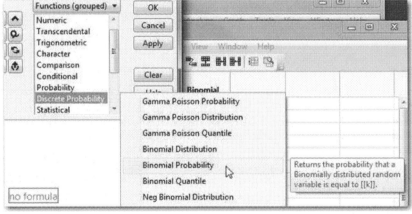

The binomial probability function (Figure 6.6) has three arguments: the probability of success on each trial, the number of trials, and the particular value of the variable to be evaluated.

Figure 6.6: The Binomial Probability Function

Binomial Probability(p, n, k)

3. In the red box labeled **p**, type the probability of success **.3** and press ENTER.

4. Select the box labeled **n**, and type the number of trials **20** and press ENTER.

5. Now select the box labeled **k**. Rather than typing a value, move the cursor to the **Table Columns** list in the upper left and click once on Value k. Click **OK** in the formula editor and **OK** in the **Column Info** dialog box.

The Binomial column now contains the probabilities corresponding to each possible value, *k* of the random variable. Reading down the list of probabilities, it becomes clear that with these particular parameters of *n* and *p*, our survey researcher is more likely to obtain, say, 5 or 6 respondents than she is to succeed with 15 or 16. Because each value of *k* is mutually exclusive with other values, we can add probabilities to determine the likelihoods of any range of specific number of successes. For example, the probability of 2 or fewer = 0.0008 + 0.0068 + 0.0278 = 0.0354. There is less than a 0.04 probability that the researcher will succeed with 2 or fewer volunteers; if that were to happen, the researcher would be surprised and disappointed. Similarly, there is only about a 0.05 probability that more than 9 volunteers will complete the survey. If 10 or more complete the survey, the researcher should also be surprised and delighted.

> Had we wanted to JMP to add up the probabilities, we would compute *cumulative probabilities*—the probability of finding *k* or fewer successes in *n* trials—and we could have used the **Binomial Distribution** function instead of the **Binomial Probability** function.

Poisson

Similar to the binomial distribution, the Poisson distribution refers to the number of times an event occurs. Unlike the binomial, the Poisson refers to the number of occurrences over a continuous span of time or distance rather than over a discrete number of trials. Because of this difference, a Poisson-distributed variable has no inherent maximum. In other respects, its characteristics are the same as the binomial; in a

Poisson distribution, the likelihood of a success is identical in all time periods, and one period is independent of the others.

A Poisson distribution has a single parameter, conventionally known as λ (lambda), which is equal to the mean number of occurrences within one unit of time or distance. For the sake of this example, we'll compute probabilities for a Poisson random variable with a mean rate of occurrence of $\lambda = 6$.

1. As you've done before set up another new (fourth) column and name it **Poisson**.

2. Click **Column Properties** and select **Formula**.

3. In the **Formula Editor** select **Poisson Probability** from among the **Discrete Probability** functions.

4. Within the function, replace **lambda** with **6**, and again select **Value k** as the argument **k**. Click **OK** in both open dialog boxes, and look at your data table.

Notice that the values with the highest probabilities are $k = 5$ and $k = 6$. Though there is no maximum possible value, the probability of 20 successes is listed as 3.7250619–6, or a tiny probability of 0.0000037. In other words, 20 successes would be nearly impossible.

The integer, binomial, and Poisson distributions are commonly included in introductory statistics courses. We have seen that the list of discrete distribution functions is quite long, but this brief introduction should give you some sense of how we can work with theoretical distributions. In the next section of this chapter, we see how JMP can be used to generate random values drawn from a specific underlying distribution. Simulation can be helpful in understanding some characteristics of random variation, and has applications in decision-modeling and design as we will see in upcoming chapters.

Simulating Random Variation with JMP

In this section we'll use the binomial and Poisson distributions to simulate random samples of 1,000 observations. We'll open a new data table and populate it with artificial simulated observations.

1. Select **File ▸ New ▸ Data Table**. Create a new data table and name it Simulated Data.

2. Click the title Column 1, right-click and choose **Column Info**. Change the Column Name to Observation. This column will just identify the observations. Refer back to Figure 6.3 to create a column of sequential integers from 1 to 1,000.

3. Create a new column and name it Integer.

4. Click the **Column Properties** button, choose **Formula**.

5. Now, from among the functions choose **Random ▶ Random Integer**.

6. Specify that *n1* = 20 click **OK** in both dialog boxes.

7. Next create a new column and name it Binomial.

8. Click the **Column Properties** button, choose **Formula**.

9. Now from among the functions choose **Random ▶ Random Binomial**.

10. Specify that *n* = 20 and *p* = 0.3, and click **OK** in both dialog boxes.

Before examining the simulated data, we'll simulate some Poisson observations. By this point, you should be able to create a new column on your own by retracing the steps you've just completed.

11. When you are ready to enter the formula, select the **Random Poisson** function with a parameter of **lambda** equal to 6.

12. Finally, we will examine the pattern of variation in the simulated data generated by these three probability models. **Analyze ▶ Distribution**. Cast the Integer, Binomial and Poisson columns as **Y**, and click **OK**. The histograms should look similar to Figure 6.7.

Figure 6.7: Comparison of Simulated Sets of Binomial and Poisson Random Data

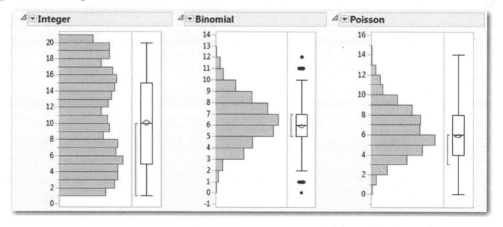

The Y-axis represented the number of successes in each simulated observational unit. Because this is a simulation, your results could look different from the graph shown here, but the overall pattern in the distribution should be similar. Look at these histograms, especially noting the similarities and differences in their shapes. These graphs illustrate the long-run nature of random variation; even though it is impossible to predict the value of any single observation, we can reliably characterize the long-run pattern for a random variable. In the Integer graph, all outcomes occur with very similar (though not identical) frequencies. In contrast, in both the binomial and Poisson distributions (for example) 4 to 8 successes occur frequently, and more than 12 successes are possible, but quite rare.

Discrete Distributions as Models of Real Processes

The Pipeline and Hazardous Materials Program (PHMSA) is an office of the U.S. Department of Transportation. PHMSA monitors all disruptions in the natural gas pipelines supplying natural gas throughout the United States, maintaining detailed records about all major incidents. We have some PHMSA data in the **Pipeline Safety** data table for the period from 2003 through 2008. Each row of the table represents an incident in which a natural gas pipeline was somehow disrupted.

The last column of the table represents the day of the year (1 through 366) on which the disruption occurred. Figure 6.8 is a histogram of the frequency with which each day of the year appears within the table; disruption events occurred on 268 different days of the year during the period covered by the data table meaning that there were 366 – 268 = 98 days of the year with no disruptions.

Figure 6.8: Pipeline Disruptions per Day of the Year (2003 through 2008)

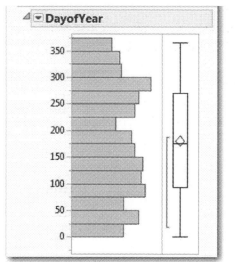 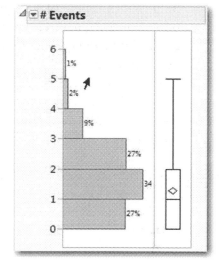

Comparing this distribution on the left to the Uniform distribution shown in the left-most panel of Figure 6.7, we could provisionally suggest that pipeline disruptions occur with almost equal likelihoods on any day of the year. Using this observed information, the sample mean of events per day is 468/368 – 1.2787 disruptions per day. Might a Poisson model be a reasonable approximation of this process?

To answer that question, we could look at the frequencies from a different perspective. Using JMP, I tabulated the number of events occurring each day during the sample period, and then counted the number of days on which 0, 1, 2… events occurred. The right-hand panel of Figure 6.8 displays that distribution. Within the sample, there were no disruptions at all on 27% of the days. If we use JMP to compute the probability of 0 successes in a Poisson distribution with $\lambda = 1.2787$, we find a theoretical probability of 0.2784, or 27.8% compared to the observed relative frequency of 27%. To this limited extent, the theory matches closely to the empirical observations.

In the next few chapters, we'll make further use of simulated data to help us visualize important statistical concepts and develop deeper understanding of the ideas, tools, and techniques of statistical investigations. Before that, though, we'll extend our thinking about probability distributions to include *continuous* variables.

Application

Now that you have completed all of the activities in this chapter, use the concepts and techniques you've learned to respond to these questions.

1. *Scenario:* In these questions, we return to the **NHANES** data table containing survey data from a large number of people in the U.S. in 2005. For this analysis, we'll focus on just the following variables:

   - **RIDRETH1:** Respondent's racial or ethnic background.

   - **DMDMARTL:** Respondent's reported marital status.

   For each of the following questions, use JMP to create a suitable contingency table, and report on the probability of sampling one person that satisfies the specified events. What is the probability that we sample an individual who reported being the following:

   a. Mexican American?

   b. Mexican American or never married?

   c. Mexican American and never married?

   d. Never married, given that they are Mexican American?

   Finally, use your responses to consider the question of independence:

   e. Are the events "Never married" and "Mexican American" independent? Explain.

2. *Scenario:* The following questions were inspired by an article on alcohol consumption and risky behaviors (Hingson *et al.*, 2009) based on responses to a large national survey. The data table **BINGE** contains summary frequency data calculated from tables provided in the article. We'll focus on the self-reported frequency of binge drinking (defined as five or more drinks in a two-hour period for males and four or more drinks in a two-hour period for females) and whether the respondents reported having ever been involved in a car accident after drinking.

For each of the following questions, use JMP to create a suitable contingency table, and report on the probability of sampling one person that satisfies the specified events. What is the probability that we sample an individual who reported:

   a.   binge drinking at least once a week?

   b.   never binge drinking?

   c.   having had an accident after drinking?

   d.   having had an accident after drinking or binge drinking at least once a week?

   e.   having had an accident after drinking given binge drinking at least once a week?

   f.   binge drinking at least once a week given having had an accident after drinking?

Finally, use your responses to consider the question of independence:

   g.   are binge drinking at least once a week and having had an accident after drinking independent events? Explain.

3.   *Scenario:* In this scenario we return to the data from the American Time Use Survey (**TimeUse**). Our data table contains observations from approximately 16,000 respondents in 2003 and 2007. Use the Data Filter to select and include just the 2007 responses. Suppose we select one 2007 respondent at random. What is the probability that we have chosen an individual who:

   a.   is not in the labor force?

   b.   is from the South?

   c.   works part-time or is female?

   d.   works part-time, given that she is female?

   e.   is not married?

Finally, use your responses to consider the question of independence:

   f.   Are the events "from the South" and "female" independent? Explain.

   g.   Are "divorced" and "not in the labor force" independent? Explain.

4.   *Scenario:* This scenario continues the analysis of the **Pipeline Safety** data table. Each row of the table represents an incident in which a natural gas pipeline was somehow disrupted. For each of the following questions, use JMP to create a suitable contingency table, and report on the probability of sampling one

disruption incident that satisfies the specified events. What is the probability that we sample an incident that:

    a. occurred in the central United States?

    b. involved a pipeline rupture?

    c. required evacuation of local residents?

    d. required evacuations, given that the pipeline ruptured?

    e. involved a rupture or an explosion?

Now we return to the issue of how well a Poisson model fits these data. Figure 6.8 reported that 1 incident occurred 34% of the time, 2 incidents 27% of the time, and 5 incidents only 1% of the time.

    f. Use JMP to compute probabilities of 1, 2, and 5 incidents assuming a Poisson distribution with $\lambda = 1.2787$. Compare the probabilities to the empirical percentages and comment on your findings.

5. *Scenario:* The next few questions refer to data from the **Hubway** data. This table contains observations of 55,230 individual bike trips using a bike rental service in Boston, Massachusetts. People can register with the program or simply rent a bike as needed. Individual trips within the data table are labeled as taken by **Registered** or **Casual** users. For each of the following questions, use JMP to create a suitable contingency table and report on the probability of sampling one trip that satisfies the specified events. What is the probability that we sample a trip:

    a. by a registered user?

    b. by a female rider?

    c. by a female who is a registered user?

    d. by a female, given that the rider is registered?

    e. that began at South Station – 700 Atlantic Avenue?

    f. that ended at Boston Public Library – 700 Boylston St?

    g. that started at South Station and ended at the Boston Public Library?

    h. the ended at the Boston Public Library given that it began at South Station?

6. *Scenario:* The data table called **NC Births** contains a random sample of 1,000 individuals drawn from a larger data set from the state of North Carolina[2]. In 2004, the stat gathered and released an anonymized data from a large number of

recent birth. The data table includes several attributes of the mother, father and baby.

For each of the following questions, use JMP to create a suitable contingency table and report on the probability of sampling one person that satisfies the specified events. What is the probability that we sample an individual who is a:

   a.   Smoker with a premature baby

   b.   Smoker with a low birth weight baby

   c.   Mature mom and smoker

Finally, use your responses to consider the question of independence:

   d.   Are the events "smoker" and "low birth weight" independent? Explain.

7. *Scenario:* Recall the FAA bird-strike data we examined in a prior chapter. In that instance, we used a subsample of data from three California airports. We have the full data set called **FAA Bird Strikes**. In this question, we focus on the extent to which two observed variables appear to follow an Integer Distribution.

   a.   The column labeled **Month** indicates the month of the bird-strike event. Create a histogram of that column. Do strikes occur with equal frequency across the twelve months of the year? Explain your thinking.

   b.   The column labeled **DayofWeek** indicates the day of the week (Sunday = 1, Saturday = 7) of the bird-strike event. Create a histogram of that column. Do strikes occur with equal frequency across the seven days of the year? Explain your thinking.

[1] For the sake of clarity, this section uses labels from the data table to identify events rather than using symbols like A and B. *Feed* refers to the event "observed a dolphin feeding."

[2] This data table comes from *OpenIntro Statistics,* 2nd Ed. (2012) by David M Diez, Christopher D Barr, and Mine Cetinkaya-Rundel, available online at http://www.openintro.org/stat/textbook.php.

The Normal Model

Overview

The prior chapter presented some probability models that can represent discrete data. The focus of this chapter is on one essential model widely applicable to continuous data: the so-called normal model. In addition to usefully representing many real-world phenomena, the normal model underlies a large number of traditional inferential techniques and can help us to understand standard approaches to inference.

Continuous Data and Probability

In Chapter 6, we developed a model of discrete random variables based on the concept of a probability value associated with individual possible values of a random variable. In this chapter, we'll consider *continuous* random variables. Continuous data has the property that there are an infinite number of possible values between any two possible values. For example, the winning time in the next Olympic marathon is a continuous variable: no matter how thinly we slice the imaginable values, it is always possible that the winner's time will fall between two values that we imagined. As a practical matter, we tend to measure continuous variables discretely—the official scores might report race times to the nearest 100th second—but that does not change the fact that time is continuous.

Because of the nature of continuous data, our discrete model of probability distributions doesn't apply to continuous data. We cannot list all possible values of a continuous random variable, and because of the infinitesimal nature of the variables, each possible value has a probability of zero. We need a different approach.

Density Functions

For a continuous random variable X, we'll think about the probability that X lies between two values instead of thinking about X assuming one specific value. In other words, rather than solving for $Pr(X = a)$ we'll solve for $Pr(a \leq X \leq b)$. Fortunately, the software handles all of the computational issues, but it's helpful to have some basic grounding in the underlying theory.

With continuous random variables, probability values are derived from the *density function* that describes the variable. From the standpoint of data analysis, density functions *per se* have little direct application. You might think of a density function as representing the concentration of probability near different possible values of X.

To help illustrate this concept, we'll take advantage of one of the graphing features of JMP and look at some histograms in a new way.

1. Open the **NHANES** data table once again.

We've worked with this table before; it contains health-related measurements and survey responses gathered in 2005 from 9,950 people in the United States. Two of the columns in the table contain the blood pressure readings of the adults in the sample. Blood pressure—the force exerted by flowing blood on the walls of arteries and veins—is a continuous variable. By standard practice, it is measured to the nearest whole number, but it is continuous nonetheless. When JMP constructs a histogram of these measurements, it groups similar values into bins, and we see a relatively small number of bars in the graph.

2. Select **Analyze ▶ Distribution**. Cast the last two columns **BPXSY1** and **BPXD1** as **Y** and click **OK**. Your results will look like Figure 7.1.

Figure 7.1: Histograms of Blood Pressure Measurements

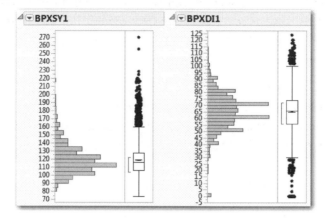

The bars in these histograms are the default choices selected by JMP. In Chapter 3, we learned to adjust the number and therefore the width of the bars. The bars are an artificial convention used to represent the shape of the distribution. JMP provides another way to depict a continuous distribution that better conveys the concept of density.

3. Click the red triangle next to **BPXSY1**. Select **Histogram Options ▶ Shadowgram**. The graph should now look like Figure 7.2.

Figure 7.2: Shadowgram of Systolic Blood Pressure Data

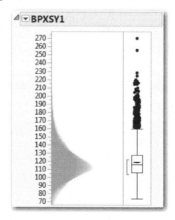

> Shadowgrams are a unique feature in JMP. When one varies the bin width in the histogram, the apparent shape of the distribution changes. The software averages the effects of many possible bin widths, and the shadowgram is the visual result.

In Figure 7.2, the jagged steps of the histogram are replaced with a smooth, shadowy density graph. In both graphs it is clear that the systolic blood pressures are most common near 110 to 120 mmHg[1] and fall off in frequency at lower and higher pressures.

4. Compare the shadowgram of systolic pressure to the histogram of diastolic pressure (**BPXD1**). Do you notice the difference in visual impressions created by the two graphs? The shadowgram should leave the impression of a variable with smoothly changing likelihoods.

With continuous values, probability is computed as the area under the density curve between two points *a* and *b*. Some density curves have infinite domains, others have minima and maxima. For all density functions, the area under the entire curve equals 1, analogous to the property that probabilities for a discrete variable sum to 1. As you advance through your study of statistics and encounter an increasing portfolio of applications, you will eventually work with several commonly used classes of density functions. In this chapter, we'll focus on the *normal* (also known at the *Gaussian*) density function, which has wide application and is particularly relevant to the analysis of data from simple random samples.

The Normal Model

As in Chapter 6, we'll rely on your principal textbook to provide an extensive introduction to normal distributions. In this section, we'll review several attributes of the normal distributions that are essential for practical data analysis. Though we can use normal distributions in many ways, one frequent application is as a model for the population or process from which we're sampling data. Specifically, we often want to ask whether a normal model suitably describes observations of a random variable.

There are an infinite number of normal distributions, each uniquely identified by its mean μ and standard deviation σ. Figure 7.3 shows the graph of a normal distribution known as the *Standard Normal Distribution* whose mean equals 0 and standard deviation equals 1. The standard normal variable is commonly represented by the letter Z and is the basis for z-scores. This curve illustrates several of the major properties of all normally distributed random variables:

- If we know that a variable follows a normal distribution with mean = μ and standard deviation = σ, we know its shape, center, and spread; in short, we know everything that we need to know to find probabilities.

- The distribution is symmetric, with its mean at the center. Half of the area lies on either side of the mean.

- The mean and median of the distribution are equal.

- Nearly the entire area of the density curve lies within three standard deviations of the mean.

- The density function has an infinite domain. The tails of the curve (the ends nearest the X axis) approach the axis asymptotically.

Figure 7.3: A Normal Probability Density Function

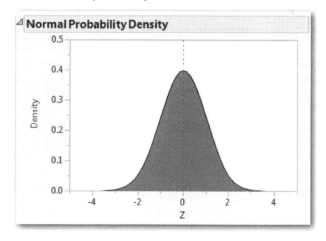

Normal Calculations

There are two classes of problems that we typically solve using a normal distribution:

- For X values a and b, we want to solve for $Pr(a \leq X \leq b)$. In these problems, we are solving for *cumulative probability*.

- For a probability value, p, we want to solve to find x^* such that $Pr(X \leq x^*) = p$. In these problems, we are solving for *inverse cumulative probability*.

For the sake of illustration, consider a normal distribution that shares the same mean and standard deviation as the sample of systolic blood pressure measurements. Figure 7.4 displays the summary statistics for our sample of 6, 668 readings. The sample mean is 119.044 mgHg and the sample standard deviation is 18.841 mmHg. For the sake of this illustration, then, we'll consider a normal distribution with $\mu = 119.044$ and $\sigma = 18.841$, which from now on we'll write as $X \sim N(119.044, 18.841)$.

Figure 7.4: Summary Statistics for the Systolic Blood Pressure Data

Summary Statistics	
Mean	119.04379
Std Dev	18.841373
Std Err Mean	0.2307357
Upper 95% Mean	119.49611
Lower 95% Mean	118.59148
N	6668

Solving Cumulative Probability Problems

Let's introduce the variable $X \sim N(119.044, 18.841)$ as a model for adult systolic blood pressure and use this model to estimate the proportion of adults whose blood pressure falls within specific intervals. For example, suppose we wanted to know the approximate percentage of adults who have systolic pressure below 110 mgHg.

Recall that cumulative probability is defined at the probability that a variable is less than or equal to a particular value. In JMP 11, there are two methods we can use to find probabilities with normal distributions. The first method uses a specialized calculator with a visual interface; in the second approach, we set up a small data table and use cumulative probability density functions to compute whatever probability values we want to find. The first method is easier to learn, but the second provides additional capability and control. We'll see both methods applied to the same problem.

The "Easy" Way

All normal distributions share the same shape but differ with respect to mean and standard deviation. JMP's calculator takes advantage of that fact to compute areas under the curve. We'll specify the two parameters of the distribution as well as the X value of interest (110 mgHg), and the calculator will do the rest.

1. Choose **Help ▶ Sample Data**. This opens the Sample Data Index.

2. In the lower right, find the label marked **Teaching Demonstrations** and click the gray disclosure arrow next to it.

3. About halfway down the list of demonstration scripts, select **UtilNormalDistribution.**

You should now see the window shown in Figure 7.5. There are three numerical inputs at the upper right (Mean, SD, and X) and the results appear in the graph and in the Probability area of the window.

Figure 7.5: The Initial State of the Normal Distribution Calculator

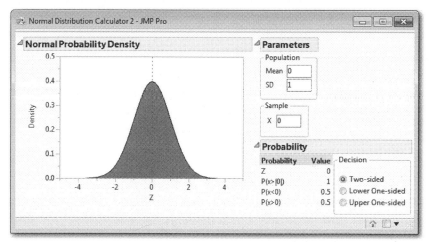

4. In this example, we are assuming a normal population distribution with the parameters cited above, namely a mean of 119.044 mgHg and a standard deviation of 18.841 mgHg. Enter those values in the corresponding boxes.

5. We are seeking the area to the left of a blood pressure of 110. This will be our **Sample** value of X.

6. Finally, we want to graph the area to the left of X = 100, so click the radio button marked **Lower One-Sided**.

> When referring to continuous distributions, the portions of the curve that approach the horizontal axis are often called "tails;" you will see references to the lower and upper tails. "Lower" indicates the Z-values furthest to the left from 0, not to the height of the density curve.

When you are finished, your screen will look like Figure 7.6. From the graph, we see that a large proportion, but fewer than half, reported systolic pressures less than 110 mgHg. The small list of probabilities shows that P(x < 110) = 0. 31561, or about 32% of adults, have such low systolic numbers.

Figure 7.6: The Problem Solved

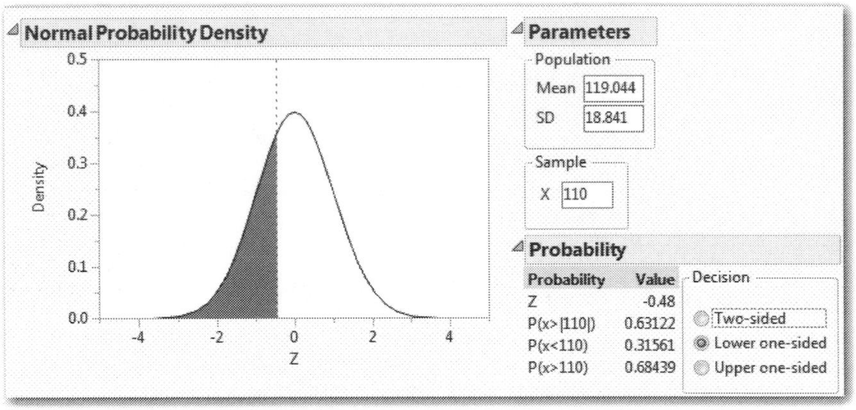

The Less-Easy, but More Powerful Way

We'll use the Normal Distribution function in JMP to find $Pr(X \leq 110)$. To do so, follow these steps:

1. Select **File ▶ New ▶ Data Table**.

2. Rename **Column 1** as Systolic and initially create the column with two empty rows.

3. Move your cursor into the first empty cell of the Systolic column and type 110.

4. Add a new column called CumProb (**Cols ▶ New Column**) and under **Column Properties** choose **Formula**.

5. From the list of **Functions** select **Probability ▶ Normal Distribution**, complete the formula as shown in Figures 7.7 and 7.8, and click **OK** twice.

> Take extra care in entering the arguments for the Normal Distribution function. For the first argument choose the column **Systolic**. Then type a comma (**,**) and enter the mean. Again, type a comma and enter the standard deviation.

In the new data table, you'll now see the value 0.31560766. This indicates that for this specific normal distribution, nearly 31.6% of the distribution lies to the left of 110 mmHg, or that according to this model, approximately 32% of adults have systolic readings less than 110 mmHg.

Figure 7.7: Creating a Column Using the Normal Probability Function

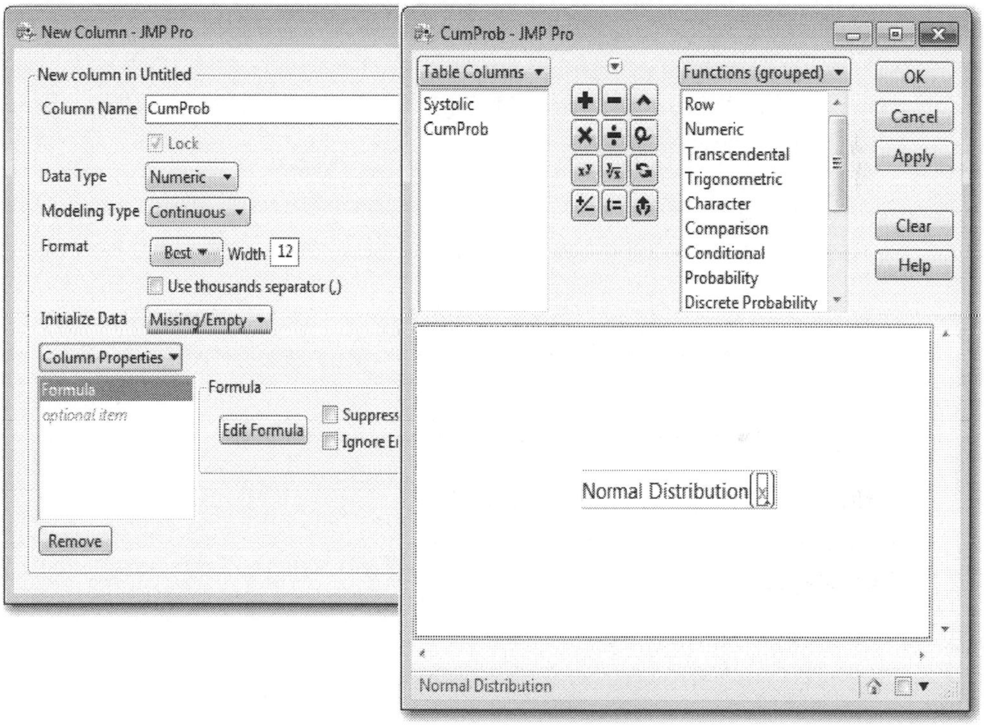

Figure 7.8 Normal Distribution Function with Arguments

Let's consider another example. Systolic pressures above 140 are generally considered high. According to this model, what proportion of the population has readings in the high range?

6.   Within the data table, enter the value 140 in the second row of the **Systolic** column.

7.   The corresponding value in the **CumProb** column is now **0.86698578**.

This is the probability of observing a reading less than 140; the probability of the complementary event is approximately $1 - 0.867 = 0.133$. In other words, in a population described by this model, approximately 13.3% of the people have high blood pressure.

As a final example, we'll find the probability between two X values. In our data set, the first and third quartiles were 106 and 128, respectively. Let's calculate the probability between these two values using our model.

8.   In the first row of the data table, replace 110 with 128.

9.   Replace 140 with 106 in the second row.

The **CumProb** column will now contain the (rounded) values 0.683 and 0.244. Because 68.3% of the distribution is less than 128 and 24.4% is less than 106, we conclude that $68.3\% - 24.4\% = 43.9\%$ of the theoretical normal distribution lies between 106 and 128 mmHg.

At this point, we should note that this model is not a perfect match to our observed data with respect to the quartiles. In the empirical data table, 75% of the observations fell below 128, but only 68.3% of the model distribution lies below 128. On the other hand, 25% of the actual observations are less than 106 mmHg, and 24.4% of the model lies below 106. Fully one-half of the sample observations are between 106 and 128 mmHg, and only about 44% of the model is between the two numbers. Soon, we'll consider the question of how closely a model corresponds to an observed table of data. First, let's see how inverse probability calculations work.

Solving Inverse Cumulative Problems

Suppose that public health officials wanted to address the group of adults with the highest systolic blood pressure, and as a first step, want to identify the top 10% of such individuals. According to our model, we already know that that about 13% of the population has blood pressure above 140 mmHg; at what pressure reading have we reduced the group to 10%?

We want to find x^* such that $Pr(X > x^*) = 0.10$. We do this with the *Normal Quantile* function in JMP. This function returns a value of x^* that satisfies the cumulative probability condition $Pr(X \le x^*) = p$ for any probability value (p) that we choose. So, our first step is to recognize that the problem we want to solve is the equivalent of solving for $Pr(X \le x^*) = 0.90$. In other words, we want to find the 90th percentile of the distribution.

1. Add a column to our data table. Name the column Prob and type the value .9 in the first row.

2. Add another new column named Inverse, and choose **Column Properties ▶ Formula**.

3. As we did earlier, choose **Probability** from the list of functions. This time, select the **Normal Quantile** function, and supply the arguments as shown in Figure 7.9.

Figure 7.9 Normal Quantile Function to Compute Inverse Probabilities

Normal Quantile(Prob , 119.044 , 18.841)

In the new column of the data table, you'll now see the value 143.1897. This value divides the distribution with 90% to the left of the value and 10% to the right. In short, in our model population 10% of the pressure readings are above 143.1897 mmHg.

Sometimes we want to use inverse probability to identify the middle p% of a normal distribution. For example, earlier we referred to the first and third quartiles of our sample data. We know that the quartiles mark off the middle half of the data. We might similarly want to find the middle 50% of this normal distribution. One way to do this is to recognize that the middle 50% is defined as the region that is 25% below the center and 25% above the center, defined by the $50 - 25 = 25$th percentile and the $50 + 25 = 75$th percentile. Just as we did for the 90th percentile, we can use our formula to find the 75th and 25th percentiles.

4. In the data table, change the value 0.9 in the **Prob** column to 0.75.

5. Enter the value 0.25 in the second row of the **Prob** column.

There should be two new values in the **Inverse** column that round to 131.75 and 106.34, corresponding to the quartiles of the theoretical normal distribution. In the observed data, the first quartile was 106 and the third was 128; here (consistent with our earlier calculations), we see a very close match between the observed and theoretical first quartile and a more disparate contrast between the third quartiles. The degree of correspondence brings us to the next topic in this chapter.

Checking Data for the Suitability of a Normal Model

We've already noted that there are times when we'll want to ask how well a normal distribution can serve as a model of a real-world population or process. Because there are often advantages to using a normal model, the practical question often boils down to asking if our actual data deviate too grossly from a normal distribution. If not, a normal model can be quite useful.

In this chapter, we'll introduce a basic approach to this question, based on the comparison of observed and theoretical quantiles (that is, the types of comparisons we've just made in the prior section). In Chapter 9, we'll return to this topic and refine the approach.

Normal Quantile Plots

In the previous section, we computed three different quantiles (percentiles) of the normal variable $X \sim N(119.044, 18.841)$. We also have a large data table containing well over 6,000 observations of a variable that shares the same mean and standard deviation as X. We also know the medians (50th percentile) for each. If we compare our computed quantiles to the observed quantiles, we see the following:

Table 7.1: Comparison of Observed and Theoretical Quantiles

Percentile Value	Observed Value BPXSY1	Computed Value of X
25	106	106.34
50	116	119.04
75	128	131.75
90	142	143.19

The observed and computed values are similar, though not identical. If **BPXSY1** were normally distributed, the values in the last two columns of Table 7.1 would match perfectly.

We could continue to calculate theoretical quantiles for other percentiles and continue to compare the values one pair at a time. Fortunately, there is a more direct and simple way to carry out the comparison—a *Normal Quantile Plot* (sometimes known as a *Normal Probability Plot,* or *NPP*).

In JMP, a normal quantile plot is a scatterplot with values of the observed data on the vertical axis and theoretical percentiles on the horizontal.[2] If the normal model were to match the observed data perfectly, the points in the graph would plot out along a 45° diagonal line. For this reason, the plot includes a red diagonal reference line. To the extent that the points deviate from the line, we see imperfections in the fit. Let's look at two examples.

1. Return to the **NHANES Distribution** report window.

2. Hold down the CTRL key and click the red triangle next to **BPXSY1**; select **Normal Quantile Plot**.

> Remember that depressing the CTRL key while clicking a red triangle applies changes or selections to all open panels.

Figure 7.10 shows the plots for both blood pressure columns. Neither shows a perfectly straight diagonal pattern, but the plot of diastolic pressure on the right more closely runs along the diagonal for most of the distribution. The normal model fits poorly in the tails of the distribution, but fits well elsewhere.

Recall that we have more than 6,600 observations here. The shadowgram, histogram, and box plots show that the diastolic distribution is much more symmetric than the systolic.

Figure 7.10: Normal Quantile Plots for Blood Pressure Data

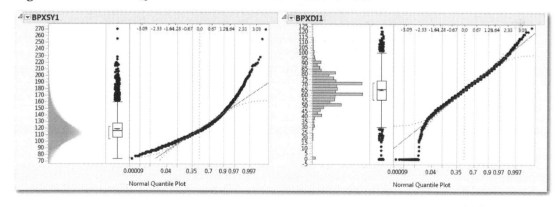

As an example of a quite good fit for a normal model, let's look at some other columns from a subset of the **NHANES** data table, selecting just the two-year-old girls in the sample.

3. Select **Rows ▸ Data Filter**. We need to specify two conditions to select the rows corresponding to two-year-old girls. Within the dialog box, click **Add** and use the CTRL key to choose **RIAGENDR** and **RIDAGEYR**. Check the **Show** and **Include** boxes (shown in Figure 7.11).

4. For gender, click the **Female** button.

5. Click the 0 to the left of **RIDAGEYR** and change it to a 2; do the same with the 85 to the right side. This will select the two-year-old girls.

Figure 7.11: Completed Data Filter Dialog

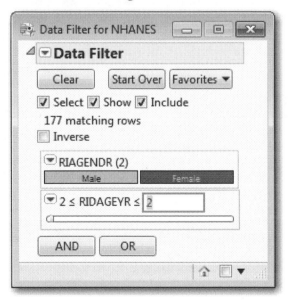

Scroll down the rows of the data table and you'll find a relatively small number of rows selected. In fact, there are just 177 two-year-old girls in this sample of almost 10,000 people.

6. Select **Analyze ▸ Distribution**. Cast **BMXRECUM** as **Y**. This column is the recumbent (reclining) height of these two-year-old girls.

7. Create a normal quantile plot for the data; it will look like Figure 7.12. Here we

find that the points are very close to the diagonal, suggesting that the normal model would be very suitable in this case.

Figure 7.12: Histogram and Normal Quantile Plot for Recumbent Length

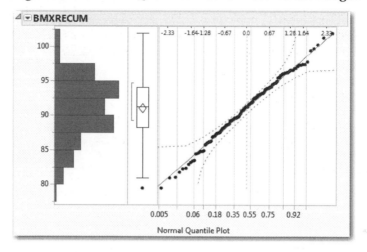

8. As a final example in this section, perform all of the steps necessary to create a normal quantile plot for **INDFMPIR**. The correct result is shown in Figure 7.13.

This column is equal to the ratio of family income relative to the federally established definition of poverty. If the ratio is less than 1, this means that the family lives in poverty. By definition, if the ratio is more than 5.0, **NHANES** records the value as equal to 5; there are 11 such families in this subsample.

Figure 7.13: Distribution of Family Income Poverty Ratio

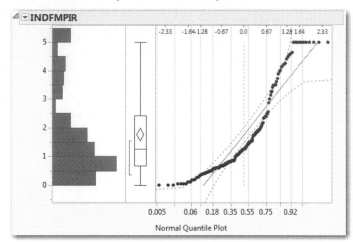

In contrast to the prior graph, we should conclude that a normal distribution forms a poor model for this variable. One could develop a normal model, but the results calculated from that model would rarely come close to matching the reality represented by the observed data from the families of two-year-old girls.

Generating Pseudo-Random Normal Data

We conclude this chapter in much the same way that we concluded Chapter 5. We may sometimes want to generate a simulated sample of random data that conforms to an approximate normal distribution. We could use the **Random** function as we did in Chapter 6, but JMP provides an even easier method.

To illustrate, we'll use the current data table and create some simulated data for a normal random variable with a mean and standard deviation matching the recumbent length measurements. Look back at the distribution of **BMXRECUM** and you'll see that the mean length was 91.06 cm and the standard deviation was 4.33.

1. Select **Cols ▸ New Column**. Name the new column **Normal**. In the dialog box, change the **Initial Data Values** from **Missing/Empty** to **Random**.

2. As shown in Figure 7.14, select **Random Normal** and enter a mean value of 91.06 and a standard deviation value of 4.33.

Figure 7.14: Filling a New Column with Normal Random Data

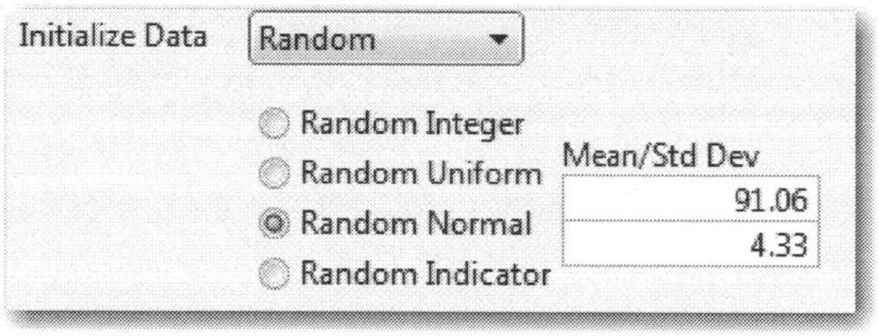

3. Analyze the distribution of this new column and generate a normal quantile plot.

Because each simulation is different, your results will be unique. However, you should find that the normal quantile plot generally, if imperfectly, follows the red diagonal line. You'll also find that the sample mean is approximately 91 and the sample standard deviation is approximately 4.3.

The summary statistics of your simulated sample will almost always deviate slightly from population parameters that you specified. The same is true of real random samples, and that simple and unavoidable fact sets the stage for the next chapter.

Application

Now that you have completed all of the activities in this chapter, use the concepts and techniques that you've learned to respond to these questions.

1. *Scenario:* We'll continue to examine the World Development Indicators data in **BirthRate 2005**. This time we'll see if we can use a normal model to represent one or more variables.

    a. Complete the steps necessary to create a normal quantile plot for the **Fertil** column. **Fertil** is the mean number of children that women have during their child-bearing years in each country. Report on what you see in the graph and what you conclude about the suitability of a normal model.

    b. Consider a random variable $X \sim N(2.993, 1.590)$. Such a variable has the same mean and standard deviation as **Fertil**. For **Fertil**, 10% of observations fell below 1.319 children. What proportion of X lies to the left of 1.319?

    c. Using the same X, find $Pr(X > 5.5)$. Compare your result to the observed data for **Fertil**.

    d. Find the third quartile for X and compare the result to the third quartile for **Fertil**.

2. *Scenario:* In the first chapter we looked at Michelson's early measurements of the speed of light. The data table is **Michelson 1879.**

   a. Complete the steps necessary to create a shadowgram of the **Velocity** data. Paste a copy of the graph in your report and comment on what you see.

   b. Construct a normal quantile plot for the **Velocity** column. Do you think that a normal model is suitable for this variable? Explain.

   c. "Measurement error" is sometimes assumed to be normally distributed. What does this set of data suggest about that assumption?

3. *Scenario:* In Chapter 2, we first looked at the **Concrete** data table containing experimental data. The point of the experiment was to develop a concrete mixture that has high compressive strength.

   a. Complete the steps necessary to create a normal quantile plot for the **Compressive Strength** column. Report on what you see in the graph and what you conclude about the suitability of a normal model.

   b. Consider a random variable $X \sim N(35.818, 16.706)$. Such a variable has nearly the same mean and standard deviation as **Compressive Strength**. Use this normal model and find its 10th, 25th, 50th, 75th, and 90th percentiles. Compare your computed values to the observed corresponding percentiles and report what you find.

4. *Scenario*: Normal models are frequently used to describe the fluctuations in stock market prices. Open the data table called **Stock Index Weekly Changes**. This table contains the weekly proportionate changes in six major international stock market indexes for all of 2008. In the fall of 2008, world financial markets suffered precipitous declines in value.

   a. Construct a normal quantile plot for each column in this table (except **Date**). Which column would you say is best described by a normal model? Explain your thinking.

   b. Which column would you say is least described by a normal model? Explain your thinking.

   c. Identify the mean and standard deviation of a normal distribution that might be suitable for modeling the HangSeng index. Use your model to estimate the percentage of weeks that the index lost value (that is, had a weekly change that was less than or equal to zero).

   d. Compare your result in the prior question to the actual percentage of weeks that the HangSeng lost value in 2008. Comment on the comparison.

5. *Scenario:* In this scenario, we return to the data from the American Time Use Survey (**TimeUse**). Our data table contains observations from approximately 16,000 respondents in 2003 and 2007. Use the Data Filter to select and include just the 2003 responses.

   a. Complete the steps necessary to create a shadowgram of the **Sleeping** data (that is, time spent sleeping). Paste a copy of the graph in your report and comment on what you see.

   b. Construct a normal quantile plot for the **Sleeping** column. Do you think that a normal model is suitable for this variable? Explain.

   c. Repeat steps a and b for the column containing the ages of the respondents.

6. *Scenario:* In this scenario, we return to the **FTSE100** data table, with values from the London Stock Exchange index. We'll focus on the daily closing value of the index, the daily percentage change, and the daily volume (number of shares traded).

   a. Use the **Distribution** platform to summarize the daily closing values and the daily percentage change. Comment on interesting features of these two distributions.

   b. The daily percentage changes are calculated from the data in the **Close** column, and clearly are much better approximated by a normal model than are the daily closing figures. In fact, computing percentage changes is one strategy that analysts use to transform time series into a nearly normal variable. Construct NPPs for both **Close** and **Change%**, and report on your findings.

   c. Would a normal model be an apt description of the daily volume variable? If so, which normal model fits this variable?

7. *Scenario:* In this scenario, we return to the World Development Indicators data (**WDI**). Use the **Data Filter** to **Show** and **Include** only data from the year 2010.

   a. Two columns measure the prevalence of telecommunications technology around the world. They are the number of mobile cellular subscriptions per 100 people (**cell**) and the number of fixed broadband Internet subscribers per 100 people (**broadband**). Construct histograms and NPPs for the two columns. Which column is better summarized as an approximate normal distribution? Explain your reasoning.

b. Using the mean and standard deviation of the more normal column, use JMP to estimate the percentage of countries with 70 or fewer subscribers per 100 people. Explain the approach you took.

c. As in part b, estimate the percentage of countries with 150 or more subscribers per 100 people.

d. Approximately 90% of countries have *at least* X subscribers per 100 people. Find X, and explain how you did so.

8. *Scenario:* The **States** data table contains measures and attributes for the 50 U.S. states plus the District of Columbia. Two of the columns are **Poverty** (percentage of individuals living below the poverty level) and **Income** (median income per capita).

a. Complete the steps necessary to create a normal quantile plot for each of these two columns.

b. Comment on what you see in the graph and what you conclude about the suitability of the normal model for these columns.

9. *Scenario:* In this scenario, we return to the Federal Aviation Administration data on wildlife strikes (**FAA Birdstrikes**).

a. Prepare normal quantile plots for height and speed. Which column is better described by a normal model? Explain your thinking.

b. We know from the **Quantiles** for **Speed** that 25% of the strikes occur at speeds of 120 mph or less. Use the normal model with mean and standard deviation equal to the **Summary Statistics** for **Speed** to estimate the percentage of the distribution below 120. Comment on the comparison.

c. From the **Quantiles**, we also know that about 10% of the bird strikes occur at speeds in excess of 210 mph. What is the area of the same normal curve above 210? How does this compare to the observed 90th percentile?

d. Compare your findings in parts b and c. How do you explain the comparisons between the observed data and the normal approximations?

---

[1] mmHg is the abbreviation for millimeters of mercury (chemical symbol Hg), based on the now old-fashioned mercury gauges used to measure blood pressure.

[2] These are the default axis settings.

Sampling and Sampling Distributions

8

Overview

Although current technology makes it possible to analyze population-level data nearly in real time, a great deal of statistical analysis begins with carefully selected samples from a population. The use of sample data introduces a layer of uncertainty for analysts who need to make conclusions about the parent population. This chapter is devoted to two topics: the relationship between the *method* of sampling and the amount of uncertainty, and also about nature of the uncertainty created by simple *random sampling*.

Why Sample?

Early in Chapter 2, we noted that we often analyze data because we want to understand the variation within a population or a process, and that more often than not, we work with just a subset of the target population or process. Such a subset is known as a *sample*, and in this chapter, we focus on the variability across possible samples from a single population.

It is usually impossible or impractical to gather data from every single individual within a population. In Chapter 2, we presented the following sequence that describes the situation in many statistical studies:

- We're interested in the variation within one or more attributes of a population.

- We cannot gather data from every single individual within the population.

- Hence, we choose some individuals from the population and gather data from them *in order to generalize about the entire population.*

We gather and analyze data from the sample of individuals instead of doing so for the whole population. The particular individuals within the sample are *not* the group we're ultimately interested in knowing about. We really want to learn about the variability within the population and to do so, we learn about the variability within the sample. Before drawing conclusions, though, we also need to know something about the variability among samples we might have taken from this population. While many potential samples probably do a very good job of representing the population, some might do a poor job. Variation among the many possible samples is apt to distort our impression of what's happening in the population itself.

Ultimately, the trustworthiness of statistical inference—the subject of the chapters following this one—depends on the degree to which any one sample might *misrepresent* the parent population. When we sample, we run the risk of unluckily drawing a misrepresentative sample, and therefore it's important to understand as much as we can about the size and nature of the risk. Our ability to analyze the nature of the variability across potential samples depends on which method we use to select a sample.

Methods of Sampling

Chapter 2 also introduced several standard approaches to sampling. In an introductory course, it is valuable to recognize that there are different reasonable methods. It is also valuable to understand that each method carries different risks. In this chapter, we'll focus the discussion on simple random samples (SRS) and the ways in which sample proportions and sample means tend to vary in different samples from one population.

This chapter should help you to understand the nature of simple random sampling. It may also help you to become a more critical consumer of statistical studies by teaching you to ask questions about the implications of sampling methods other than simple random sampling. So-called complex sampling methods, such as stratification and clustering, can also lead to excellent representative samples, but they are generally considered to be beyond the scope of an introductory course. Non-random methods do not reliably yield representative samples, despite the fact that they are widely used. We'll concentrate on the SRS in this chapter to develop a good foundation in the logic of statistical inference, but do discuss complex methods briefly as well.

In the next part of this chapter, we'll use JMP to select random samples from a *sampling frame*, a complete list of individuals within a population. Then, we'll make some visual and numerical comparisons between the known population distribution and the distribution of a random sample. We will then extend our examination of sampling by using the simulation capabilities of JMP to visualize the variability across a large number of possible random samples. We'll begin with an artificially small population from which we'll select some simple random samples. Our goal is twofold: first, simply to learn how JMP can help us select a random sample, and second, to develop an understanding of the ways in which a sample might mislead a researcher. If we were truly working with such a small population, we'd clearly include every observation in the analysis. Our second goal here is subtle but important—we're starting small so that we can actually examine all possible samples from this population.

Using JMP to Select a Simple Random Sample

The concept of a random sample is straightforward. Choose a subset of a population, taking steps to ensure that all equally sized subsets share the same chance of selection. In practice, there are two common obstacles to random sampling: we often don't know the precise size or the identities of all members of the population. While common, these obstacles are easily overcome, but the resulting samples are not simple random samples. In this section, we'll confine our attention to two ways in which we can use JMP to select a simple random sample.

To choose *n* observations as a simple random sample, we need a way to gather or list all elements in the population. This is known as a *sampling frame*—the set from which our random subset is drawn. We also need some type of randomizing algorithm or method.

By way of illustration, let's revisit the **Military**[1] data once again. Recall that this data table includes some categorical data concerning the 1,048,575 active-duty personnel from four branches of the U.S. military, excluding the Coast Guard, in April 2010. We'll treat this group as the sampling frame. For the sake of this example, we will suppose that we want to select a 100-person random sample from the population for further follow-up study.

1. Open the **Military** data table.

2. Select **Analyze ▸ Distribution**. Examine the distributions of the first three columns in the table: **grade**, **branch**, and **gender**.

We see that approximately 78% of the individuals are enlisted personnel, that 6% are in the Air Force, and that 13% are female. We'll keep these population proportions in mind as we examine some samples. Recall that our goal here is to gain insight in the extent to which different methods of sampling produce samples that mirror the population.

3. Choose **Tables ▸ Subset**. Complete the dialog box as shown in Figure 8.1, requesting 100 randomly selected rows. Check **Keep dialog open**, and click **OK**.

Figure 8.1: Selecting a Simple Random Sample

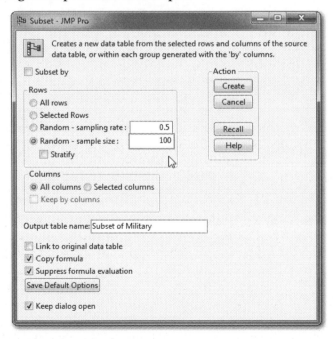

> If we did not already have the full data table available to us in electronic form (for Military or some other population, like all of the names listed in the local city telephone directory), or if we didn't intend to use the current data columns but want to contact the respondents with other questions, we might use JMP to generate a list of random integers, and then use those numbers to select our sample.

You should now see a new data table titled **Subset of Military**, with 100 rows randomly selected from the sampling frame. The new table contains all columns from the main table, but includes only 100 rows randomly selected from the population.

Because this is a random process, your list of 100 personnel will be (nearly) unique. It will differ from the samples other readers obtain and from the one used as an illustration here. Now let's compare the descriptive statistics for **grade**, **branch**, and **gender** in the random sample to the actual population parameters:

4. Create a distribution for the columns **grade**, **branch**, and **gender**.

5. Compare the proportion of personnel who are enlisted, who are in the Air Force, and who are female to the population proportions.

It is very likely that you will notice first that the proportions won't match exactly, but that they are similar in magnitude. The author's subset contained 73 enlisted personnel, 7 Air Force, and 15 women. Perhaps you have 70% or 85% enlisted, but you probably don't have 50% or 100% enlisted. Let's try it again with a different subset.

6. Earlier we left the **Subset** dialog open. Return to it now and click **OK** to choose a different subset. Repeat steps 4 and 5.

Table 8.1 summarizes the author's findings in this exercise. Your results will yield different proportions, but they should be similar to these:

Table 8.1: Comparison of the Population and Random Samples for Selected Categories

Data Table	% Enlisted	% Air Force	% Female
Military (n=1,048,575	78	6	13
Subset 1 (n=100)	73	7	15
Subset 2 (n=100)	75	9	15

In practice, we typically draw one random sample, collect and summarize the sample data, and then attempt to draw reasonable conclusions about a population. Here we see that sample data is apt to be "close" to an accurate match with a population, but not

exact. We want to ask just how much difference it makes to look at one sample rather than a different one.

With more than one million rows in this Military table, there are a vast number of different 100-person samples that we might have randomly selected. In the next section, we'll begin to think about how the variability across those many samples can blur our perception of the patterns of variation within the population. We'll start by simulating repeated subsets of a particular size from one population. Because the number of possible samples can truly be enormous with even modest-sized populations, we'll automate the process until we have a large enough group of subsets to see a pattern.

Variability Across Samples: Sampling Distributions

All samples vary, and the relevant variation is different depending on whether we're looking at quantitative continuous data or qualitative ordinal or nominal data. In this section, we'll look more closely at how we can gauge and characterize the *sampling distribution* of the sample proportion for qualitative, categorical data and the sampling distribution of the sample mean for continuous data.

> Don't be thrown by the terminology here. We want to describe the variability of a sample statistic (like the proportion of Air Force personnel) across all possible samples. Statisticians call such a distribution a *sampling distribution* to distinguish it from the distribution within one sample or the distribution in an entire population.

Sampling Distribution of the Sample Proportion

To illustrate how different samples will yield different results, we'll focus just on the gender column in the data. We know that 13.221% of the more than one million personnel are women. That value—13.221%--is a population parameter. We want to think about what would happen if we did not have the entire population data set, and needed to use a simple random sample in order to estimate the parameter. How much might our estimates deviate from the actual value?

In this instance, there are far too many possible outcomes for us to list them all. Instead we'll use a JMP *script* that has been written to simulate situations just like this.

A script is a file containing a sequence of JMP commands written in the JMP Scripting Language (JSL). Scripts are particularly useful when we want to run through a series of commands repeatedly, as in this case, where we'll generate numerous random samples and tabulate the results. You can extend your JMP capabilities by learning to use and write scripts, but that's a subject for another book. In this chapter, we'll just run two scripts to illustrate patterns of sampling variability. These scripts were developed by Amy Froelich of Iowa State University and Bill Duckworth of Creighton University. They were programmed by Wayne Levin and Brian McFarlane of Predictum Inc. and are available for download at http://www.jmp.com/modules.

1. Select **File ▸ Open**. Look in the **Scripts** folder within our **Data Tables** directory and choose the file called **03_Dist_Sample_Proportions.**

Choosing this file opens the interactive simulator. (See Figure 8.2.) With this tool, we can simulate repeated sampling to see how samples vary. In this way, we can develop an approximate picture of the sampling distribution. For our first simulations, we'll simulate the process of selecting 100-person SRSs.

Figure 8.2: The Simulator Window for Sample Proportions

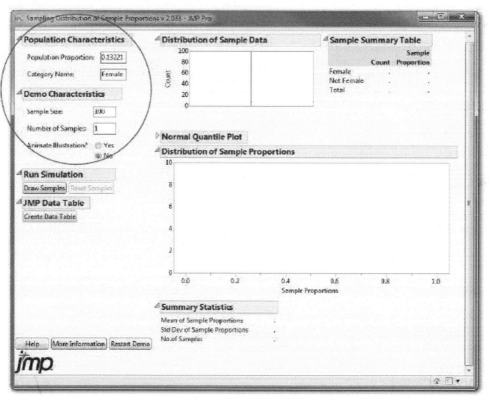

2. In the upper left of the window, find the **Population Characteristics** (circled in Figure 8.2) and change the **Population Proportion** from 0.2 to 0.13221.

3. Now, change the **Category Name** to Female.

4. Change the **Sample Size** to 100.

5. Now, click the **Draw Samples** button at the lower left to generate a single random sample of 100 military personnel from this population with a proportion of 13.221 % females.

Now, look at the upper right of the window to see the summary for your first simulated sample. Each time we draw a sample, we'll see a different result. Figure 8.3 displays one typical sample result. In this instance, our sample of 100 people contained 21 women for a sample proportion of 0.21 (21 %). This particular sample yields a reasonably high estimate of the population proportion, but yours may be closer to 13%.

Figure 8.3: Results for One Simulated Sample

6. Click **Draw Additional Samples** nineteen more times, keeping your eyes on the bar chart labeled **Distribution of Sample Data**. Notice how the bars oscillate up and down indicating the variability across different samples.

By now, you may also have noticed that the large region titled **Distribution of Sample Proportions** is starting to take shape. Figure 8.4 shows the results after one set of 20 samples; your graph will not match this one exactly, but will very likely be similar in center, shape, and spread. At this point in this simulation, all of the sample proportions have fallen between approximately 0.08 and 0.22, with a concentration at around 0.12–0.14. We can see in the summary statistics below the graph that the mean of the sample proportions from all 20 samples is 0.142 – just slightly higher than that actual population proportion. Again, your simulation results will be different from these.

Let's keep adding more simulated samples, but pick up the pace.

7. In the center left of the simulator window under **Demo Characteristics**, choose **Number of Samples ▸ 100**, click **Draw Additional Samples**, and watch the histogram of Figure 8.4 morph into a different picture.

Figure 8.4: Distribution of Simulated Sample Proportions after 20 Repetitions

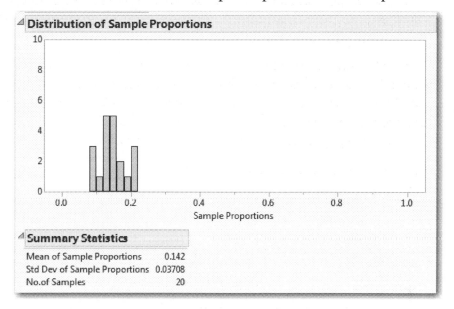

8. Repeat this step once more.

9. Now, let's turn off the animation and generate several thousand samples. Select **Number of Samples ▸ 1000** and click the radio button **No** just below the number of samples. Click **Draw Additional Samples** until you've drawn more than 6,000 samples.

The sampling distribution graph should now look very much like Figure 8.5. There are several things to notice in these results. The distribution of observed sample proportions is symmetric with a strong central concentration near 0.132. The vast majority of sample proportions fell between 0.08 and 0.2. The summary statistics inform us that the mean proportion after 6,220 repetitions was 0.132—very nearly matching the population proportion. In a single sample of 100 military personnel we will never have exactly 13.221% females because that is impossible, but out of all possible samples the mean sample proportion is 13.221%.

Figure 8.5: Results After More than 6,000 Replications

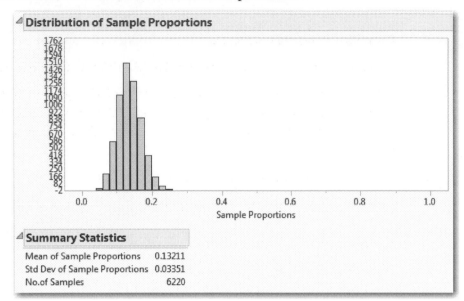

From Simulation to Generalization

It's easy to see that the distribution shown in Figure 8.5 might be reasonably approximated by a normal distribution. In fact, the normal approximation has long been the traditional approach in analyzing sampling distributions. With inexpensive computers and software like JMP, it is now equally practical to take a more direct and more accurate approach. If we take genuine 100-person random samples from a very large population with a population proportion of 0.13221, then we can use the binomial distribution to describe the theoretical sampling distribution[2]. If we define $\hat{p}$ as the sample proportion of households in a 50-person random sample with low or very low income security, then the distribution of $\hat{p}$ is described by a binomial distribution with parameters $n = 100$ and $p = 0.13221$. In this way, we can compute the likelihood of drawing a random sample that lies a specific distance from the population proportion.

For example, the simulation showed that although the population proportion is 13.2%, it is possible to draw a simple random sample that could mislead us into estimating that the population proportion is 10% or less. But how likely is that? In other words, how likely is it that an investigator would observe 10 or fewer women in a sample of 100? Alternatively, we might ask the question this way: "If the population actually contains

13.2% women, how surprised should we be if our sample of 100 contains fewer than 10% women?"

1. Create a new empty data table.

2. Rename the first column x, and enter the value 10 into the top row.

3. Add a new column containing the **Discrete Probability** formula Binomial Distribution (0.13221,100,x). Figure 8.6 shows the formula editor dialog box. You might want to refer back to Chapter 6 to refresh your memory about how to create a column like this.

Figure 8.6: Computing a Cumulative Binomial Probability

After you click **OK** to set up the column, you should find the value 0.2146 (rounded) in your data table. This is the probability of observing ten or fewer women in a sample of 100 according to the binomial model. So, in this example, more than 21%--or about one-fifth—of all possible samples from this population have 10 or fewer women.

Sampling Distribution of the Sample Mean

Thus far in the chapter, all of the examples have focused on categorical data. What happens when the variable of interest is continuous? Let's consider the work of the PHMSA, the Pipeline and Hazardous Materials Program, an office of the U.S. Department of Transportation. PHMSA monitors all disruptions in the pipelines supplying natural gas throughout the United States, maintaining detailed records about all major incidents.

For this example, we'll focus on one of the many variables that PHMSA tracks, namely the total elapsed time from the first reporting of a disruption until the area was made safe. Time is a continuous variable, measured here in minutes.

1. From our set of data tables, open the data table called **Pipeline Safety**. This table lists all major pipeline disruptions in the U.S. from 2003 to 2008.

2. Analyze the distribution of the column **STMin**.

This column is strongly skewed, which makes sense. In most cases, the areas are secured by local public safety and utility officials in a few hours, but sometimes the disruptions can be quite lengthy.

As we did earlier, let's initially assume that this data table represents an entire population of incidents. Within the population, the mean elapsed time is 149.02 minutes with a standard deviation of 212.64 minutes. We will use simulation to examine the sampling variation from this population.

Specifically, we'll look at variation in the *sample means* among many simulated samples from this population. This data table contains 468 rows and we can use those rows to seed another simulation script. We'll arbitrarily sample fewer than 10% of the population each time by setting our sample size at 45.

3. From the **Scripts** folder again, open the file **02_Dist_Sample_Means**. This simulator (shown on the next page in Figure 8.7) works very much like the one for proportions, but this one works with the continuous data and the sample mean. As before, the goal is to investigate the variability in sample means across different possible samples.

4. We first need to specify the **Population Characteristics**. Unlike the prior example, we'll rely on the empirical data in the **Pipeline Safety** data table to define the population attributes. From the drop-down list next to **Population Shape**, choose **My Data**.

Figure 8.7: Sampling Distribution Simulator for a Sample Mean

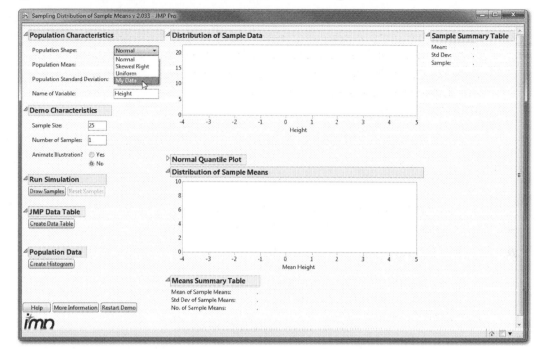

5. This opens a dialog box from which we can choose one of the open data tables or select an unopened table. Choose **Pipeline Safety**.

6. After choosing the data table, a standard dialog box opens to select one column (not shown here). Just select **STMin** near the bottom of the list of available columns and click **OK**. After you do this, notice that the simulator has automatically filled in the three boxes for the mean, standard deviation, and variable name.

7. Now type 45 into the **Sample Size** box, indicating that we want to simulate drawing a sample of 45 pipeline disruption incidents.

8. This simulator operates like the one we used for proportions. Click the **Draw Samples** button a few times and look at the histogram in the upper portion of your window.

Within each sample, we see a right-skewed distribution of values, reflecting the shape back in the parent population. When we look at the **Distribution of Sample Means**, we find that the sample means are starting to cluster to the left of 200 minutes.

9. Increase the number of samples from 1 to 5,000. Generate about 20,000 samples (depending on your computer's speed, this may take a few moments).

Each group of 20,000 simulated samples is different, but your results probably look very much like those shown below in Figure 8.8. Despite the very strongly skewed population and samples, the means of the samples follow a much more symmetric shape.

Figure 8.8: Typical Results of More than 20,000 Simulations

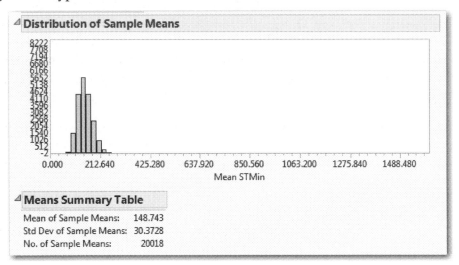

Moreover, we notice in the **Means Summary Table** that the mean of the sample means is just a little less than 149 minutes. In other words, samples from this population yield sample means whose values cluster rather closely near the actual population mean. This is very important; not only do samples tend to have means close to the true population mean, they vary in a predictable way. So even though we can't whether one particular sample will perfectly match the population mean, we can describe the pattern and extent of variation across possible samples.

The Central Limit Theorem

The symmetric shape of the distribution and the clustering near the population mean is actually predictable. The *Central Limit Theorem* (CLT) applies to the variation among sample means for continuous data. The theorem essentially says that regardless of the shape of the parent population, for sufficiently large samples the distribution of all possible sample means is approximately normal in shape, with a mean value of μ (the population mean) and a standard deviation of $\sigma/\sqrt{n}$. So as to distinguish the standard deviation of the sampling distribution from the standard deviation of the population in this book, we'll restrict the term standard deviation to refer to the variation among individual cases in a sample or population, using the symbols s and σ respectively. In reference to a sampling distribution, we'll use the term *standard error*. In this section, we're asking about variability of the sample mean, so we'll refer to the *standard error of the mean*, denoted as $\sigma_{\bar{x}}$.

More precisely, the sampling distribution is described by a t *distribution*. The t distributions are another family of mound-shaped symmetrical distributions, similar in shape to normal curves but with proportionately more area in the tails. Similar to the normal distributions, there are a large number of t distributions, each uniquely and completely identified by one parameter known as *degrees of freedom* (or *df*). For a t distribution, we calculate degrees of freedom as $df = n - 1$.

According to the CLT the means of all possible 45-observation random samples from this population will follow a t distribution with 44 degrees of freedom. Their distribution will have a mean of 149.02 and a standard error of $212.64/\sqrt{45} = 31.7$. We've already seen in the simulation that the mean of more than 20,000 sample means is close to 149. Now let's check the standard error.

Now, look at your screen just below the histogram of the distribution of sample means (refer back to Figure 8.8). Using the CLT, we just calculated the standard error of the mean to be 31.7 minutes. This is the spread of the theoretical distribution of all possible sample means. In the simulation illustrated here, the standard deviation of the collection of 20,018 simulated sample means was about 30.37 minutes. Although not perfect, over the span of just 20,000 samples, the CLT does an impressive job of informing us about the ways in which sampling distorts our estimates of the population mean. If we were to continue simulating samples, the CLT predicts that three things will happen:

- the mound shape of the histogram will approach the normal distribution shape,
- the mean of the sample means will approach 149.0,
- the standard deviation of the sample means will approach 31.7

Armed with the CLT and an understanding of t distributions, we can estimate how often and how far we are likely to be misled by our samples. With a normal model, we could confidently say that 95% of all observations lie within almost two standard deviations of the mean. Similarly, with a normal model, we could say that 95% of all possible samples have means within about two standard *errors* of the mean.

JMP has a *t* distribution function that we can use to calculate the proportion of sample means lying more than two standard errors from the mean, assuming a sample size of 45.

1. In JMP, return to the data table we created to make the binomial computation, and type –2 into the first row of the **x** column and 2 into the second row of the **x** column.

2. Set up a formula in **Column 2** as illustrated in Figure 8.9. The **t Distribution** function is listed under the **Probability** group. Complete the formula as shown.

Figure 8.9: Using the *t* Distribution Function

The results of this command appear in your new data table (see Figure 8.10.)

Figure 8.10: Results of *t* Distribution Calculations

	x	Column 2
1	-2	0.0258483015
2	2	0.9741516985

This shows that about 2.6% of the *t* distribution with 44 *df* lies to the left of –2 and the complementary 97.4% lies below +2, meaning that another 2.6% is to the right of +2. Therefore, we can conclude that approximately 5.2% of all possible samples of 44 incidents would have sample means more than two standard errors from the true population mean. In other words, any one sample could mislead us, but about 95% of all possible samples would mislead us by no more than 63.4 minutes.

Stratification, Clustering, and Complex Sampling (optional)

As noted previously, simple random sampling requires that we can identify and access all N elements within a population. Sometimes this is not practical, and there are several alternative strategies available. It's well beyond the scope of this book to discuss these strategies at length, but many of the JMP analysis platforms anticipate the use of these methods, so a short introduction is in order.

Two common alternatives break a population up into groups, and then use random sampling methods to gather a probability sample based on the groups. Fundamentally, we group in one of two ways:

Stratification: Strata (the singular form is stratum) are logical or natural subgroups within a population such that individuals within a stratum tend to share similar characteristics, but those characteristics are likely to vary across strata. In a *stratified sample,* we identify strata within the population and then randomly sample from each stratum. For example, imagine a public health study focusing on diet and blood pressure among adults in the U.S. We might reasonably stratify the population by gender and age group.

Clustering: Clusters are often geographical or arbitrary groupings within a population such that there is no particular reason to expect individuals within a cluster to be more or less homogenous than individuals across clusters. In a *cluster sample,* we identify clusters within a population, randomly sample a small number of clusters, and then proceed to sample individuals from those clusters.

For example, many telephone surveys in the United States cluster the population by area code and randomly select a group of area codes, and subsequently use computers to randomly dial seven-digit telephone numbers. In such a process, where the sample is built up in stages, the area codes are referred to as the Primary Sampling Units (PSUs).

Complex sampling refers to methods that combine stratification and clustering. Whether a sample is complex, stratified, or clustered, analysis of the sample data generally will require *weighting* if we want to generalize from the sample to the entire population. In a simple random sample of n individuals from a population with N members, each of the individuals represents the same fraction of the total population (n/N). However, consider what happens when a pollster randomly calls households within two different area codes. The entire state of Wyoming (population of approximately 0.5 million) is covered by area code 307. In New York City, area code 212 covers part of the borough of Manhattan that is home to about 1.6 million people. If the pollster were to phone 10 people in each of these two area codes, the Wyoming subsample would be representing about 50,000 people apiece while the Manhattan subsample would be representing nearly three times that number.

When we use a data table assembled from a complex sample, we'll typically find one or more weighting variables—sometimes identified as *post-stratification weights*—and a PSU column. Most of the analysis platforms in JMP provide the option of specifying a **weight** column, to be used if our data came from a complex sample.

> Several of the data tables used in this book come from survey organizations that use complex sampling methods. Among these examples are the NHANES data and the American Time Use Survey data tables.

Clustering and stratification both rely on random selection. Looking back at our military personnel table, you can understand how we might proceed if we wanted, say, to cluster by military branch. We would establish a data table listing the branches, randomly subset those tables, and proceed through the process.

The Subset platform enables us to stratify using columns already within the data table. Let's stratify by grade; there are three grades in this table. To illustrate the concept of sampling weights, we'll choose 100 individuals from each grade, recognizing that with varying numbers of personnel in each grade, a sample with an equal number of rows from each grade will misrepresent the military.

1. Select the **Military** data table.

2. Choose **Tables ▶ Subset**. Specify a random sample size of 100 and select the **Stratify** check box, indicating we want 100 observations from each stratum.

Figure 8.11: Starting to Create a Stratified Random Sample

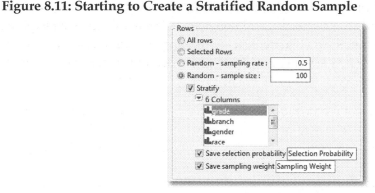

3. As shown above in Figure 8.11, select grade as the stratification column.

4. Select the two options to save the selection probabilities and sampling weights, and click **OK**. You will see a new data grid containing the stratified sample.

Notice the two new columns to the far right of the table. In this illustration, the first 100 rows are officers; they all have the sample selection probability and sampling weights. If you scroll down, the rows you should notice that the values in the two new columns change when you come to the warrant officers, and then change again when you come to the enlisted personnel. In the full data table, there are 211.525 officers or approximately 20% of the population. In contrast, there are only 19,179 warrant officers, or just fewer than 2% of the population. Our sample contains 100 officers out of the 211,525 in the population, and the **Sampling Probability** equals exactly that: 100/21,525 = 0.000473. For warrant officers, the sampling probability is just about ten times higher. Inversely, in our sample 100 officers will represent all 211,525 officers in the population, so each officer represents 211,525/100 = 2115.25 officers, which is what the **Sampling Weight** indicates. In round numbers, each sampled officer is standing in for more than 2000 peers, but each warrant officer is standing in for only about 200 peers. Hence, in later analysis, JMP can adjust its calculations to reflect that imbalance.

Application

Now that you have completed all of the activities in this chapter, use the concepts and techniques that you've learned to respond to these questions.

1. *Scenario:* You live in a community of 23,847 households. You want to select a simple random sample of 250 of these households from a printed list of addresses.

   a. Use JMP to generate a random list of 250 integers; report the first 10 and last 10 integers on the list.

   b. Suppose that, from prior figures, you estimate that 18% of these households (approximately 4,300) have no Internet service. Use what you've learned in this chapter to estimate the probability that a random sample of 250 households would contain only households with Internet service.

   c. Estimate the probability that an SRS of 250 households would include 25 or fewer homes without Internet service.

   d. Is there any realistic chance that an SRS of 250 households would be made up entirely of homes without Internet service? Explain your thinking.

2. *Scenario:* Recall the World Development Indicators data in **BirthRate 2005**.

   a. First, analyze the full table to determine the actual proportion of all countries that are in Sub-Saharan Africa. What is that proportion?

   b. Now find and report the mean and standard deviation of the infant mortality rate for all countries.

c. Select a random sample of 30 countries. In your sample, what was the proportion of Sub-Saharan African countries? What was the mean infant mortality rate? How do your sample statistics compare to the values that you found in parts a and b? In your own words, explain the discrepancies.

3. *Scenario*: You want to develop a deeper understanding of sampling variability for categorical data by making further use of the sampling distribution of the sample proportion simulator script. We'll imagine that you want to conduct a large-scale survey to learn what proportion of the population favors a particular issue. Complete the following activities, and report briefly on what you find.

   a. Set the population characteristics as follows: the Population Proportion is 0.4, and the sample size is 1,000. Set the Category Name to Favor. Generate 5,000 samples by repeatedly running the simulation and describe the distribution of sample proportions.

   b. Reset the simulator, and keep the population proportion set at 0.4, but change the sample size to 250. Again generate 5,000 simulated samples and report on what you find. If we reduce the sample size by a factor of ¼, what happens to the center, shape, and spread of the sampling distribution?

   c. Reset once again, and change the population proportion to 0.95, keeping the sample size at 250. Generate 5,000 samples, and compare your results to those obtained in part b.

   d. In major public opinion polls, pollsters often choose sample sizes around 250 or around 1,000. What difference does the size of the sample make?

   e. If you were conducting an important public opinion survey and were concerned about the potential adverse impacts of sampling variation, would you be better off sampling from a population in which there was considerable uniformity (say, in which 95% of the population agreed on an issue) or a population that was split on the issue? Explain how this exercise shapes your response.

4. *Scenario*: You want to develop a deeper understanding of sampling variability for continuous data by making further use of the sampling distribution of the sample means simulator script. In this scenario, you are investigating the number of seconds required for a particular Web site to respond after a user issues a request. Complete the following activities, and report briefly on what you find.

   a. Set the population characteristics as follows: the **Population Shape** is **Normal**, the **Population Mean** is **15** seconds, the **Standard Deviation** is **3.3** seconds, and the **Name of the Variable** is **Response Time**. The **Sample Size** is **1,000**. Generate 10,000 samples, and describe the distribution of sample means.

   b. Reset the simulator, and keep the other settings the same as in part a, but change the sample size to 250. Again, generate 10,000 simulated samples, and report on what you find. If we reduce the sample size by a factor of ¼, what happens to the center, shape, and spread of the sampling distribution?

   c. Reset once again, and change the population standard deviation to 6.6 seconds, keeping the sample size at 250. Generate 10,000 samples, and compare your results to those obtained in part b.

   d. Now reset the simulator, change the **Population Shape** to **Right-Skewed**, and repeat the other settings from part a. How do your results compare to those obtained in part a?

   e. Once again, reset and change the shape to Uniform to model a population in which all values have the same probability.

   f. In the prior questions, we've varied the shape of the parent population, the size of our samples, and the amount of variability within the population (standard deviation). Write a few sentences summarizing how each of these changes impacts the nature of sampling variability.

5. *Scenario:* Recall the **FAA BirdStrikes** data.

   a. First, analyze the full table to determine the actual proportion of all strikes that occurred for California strikes. What is that proportion?

   b. Now find and report the mean and standard deviation of the cost of repairs for all strikes.

   c. Select a random sample of 100 observations. In your sample, what was the proportion of California flights? What was the mean cost of repairs? How do your sample statistics compare to the values that you found in parts a and b? In your own words, explain the discrepancies.

6. *Scenario:* The city of Boston, Massachusetts introduced a bike sharing program in 2007. Using technology installed in the bike racks and bicycles, the Hubway system can track usage patterns and habits. The data table titled **Hubway** contains data drawn from more than 55,000 actual bike trips between and among nearly 100 different rack locations in the city.

   a. First, analyze the full table to determine the actual mean **rider_age** (third to last column) for all trips. What is the population mean?

   b. How would you describe the shape of the distribution of ages?

   c. Using the Central Limit Theorem, what do we expect the sampling distribution of the mean to be? (center, shape, spread)

   d. Use the sampling distribution of the mean simulator to simulate samples of size $n = 50$, using the **Hubway** data table to define the population characteristics. Run 10,000 simulations

   e. Describe the pattern of variation among the 10,000 simulated sample means noting the center, shape and spread of the sampling distribution.

   f. Using the lower histogram in the simulator window, and without making any calculations, how likely do you think it would be to select a random sample in which the mean rider age was more than 35 years old? Less than 30 years? More than 45 years? Explain your thinking.

---

[1] This section is inspired by an example in *OpenIntro Statistics, 2nd Ed.* (2012) by David M Diez, Christopher D Barr, and Mine Cetinkaya-Rundel, available online at http://www.openintro.org/stat/textbook.php.

[2] Because the population is finite in size, one more properly should use the *hypergeometric* distribution to compute the exact probabilities. This is because the probability of success changes after each individual is selected, since the size of the remaining pool from the population is decreased by 1. With a population this large in comparison to the sample size, the impact is negligible. We studied the binomial in an earlier chapter, so we will use it.

Review of Probability and Probabilistic Sampling

Overview

The past three chapters have focused on probability—a subject that forms the crucial link between descriptive statistics and statistical inference. From the standpoint of statistical inference, probability theory provides a valuable framework for understanding the variability of probability-based samples. Before moving forward with inference, this short chapter reviews several of the major ideas presented in Chapters 6 through 8, emphasizing two central concepts: (a) probability models are often very useful summaries of empirical phenomena, and (b) sometimes a probability model can provide a framework for deciding if an empirical finding is unexpected. In addition, we'll see another JMP script to help visualize probability models. As we did in Chapter 5, we'll use the **WDI** data table to support the discussion.

Probability Distributions and Density Functions

We began the coverage of probability in Chapter 6 with elementary rules concerning categorical events, then moved on to numerical outcomes of random processes. In this chapter, we'll start the review by remembering that we can use discrete probability distributions to describe the variability of random outcomes of a discrete random variable. A discrete distribution essentially specifies all possible numerical values for a variable as well as the probability of each distinct value. For example, a rural physician might track the number of patients reporting flu-like symptoms to her each day during flu season. The number of patients will vary from day to day, but the number will be a comparatively small integer.

In some settings, we know that a random process will generate observed values that are described by a formula, such as in the case of a binomial or Poisson process. For example, if we intend to draw a random sample from a large population and count the number of times a single categorical outcome is observed, we can reliably use a distribution to assess the likelihood of each possible outcome. The only requirement is that we know the size of the sample and we know the proportion of the population sharing the categorical outcome. As we'll see later, we know that approximately one-third of the countries in the WDI table are classified as "high income" nations. The binomial distribution can predict the likelihood of selecting a random sample of, say, 12 nations and observing a dozen high-income countries.

The Normal and *t* Distributions

When the variable of interest is continuous, we use a *density function* to describe its variation. In conventional parlance, a continuous density function is said to specify the distribution of a variable or family of variables. Chapter 7 presented an in-depth treatment of the normal model and some of its uses. The normal model is one notable example of a theoretical continuous distribution, i.e. a distribution that can be characterized mathematically. In Chapter 8 we encountered Gosset's *t* distribution, similar to the normal in that it is symmetric and mound-shaped. These are two continuous distributions that underlie many of the techniques of inference, and that there are yet many more distributions used regularly by statisticians and data analysts.

For a quick visual refresher on the distributions do the following:

1. **File** ▸ **Open**. Within the directory of this book's JMP files, open the **Scripts** folder and open **JMP_Distribution_Calculator_V0_E**. Figure 9.1 shows the initial view of this distribution calculator.

Figure 9.1: An Enhanced Distribution Calculator

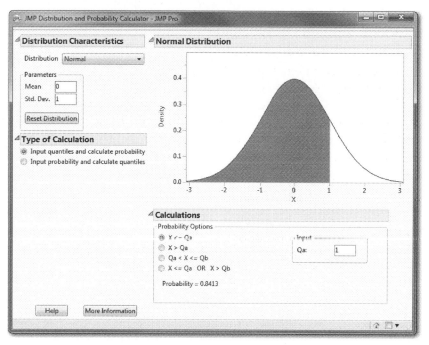

Initially the calculator displays the standard normal distribution. As you recall, all normal distributions are specified by two parameters—a mean and a standard deviation. There are two types of calculations we might want to perform. Either we want to know the probability corresponding to values of the random variable (X), or we want to know the values of X that correspond to particular probabilities. The left side of the dialog box allows the user to specify the parameters and type of calculation.

Below the graph is an input panel to specify regions of interest (less than or equal to a value, between two values, etc.) and numerical values either for X or for probability, depending on the type of calculation. Spend a few moments exploring this calculator for normal distributions.

This calculator can perform similar functions for many discrete and continuous distributions, including the few that we have studied. To review and rediscover those distributions, try this:

2. In the upper left of the dialog, click the button next to **Distribution**. This opens a dropdown list of all the distributions available for the calculator, as shown in Figure 9.2.

Figure 9.2: Available Distributions

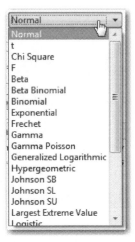

As you scroll through the list, you will find the t, normal, Poisson, and uniform distributions as well as others that are unfamiliar. Select the familiar ones and notice the parameters of each as well as their characteristic shapes. Curious readers may wish to investigate other distributions in the list as well.

The Usefulness of Theoretical Models

In statistical applications and in introductory statistics courses, we tend to put these theoretical models to work in two ways. In some instances, we may have a model that reliably predicts how random values are generated. For example, we know how to recognize a binomial or Poisson process and we know when the Central Limit Theorem is relevant. In other instances, we may simply observe that sample data are well-approximated by a known family of distributions (e.g. Normal) and we may have sound reasons to accept the risks from using a single formula rather than using a set of data to approximate the population. In such instances, using the model instead of a set of data can be a more efficient way to draw conclusions or make predictions. In the context of statistical inference, we will often use models to approximate sampling distributions; that is, the models will help us to characterize the nature of sampling variability and therefore, sampling error.

Let's consider one illustration of using a model to approximate a population, and refer back to Application Scenario 7 of Chapter 7 which in part worked with the cellphone use rates among countries in the **WDI** data table. We'll elaborate on that example here.

3. Open the **WDI** data table.

4. Select **Analyze ▶ Distribution**. Select the column **cell** (number of mobile cellular subscriptions per 100 people).

Because cell phone usage has changed considerably over the 22-year period, we'll focus on just one year, 2011. We can apply a data filter within the **Distribution** report to accomplish this.

5. Click the red triangle next to **Distributions** and choose **Script ▶ Local Data Filter** to **Show** and **Include** only data from the year **2011**.

6. Finally, add a normal quantile plot. The comparatively straight diagonal pattern of the points suggests that a normal distribution is an imperfect but reasonable model for this set of 191 observations.

The **Distribution** command reports selected quantiles for this column. We can see, for example, that 75% of the reporting countries had 122.984 subscriptions or fewer per 100 people. Suppose we wanted to know the approximate value of the 80th percentile. We might use a JMP command to report the 80th percentile from the sample, but we can quickly approximate that value using the normal distribution calculator. From the **Distribution** report, we know that the sample mean = 95.544 subscriptions, and the sample standard deviation = 43.236 subscriptions.

Follow these steps to calculate the 80th percentile of a theoretical normal population with the same mean and standard deviation.

7. Return to the **Distribution and Probability Calculator**, and re-select **Normal** distribution in the upper left.

8. In the **Parameters** panel, enter 95.544 and 43.236 as the mean and standard deviation.

9. Under **Type of Calculation**, select **Input probability and calculate quantiles**.

10. Finally, under **Calculations**, select **Left tail probability**, and type .8 in the **Probability** box.

The completed dialog is shown in Figure 9.3. In the lower right panel, we find that the theoretical 80th percentile is 131.9 subscriptions. If you repeat this exercise to locate the 75th and 90th percentiles, you will see that the theoretical normal model is a good approximation of this set of empirical data.

Figure 9.3: Using a Normal Model to Approximate a Set of Data

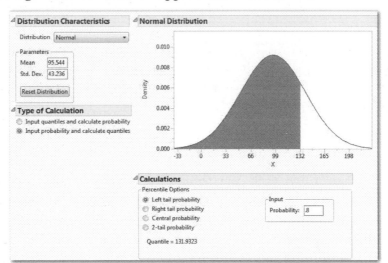

When Samples Surprise: Ordinary and Extraordinary Sampling Variability

Chapter 8 was devoted to probabilistic sampling and to the idea that samples from a population will inevitably differ from one another. Regardless of the care taken by the investigator, there is an inherent risk that one particular sample may poorly represent a parent population or process. Hence, it is important to consider the extent to which samples will tend to vary. The variability of a sample statistic is described by a *sampling distribution.*

In future chapters, sampling distributions will be crucial underpinnings of the techniques to be covered, but will slip into the background. They form the foundation of the analyses, but like the foundations of buildings or bridges, they will be metaphorically underground. We will rely on them, but often we won't see them or spend much time with them.

As we move forward into statistical inference, we will shift our objective and thought processes away from summarization towards judgment. In inferential analysis, we frequently look at a sample as *evidence* from a somewhat mysterious population, and we will want to think about the kind of population that would have yielded such a sample. As such, we'll begin to think about samples that would be typical of a particular population and those that would be unusual, surprising, or implausible.

All samples vary, and the shape, center, and variability depend on whether we're looking at quantitative continuous data or qualitative ordinal or nominal data. Recall that we model the sampling variation differently when dealing with a sample proportion for qualitative, categorical data and the sample mean for continuous data. Let's consider each case separately.

Case 1: Sample Observations of a Categorical Variable

To illustrate, consider the WDI data table once again. The table contains 40 different columns of data gathered over 22 years. Of the 214 countries, approximately one-third are classified as high income (70 of 214 nations – 0.327). If we were to draw a random sample of just twelve observations from the 4,708 rows, we would anticipate that approximately four of the twelve sample observations would be high income nations. We should also anticipate some variation in that count, though. Some samples might have four high income countries, but some might have three, or five, or perhaps even two. Is it likely that a single random sample of twelve would not have any high-income countries? Could another sample have twelve? Both are certainly possible, but are they plausible outcomes of a truly random process?

The process of selecting twelve nations from a set of 4,708 observations and counting the number of high-income nations is approximately a binomial process. It is only approximately binomial because the probability of success changes very slightly after each successive country is chosen; for the purposes of this discussion, the approximation is sufficient[1]. As such, we can use the binomial distribution to describe the likely extent of sampling variation among samples of twelve countries.

1. In the probability calculator, select the **Binomial** distribution.

2. Specify that the probability of success is 0.327 and a sample size, **N**, of 12.

Your screen will look like Figure 9.4. The binomial histogram (reflecting the discrete number of successes in twelve trials) is initially set to show the probability of obtaining one or fewer high-income countries. This probability is represented by the blue shading, and at the bottom of the dialog, we find that it is 0.0590. By selecting other options within the dialog, we can discover that it is extremely unlikely that a random sample from this data table will contain more than seven high-income countries.

Figure 9.4: A Binomial Distribution

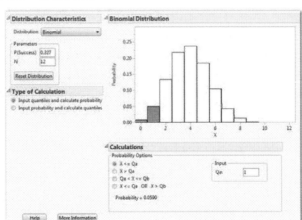

3. Below the histogram, click the radio button marked **X > Qa**, and enter a 6 into the **Input** panel box **Qa**.

4. The probability that a twelve-country sample will contain more than 6 high income countries is 0.0603. Repeat step 3 for 7, 8, and 9 countries; notice that the probability becomes vanishingly small.

Such random samples may be logically possible, but are so unlikely as to be considered nearly impossible. By far, it is most likely that this population would generate samples with between (say) 2 and 7 high-income countries.

Case 2: Sample Observations of a Continuous Variable

For this example, we'll move away from the **WDI** data and return to the **Pipeline Safety** data that we recently investigated in Chapter 8. As we did then, we'll work with the column called **STMin**, which is the number of minutes between the first report of a pipeline disruption until the area was made safe. We will recall the exercise we did then with the **Sampling Distribution** simulator and look into the results more deeply than before.

1. Open the **Pipeline Safety** data table.

2. Create a distribution of the column **STMin**.

   As we found in Chapter 8, this column is strongly skewed because most disruptions are secured rather quickly, but sometimes the disruptions can be quite lengthy.

Usually we think of samples as coming from a very large population or from an ongoing process. This data table has a sample of 468 reported disruptions over a period of time. We can think of pipeline operations as a continuous process that periodically is interrupted. For the sake of this review, imagine that this sample is representative of the ongoing process.

If all pipeline disruptions required the same amount of time to repair, then any sample we might draw would provide the same evidence about the process. In fact, a sample of n=1 would be sufficient to represent the process. Our challenge arises from the very fact that individual disruptions vary; because of those individual differences, each sample will be different.

In Chapter 8, we used the **Sampling Distribution** simulator to simulate thousands of random samples of size n = 45 from the data table. This process is sometimes known as *bootstrap sampling* or *resampling*, in which we repeatedly sample and replace values from a single sample until we have generated an empirical sampling distribution.

Readers who wish to repeat the simulation should go back to the section of Chapter 8 entitled **"Sampling Distribution of the Sample Mean"**, and regenerate perhaps 50,000 iterations until the histogram is relatively smooth and mound-shaped. The results of one such simulation appear in Figure 9.5; the histogram has been revised to more clearly show the horizontal axis. The blue vertical line at 149.018 minutes is the location of the original sample mean.

Figure 9.5: A Bootstrap Sampling Distribution

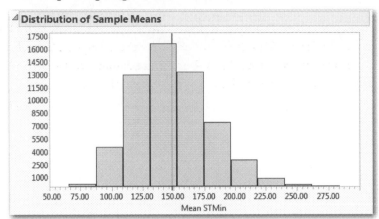

So what message are we supposed to take away from this graph and the resampling exercise? Consider these messages for a start:

- Different 45-observations samples from a given population have different sample means;

- Though different, sample means tend to be "near" the mean of the population;

- We can model the shape, center, and variability of possible sample means;

- Despite the fact that **STMin** in the original sample is very strongly skewed to the right, repeated samples consistently generate sample means that vary symmetrically.

Very importantly, the histogram establishes a framework for judging the rarity of specific sample mean values. Samples of 45 pipeline disruptions commonly have means between approximately 108 and 175 minutes. Sample means will very rarely be smaller than 80 minutes or longer than 240 minutes. It would be stunning to take a sample of 45 incidents and find a sample mean of more than 275 minutes.

We cannot conclude that a result is surprising or rare or implausible without first having a clear sense of what is *not rare*, or what is typical. Sampling distributions tell us what is common.

Conclusion

Probability is an interesting and important subject in its own right. For the practice of data analysis, it provides the foundation for the interpretation of data gathered by random sampling methods. This chapter has reviewed the content of three earlier chapters that introduce readers to a few essential elements of probability theory. Taken together, these chapters form the bridge between the two statistical realms of description and inference.

---

[1] The correct distribution to apply here is the *Hypergeometric* distribution, another discrete distribution that applies when one samples without replacement. The probability of success changes following each observation. In this particular case where the population size is so much larger than the sample size, the impact of the change in probabilities is miniscule. Inasmuch as this is a review chapter, I decided to resist the introduction of a new distribution.

Inference for a Single Categorical Variable

Overview

Beginning with this chapter, we enter the realm of inferential statistics, where we generalize from a single sample to draw conclusions about an entire population or process. When statisticians use the term *inference*, it refers to two basic activities. Sometimes we want to ask *whether the parameter is or is not equal to a specific value*, and sometimes we want to *estimate the value of a population parameter*. In this chapter, we'll restrict our attention to categorical dichotomous data and focus on the proportion of a population that falls in one category.

Two Inferential Tasks

We'll discuss the ideas of *significance testing* (often called *hypothesis testing*) and of *confidence interval* estimation. Along the way, we'll illustrate these techniques and ideas using data tables with two formats common to categorical data. We often find categorical data summarized in a frequency table, but sometimes we have detailed case-by-case data.

The main message of the previous chapter is that sampling always carries the risk of inaccurately representing the larger group. We can't eliminate the possibility that one particular sample will lead us to an incorrect conclusion about a population, but we can at least calibrate the risk that it will. To do so, it's important to consider the form and origin of our data before rushing ahead to generalize from a sample. There are also other considerations to take into account and these are the topic of the next section.

Statistical Inference is Always Conditional

Under a wide range of common conditions, good data generally leads to good conclusions. Learning to reason statistically involves knowing what those conditions are, knowing what constitutes good data, and knowing how to move logically from data to conclusions. If we are working with sample data, we need to recognize that we simply cannot infer one characteristic (parameter) of a population without making some assumptions about the nature of the sampling process and at least some of the other attributes of the population.

As we introduce the techniques of inference for a population proportion, we start with some basic conditions. If these conditions accurately describe the study and the data, the conclusions we draw from the techniques are most likely trustworthy. If the sample departs substantially from the conditions, we should avoid generalizing from the sample to the entire population.

At a minimum, we should be able to assume the following about our sample data:

- Each observation is independent of the others. Simple random sampling assures this, and it is possible that in a non-SRS we preserve independence.

- Our sample is no more than about 10% of the population. When we sample too large a segment of the population, essentially, we no longer preserve the condition of independent observations.

We'll take note of other conditions along the way, but these form a starting point. In our first example, we will look again at the pipeline safety incident data. This is not a simple random sample, even though it may be reasonable to think that one major pipeline interruption is independent of another. Because we are sampling from an ongoing nationwide process of having pipelines carry natural gas, it is also reasonable to assume that our data table is less than 10% of all disruptions that will ever occur. We'll go ahead and use this data table to represent pipeline disruptions in general.

Using JMP to Conduct a Significance Test

The goal of significance testing is to make a judgment based on the evidence provided by sample data. Often the judgment implies action: is this new medication or dosage effective? Does this strategy lead to success with clients or on the field of play? With respect to pipeline disruptions, public safety authorities may ask "Do more than 75% of disruptions involve ignition?"

The process of significance testing is bound up with the scientific method. Someone has a hunch, idea, suspicion, or concern and decides to design an empirical investigation to put the hunch to the test. As a matter of rigor and in consideration of the risk of erroneous conclusions, the test approximates the logical structure of a proof by contradiction. The investigator initially assumes the hunch to be incorrect and asks if the empirical evidence makes sense or is credible given the contrary assumption.

The hunch or idea that motivates the study becomes known as the *Alternative Hypothesis* usually denoted as H_a or H_1. With categorical data, the alternative hypothesis is an assertion about possible values of the population proportion, p. There are two formats for an alternative hypothesis, summarized in Table 10.1.

Table 10.1: Forms of the Alternative Hypothesis

Name	Example	Usage
One-sided	H_a: p > .75	Investigator suspects the population proportion is more than a specific value.
	H_a: p < .75	Investigator suspects the population proportion is less than a specific value.
Two-sided	H_a: p ≠ .75	Investigator suspects the population proportion is not equal to a specific value.

The alternative hypothesis represents the reason for the investigation, and it is tested against the *null hypothesis (H₀)*. The null hypothesis is the contrarian proposition about p asserting that the alternative hypothesis is untrue.

To summarize: we suspect, for example, that more than 75% of gas pipeline disruptions ignite. We plan to collect data about actual major pipeline incidents and count the number of fires. To test our suspicion, we initially assume that *no more than 75%* of the disruptions burn. Why would we begin the test in this way?

We can't look at all incidents, but rather can observe only a sample. Samples vary, and our sample will imperfectly represent the entire population. So we reason in this fashion:

- Assume initially that the null hypothesis is true.

- Gather our sample data and compute the sample proportion. Take note of the discrepancy between our sample proportion and the null hypothesis statement about the population.

- Knowing what we know about sampling variation, estimate the probability that a parent population like the one described in the null hypothesis would yield up a sample similar to the one we've observed.

- Decision time:

  o If a sample like ours would be extraordinary under the conditions described, then conclude that the null hypothesis is not credible and instead believe the alternative.

  o If a sample like ours would not be extraordinary, then conclude that the discrepancy we've witnessed is plausibly consistent with ordinary sampling variation.

> NOTE: if we decide against rejecting the null hypothesis, the sample data do *not* "prove" that the null hypothesis is true. The sample data simply are insufficient to conclude that the null hypothesis is false. At first, this difference might appear to be one of semantics, but it is essential to the logic of statistical inference.

In this example, suppose that our investigator wonders if the proportion of disruptions that ignite exceeds 0.75. Formally, then, we're testing these competing propositions:

| Null hypothesis: | $H_0: p \leq .75$ | (no more than 75% of disruptions ignite) |
| Alternative hypothesis: | $H_a: p > .75$ | (more than 75% of disruptions ignite) |

1. Open the data table **Pipeline Safety.**

2. Select **Analyze ▶ Distribution** to summarize the **IGNITE** column. Your results look like Figure 10.1, which was prepared in **Stacked** orientation to save space.

Figure 10.1: Distribution of IGNITE Column

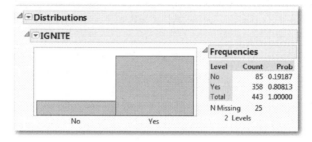

For 25 incidents in the table, we don't have information about whether or not the gas ignited. In the remaining 443 incidents, nearly 81% involved ignition. It is fairly common to find gaps like this in real observational data sets.

3. Click the red triangle beside **IGNITE** and choose **Test Probabilities**. This opens a new panel within the report, as shown in Figure 10.2.

Figure 10.2: Specifying a Significance Test

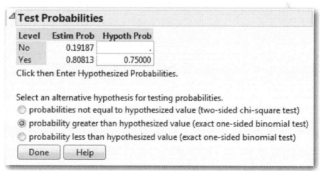

4. The upper portion of this panel displays the observed proportions of **Yes** and **No** within the sample, and leaves space for us to enter the proportions as stated in the null hypothesis. Just type .75 in the box in the second row.

5. The lower portion of the panel lists the three possible formulations of the alternative hypothesis. Click the middle option because our alternative

hypothesis asserts that the population proportion is greater than 0.75, and then click **Done**.

There are several common ways to estimate the size of sampling error in significance tests. Most introductory texts teach a "by-hand" method based on the normal distribution because of its computational ease. Because software can do computation more easily than people can, JMP does not use the normal method, but rather selects one of two methods, depending on the alternative hypothesis. For a two-sided hypothesis, it uses a *chi-square* distribution, which we'll study in Chapter 12. For one-sided hypotheses, it uses the discrete binomial distribution rather than the continuous normal. With categorical data, we always count the number of times an event is observed (like ignition of gas leaks). The binomial distribution is the appropriate model, and the normal distribution is a good approximation that is computationally convenient when there is no software available.

Your results should look like those in Figure 10.3. Look first to the bottom row of the results. There you'll see the alternative hypothesis (**p1** refers to the number that you typed in the white box above), the proportion based on the **Yes** values, and the **p1** value that you entered. The right-most number—the *p-Value*—is the one that is informative for the test.

Figure 10.3: Results of a Significance Test

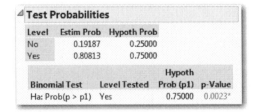

A *p-value* is a conditional probability. The assumed condition is that the null hypothesis is true, and the probability refers to the chance that an SRS would produce a result at least this far away from the hypothesized value. In this example, we interpret the *p*-value by following this logic: Our sample proportion was 0.808, which is more than the hypothesized value of 0.75 by nearly 0.6. Assuming that the null hypothesis were true (that is, at most 75% of all gas-line breaks ignite) and assuming we take an SRS of 443 incidents, the probability that we'd find a discrepancy of 0.6 or more is just 0.0023. This probability equates to an event that only occurs 23 times out of 10,000 observations. Yet that is the discrepancy we have found. In this case, it is hard to accept the idea that the null is true. Another way to express this is to say that this particular sample result is very surprising if we were expecting an ignition rate of 75%.

What do we do with that probability of 0.0023? How surprising is it? In practice, there is a conventional limit used in significance testing. The limit goes by the symbol alpha (α) and α represents our working definition of credibility. By custom the default value of α is 0.05, and there is a simple decision rule. If the *p*-value is greater than or equal to α, we conclude that the null hypothesis is credible. If the *p*-value is less than α, we reject the null hypothesis as being incredible. When we reject the null hypothesis because of a very small *p*-value, we say that the result is *statistically significant*.

> JMP places an asterisk (*) beside *p*-values that are less than 0.05 and colors them for emphasis. This convention applies to *p*-values in all types of significance tests. The rule is simple, but it can be hard to remember or interpret at first. The designers of JMP know this and have incorporated a reminder, as shown in this next step.

6. Move your cursor in a very small circle over the value 0.0023. Closely read the message that pops up on the screen. (See Figure 10.4.)

Figure 10.4: Pop-up Explanation of P-Value

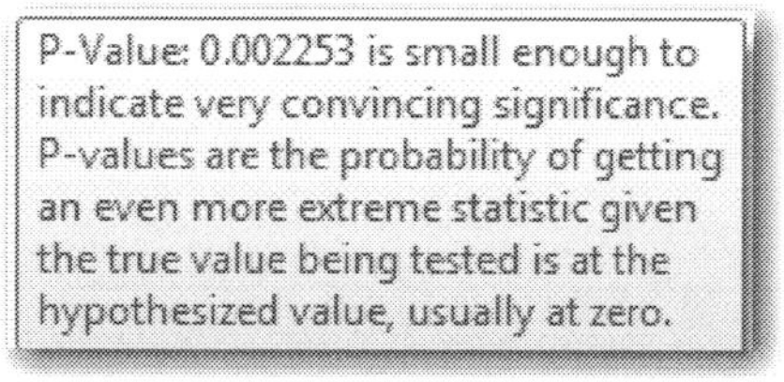

P-Value: 0.002253 is small enough to indicate very convincing significance. P-values are the probability of getting an even more extreme statistic given the true value being tested is at the hypothesized value, usually at zero.

Confidence Intervals

A significance test provides a yes/no response to an inferential question. For some applications, that is sufficient to the purpose. In other instances, though, the decision-maker may need to estimate a proportion rather than simply determine whether it exceeds a specified limit.

We already know how to calculate the proportion of ignition incidents within our sample (about 0.81 or 81%), but that only describes the sample rather than the entire population. We shouldn't say that 81% of all disruptions involve fires, because that is overstating what we know. What we want to say is something like this: "Because our sample proportion was 0.81 and because we know a bit about sampling variability, we can confidently conclude that the population proportion is between ___ and ___."

We will use a *confidence interval* to express an inferential estimate. A confidence interval estimates the parameter, while acknowledging the uncertainty that unavoidably comes with sampling. The technique is rooted in the sampling distributions that we studied in the prior chapter. Simply put, rather than estimating the parameter as a single value between 0 and 1, we estimate lower and upper bounds for the parameter and tag the bounded region with a specific degree of confidence. In most fields of endeavor, we customarily use confidence levels of 90%, 95%, or 99%. There is nothing sacred about these values (although you might think so after reading statistics texts); they are conventional for sensible reasons. Like most statistical software, JMP assumes we'll want to use one of these confidence levels, but it also offers the ability to choose any level.

Using JMP to Estimate a Population Proportion

We already have a data table of individual incident reports. In this section, we'll analyze the data in two ways. First, we'll see how JMP creates a confidence interval from raw data in a data table. Next, we'll see how we might conduct the same analysis if all we have is a summary of the full data table.

Working with Casewise Data

The Pipeline Safety data table contains raw casewise data—that is, each row is a unique observation. We'll use this data table to illustrate the first method.

1. In the **Distribution** report, click the red triangle beside **IGNITE** and choose **Confidence Intervals ▶ 0.95.**

A new panel will open in the results report. The confidence interval estimate for the entire population appears in the table row beginning with the word **Yes**, as highlighted in Figure 10.5. We find the sample proportion in the **Prob**(ability) column and the endpoints of the confidence interval in the **Lower CI** and **Upper CI**. Finally, the confidence level[1] of 0.95 appears in the column labeled **1-Alpha**. Assuming that our sample meets the conditions noted earlier, and given that 80.8% of the observed disruptions involved ignition, we can be 95% confident that the population-wide proportion of disruptions with ignition is between about 0.769 and 0.842.

Figure 10.5: A Confidence Interval for a Proportion

Level	Count	Prob	Lower CI	Upper CI	1-Alpha
No	85	0.19187	0.157916	0.231129	0.950
Yes	358	0.80813	0.768871	0.842084	0.950
Total	443				

Note: Computed using score confidence intervals.

This interval estimate acknowledges our uncertainty in two respects: the width of the interval says "this interval probably captures the actual parameter" and the confidence level says "there's about a 5% chance that the sample we drew led us to an interval that excludes the actual parameter."

Notice that the sample proportion of .808 is just about midway (but not exactly) between the upper and lower CI boundaries. That is not an accident; confidence intervals are computed by adding and subtracting a small *margin of error* to the sample proportion. There are several common methods for estimating both the center of the interval and the margin of error. One common elementary method takes the number of observed occurrences divided by the sample size ($\hat{p}$) as the center, and computes the margin of error based on percentiles of the Normal distribution. JMP uses an alternative method, as noted beneath the intervals. JMP uses "score confidence intervals," which are sometimes called the *Wilson Estimator*. The Wilson estimator doesn't center the interval at $\hat{p}$ but at a slightly adjusted value.

For most practical purposes, the computational details behind the interval are less important than the assumptions required in connection with interpreting the interval. If our sample meets the two earlier conditions, we can rely on this interval estimate with 95% confidence if our sample size is large enough. Specifically, this estimation method captures the population proportion within the interval about 95% of the time if:

The sample size, n, is sufficiently large such that $n\hat{p} > 5$ and $n(1 - \hat{p}) > 5$.

Working with Summary Data

Now suppose that we did not have the complete table of data, but had read a report or article that summarized this aspect of pipeline disruptions. Such a report might have stated that in the period under study, 358 pipeline disruptions ignited out of 443 reported incidents. So, we don't know which 358 incidents involved fires, but we do know the total number of ignitions and the total number of incidents.

1. Create a new data table that matches Figure 10.6.

Figure 10.6: A Table for Summary Data

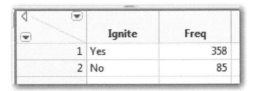

2. Select **Analyze ▶ Distribution**. As shown in Figure 10.7, cast **Ignite** into the **Y** role and identify **Freq** as containing the frequency counts.

Figure 10.7: Analyzing a Distribution with Summary Data

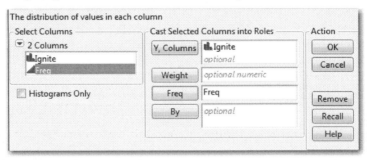

The rest of the analysis is exactly the same as described earlier. Clicking the red triangle opens a menu to generate the confidence interval or conduct a significance test.

A Few Words About Error

Whatever we conclude, there is some chance that we've come to the wrong conclusion about the population proportion. This has nothing to do with computational mistakes or data entry errors (although it's always wise to double check) and everything to do with sampling variability. When we construct a 95% confidence interval, we're gambling that our one sample is from among the 95% whose sample proportions lie a modest distance from the actual parameter. By bad luck, we might draw a sample from among the other 5% of possible samples.

With hypothesis tests, the situation is slightly different because we can go astray whether we reject the null hypothesis or not. If our *p*-value is very small, there is still a chance that we're rejecting the null hypothesis when we should not. This is called a *Type I* Error, and hypothesis testing is really designed to minimize the risk of encountering this situation.

Alternatively, a larger *p*-value (which leads us to persist in believing the null hypothesis) also carries the risk of a different (*Type II*) error whereby we should reject the null hypothesis, but do not because our evidence doesn't convince us to do so. If you can come to understand what these errors imply and why they are important to decision-making, you'll more quickly come to incorporate the logic of significance testing in your own thinking.

Application

Now that you have completed all of the activities in this chapter, use the concepts and techniques that you've learned to respond to these questions.

1. *Scenario:* We'll continue our analysis of the pipeline disruption data.

   a. Construct and interpret a 95% confidence interval for the proportion of pipeline disruptions that are classified as ruptures (rather than leaks or something else).

   b. Now construct and interpret a 90% confidence interval for the same parameter (that is, proportion of ruptures).

   c. In a sentence or two, explain what happens when we lower the confidence level.

   d. Based on your confidence intervals, would it be credible to conclude that only 15% of all disruptions are caused by ruptures? Explain your reasoning.

2. *Scenario:* You live in a community of 23,847 households. You have selected a simple random sample of 250 of these households from a printed list of addresses. Suppose that, from prior figures, you previously believed that 18% of these households have no Internet service. You suspect that more homes have recently acquired service, so that the number is now less than 18%. In your sample of 250 households, you find 35 with no service.

    a. Does this situation satisfy the conditions required for statistical inference? Explain.

    b. Construct a 95% confidence interval for the proportion of the community without Internet service.

    c. Perform a significance test using the alternative hypothesis that is appropriate to this situation. Report on your findings, and explain your conclusion.

    d. Now suppose that your sample had been four times larger, but you found the same sample proportion. In other words, you surveyed 1,000 homes and found 140 with no Internet service. Repeat parts a and b, and report on your new findings. In general, what difference does a larger sample seem to make?

3. *Scenario:* Recall the NHANES food security data in **NHANES Expanded**.

    a. Does the NHANES data set satisfy the conditions for inference? Explain.

    b. Construct and interpret a 95% confidence interval for the population proportion of all households that often worried that their food would run out before they got money to buy more (column **FSD032A**).

    c. Construct and interpret a 99% confidence interval for the same proportion using the same column. Explain fully how this interval compares to the one in part a.

    d. Repeat part b using a 90% confidence level.

    e. What can you say in general about the effect of the confidence level?

4. *Scenario:* Recall the study of traffic accidents and binge drinking introduced in Chapter 5. The data from the study were in the data table called **Binge**. Respondents were people ages 18 to 39 who drink alcohol.

   a. Assume this is an SRS. Does this set of data otherwise satisfy our conditions for inference? Explain.

   b. In that study, 213 of 3,801 respondents reported having ever been involved in a car accident after drinking. If we can assume that this is a random sample, construct and interpret a 95% confidence interval for the population proportion of all individuals who have ever been in a car accident after drinking.

   c. In the same study, 485 respondents reported that they binge drink at least once a week. Based on this sample, is it credible to think that approximately 10% of the general population binge drinks at least once a week? Explain what analysis you chose to do, and describe the reasoning that leads to your conclusion.

5. *Scenario:* Recall the study of dolphin behavior introduced in Chapter 5; this summary data is in the data table called **Dolphins**.

   a. Assume this is an SRS. Does this set of data otherwise satisfy our conditions for inference? Explain.

   b. Assuming that this is a random sample, construct and interpret a 95% confidence interval for the population proportion of dolphins who feed at a given time in the evening.

   c. Based on the data in this sample, would it be reasonable to conclude that if a dolphin is observed in the morning, there is greater than a 0.5 probability that the dolphin is socializing?

6. *Scenario:* The data table **MCD** contains daily stock values for the McDonald's Corporation for the first six months of 2009. One column in the table is called Increase, and it is a categorical variable indicating whether the closing price of the stock increased from the day before.

   a. Using the data in the table, develop a 90% confidence interval for the proportion of days that McDonald's stock price increases. Report on what you find, and explain what this tells you about daily changes in McDonald's stock.

b. Now construct a 95% confidence interval for the proportion of days that the price increases. How, specifically, does this interval compare to the one you constructed in part a?

7. *Scenario:* The data table **TimeUse** contains individual responses to a survey conducted by the American Time Use Survey in 2003 and 2007. The survey used complex sampling methods as described in Chapter 8. In all, there were 32,968 respondents, including 12,248 interviewed in 2007. For these questions, use Data Filter to restrict your analysis to the 2007 respondents. Respondents' ages ranged from 15 to 85 years.

a. Does this data table satisfy the conditions for inference? Explain.

b. Construct and interpret a 95% confidence interval for the population proportion of "Never married" people. (marst)

c. Construct and interpret a 99% confidence interval for the same proportion using the same column. Explain fully how this interval compares to the one in part a. In general, what can you say about the effect of the confidence level?

d. Using the 95% confidence interval, can we conclude that less than one-fourth (0.25) of Americans in 2007 between the ages of 15 and 85 had never been married? Explain your reasoning.

e. (Advanced) As noted above, this sample was not a simple random sample, but rather a complex sample. As such, we should properly apply the sampling weight located in the second column of the table. Reopen the **Distribution** dialog, cast marst into the Y role, and select wt06 as the **Weight**. Re-construct the 95% CI for marst. Comment on how these results compare to your earlier results.

8. *Scenario:* In this exercise, we return to the **FAA Bird Strikes** data table. Recall that this set of data is a record of events in which commercial aircraft collided with birds in the United States.

a. Does this dataset satisfy the conditions for inference? Explain.

b. Construct and interpret a 95% confidence interval for the population proportion of flights where just one bird was struck.

c. Construct and interpret a 99% confidence interval for the same proportion using the same column. Explain fully how this interval compares to the one in part a. What can you say in general about the effect of the confidence level?

   d. Based on this sample, is it credible to think that approximately 85% of the time an airplane strikes at least one bird? Explain what analysis you chose to do, and describe the reasoning that leads to your conclusion.

9. *Scenario:* The data table **Hubway** contains observations of 55,230 individual bike trips using the short-term rental bike-sharing system in Boston, Massachusetts. People can register with the program or simply rent a bike as needed; within the data table, individual trips are either by Registered or Casual users.

   a. Does this data table satisfy the conditions for inference? Explain.

   b. Estimate (90% CI) the proportion of all Hubway trips taken by Casual users.

   c. Of the 55,230 trips in the sample 2,136 of them started at the Boston Public Library. Construct and interpret a 95% confidence interval for the proportion of all Hubway trips originating at the Library.

   d. Suppose that the data table had only 5,523 rows (one-tenth the sample size) and 214 trips starting at the Library. Using this summary data, construct and interpret a 95% confidence interval for the proportion of all Hubway trips originating at the Library.

   e. Compare the intervals in parts c and d. How does the size of the sample effect the confidence interval?

---

[1] "1-Alpha" is admittedly an obscure synonym for "confidence level." It is the complement of the significance level discussed earlier (α).

Inference for a Single Continuous Variable

Overview

Chapter 10 introduced inference for categorical data and this chapter does the same for continuous quantitative data. As in Chapter 10, we'll learn to conduct hypothesis tests and to estimate a population parameter using a confidence interval. We already know that every continuous variable can be summarized in terms of its center, shape, and spread. There are formal methods of inference for a population's center, shape, and spread based on a single sample. In this chapter we'll just deal with the mean.

Conditions for Inference

Just as with categorical data, we rely on the reasonableness of specific assumptions about the conditions of inference. Because the conditions are so fundamental to the process of generalizing from sample data, we begin each of the next several chapters by asking "what are we assuming?"

Our first two assumptions are the same as before: observations in the sample are independent of one another and the sample is small in comparison to the entire population (usually 10% or less). We should also be satisfied that the sample is probabilistic and therefore can reasonably represent the population. If we know that our sample data violates one or more of these conditions, then we should also understand that we'll have little ability to make reliable inferences about the population mean.

The first set of procedures that we'll encounter in this chapter relies on the sampling distribution of the mean, which we introduced in Chapter 8. We can apply the *t*-distribution reliably if the variable of interest is normally distributed in the parent population, or if we can invoke the Central Limit Theorem because we have ample data. If the parent population is highly skewed, we might need quite a large *n* to approach the symmetry of a *t*-distribution; the more severe the skewness of the population, the larger the sample required.

Using JMP to Conduct a Significance Test

We'll return to the pipeline safety data table and reanalyze the measurements of elapsed time until the area around a disruption was made safe. Specifically, suppose that there were a national goal of restoring the areas to safe conditions in less than three hours (180 minutes) on average, and we wanted to ask if this goal has been met. We could use this set of data to test the following hypotheses:

H₀: $\mu \geq 180$ [null hypothesis: the goal has not been met]
Hₐ: $\mu < 180$ [alternative hypothesis: the goal has been met]

As before, we'll assume that the observations in the table are a representative sample of all disruptions, that the observations are independent of one another, and that we're sampling from an infinite process.

1. Open the **Pipeline Safety** data table.

2. Select **Analyze ▶ Distribution**. Select **STMin** as **Y**, and click **OK**. You've seen this histogram before.

3. Click the red triangle next to **Distributions** and **Stack** the results (shown in Figure 11.1 on the next page).

Figure 11.1: Distribution of Time Until Area Made Safe (STMin)

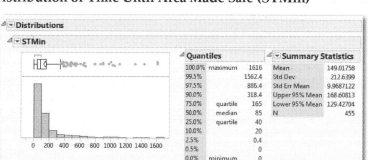

Before going further, it is important to consider the remaining conditions for inference. The histogram reveals a strongly right-skewed distribution, most likely revealing a sample that comes from a similarly skewed population. Because our sample has 455 observations, we'll rely on the Central Limit Theorem to justify the use of the *t*-distribution.

If we had a smaller sample and wanted to investigate the normality condition, we would construct a normal quantile plot by clicking on the red triangle next to **STMin** and selecting that option.

4.   Click the red triangle next to **STMin**, and select **Test Mean**. This opens the dialog box shown in Figure 11.2. Type 180 as shown and click **OK**.

Figure 11.2: Testing a Mean

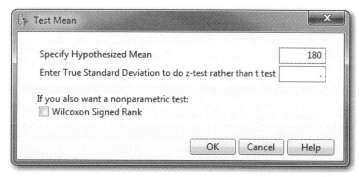

Notice that the dialog box has two optional inputs. First, there is a box into which you can **"Enter True Standard Deviation to do z-test rather than t-test."** In an introductory statistics course, one might encounter a textbook problem that provides the population standard deviation, and therefore one could base a test on the Normal distribution rather than the *t*-distribution (that is, conduct a z-test). Such problems might have instructional use, but realistically, one never knows the population standard deviation. The option to use the z-test has virtually no practical applications except possibly when one uses simulated data.

The second option in the dialog box is for the **Wilcoxon Signed Rank** test; we'll discuss that test briefly in the next section. First, let's look at the results of the *t*-test in Figure 11.3.

Figure 11.3: T-Test Results

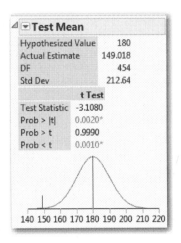

In the test report, we find that this test compares a hypothesized value of 180 minutes to an observed sample mean of 149.018, using a *t*-distribution with 454 degrees of freedom. The sample standard deviation is 212.64 minutes.

We then find the **Test Statistic**, which JMP computes using the following formula:

$$t = \frac{(\bar{x} - \mu_0)}{s_{\bar{x}}} = \frac{(149.018 - 180)}{\left(212.64 \middle/ \sqrt{455}\right)} = -3.108$$

Below the test statistic are three *P*-values, corresponding to the three possible formulations of the alternative hypothesis. We are doing a less-than test, so we're interested in the last value shown in red: 0.0010. This probability value is so small that we should reject the null hypothesis. In view of the observed sample mean, it is not plausible to believe that the population mean is 180 or higher. The graph at the bottom of the panel depicts the situation. Under the null hypothesis that the population mean is truly 180 minutes, the blue curve shows the likely values of sample means. Our sample mean is marked by the red line, which sits –3.1 standard errors to the left of 180. If the null hypothesis were true, we're very unlikely to see a sample mean this low.

More About *P*-Values

P-values are fundamental to the significance tests we've seen thus far, and they are central to all statistical testing. You'll work with them in nearly every remaining chapter of this book, so it is worth a few moments to explore another JMP feature at this point.

1. Click the red triangle next to **Test Mean** and select **PValue animation**. This opens an interactive graph (Figure 11.4) intended to deepen your understanding of *P*-values.

Figure 11.4: Interactive P-Value Visualization Tool

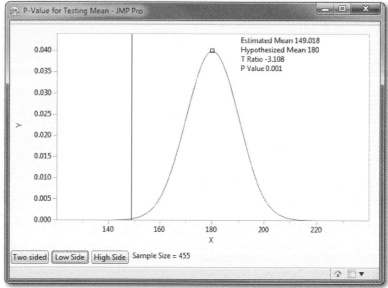

The red curve is the theoretical sampling distribution assuming that the population mean equals 180, and the vertical black line marks our sample mean. In the current example, our alternative hypothesis was $\mu < 180$, and because our sample mean was so much smaller than 180, we concluded that the null hypothesis was far less credible than this alternative. We can use this interactive graph to visualize the connection between the alternative hypothesis and the *P*-value.

2.   Click the **Low Side** button, corresponding to the less-than alternative.

This illustrates the reported *P*-value of 0.001. Clicking the other two buttons in turn shows where the probability would lie in either of the other two forms of the alternative.

Suppose that our initial hypothesized value had not been three hours (180 minutes) but a lower value. How would that have changed things?

3.   Notice that there is a small square at the peak of the curve. Place your cursor over that peak, click and drag the cursor slowly to the left, essentially decreasing the hypothesized value from 180 minutes to a smaller amount.

As you do so, the shaded area grows. Using the initial value of 180, the results of this test cast considerable doubt on the null hypothesis. Had the null hypothesis referred to a smaller number, then our results would have been less closely associated with the low side alternative.

4.   Continue to move the center of the curve until the *P*-value decreases to just about 0.05—the conventional boundary of statistical significance.

From this interaction we can see that our sample statistic of 149.018 would no longer be statistically significant in a low side test (that is, with a less-than alternative) if the hypothesized mean is about 165 or less.

What if our sample had been a different size?

5.   Click the number **455** next to **Sample Size** in the lower center of the graph. Change it to 45 and observe the impact on the graph and the *P*-value.

6.   Change the sample size to 4,500 and observe the impact.

The magnitude of the *P*-value depends on the shape of the sampling distribution (assumed to be a *t*-distribution here), the size of the sample, the value of the hypothesized mean, and the results of the sample. The size of the *P*-value is important to us because it is our measure of the risk of Type I errors in our tests.

Ideally, we want to design statistical tests that lead to correct decisions. We want to reject the null hypothesis when the alternative is true, and we want to retain the null hypothesis when the alternative is false. Hypothesis testing is driven by the desire to minimize the risk of a Type I error (rejecting a true null hypothesis). We also do well to think about the capacity of a particular test to discern a true alternative hypothesis. That capacity is called the *power of the test*.

The Power of a Test

Alpha (α) represents the investigator's tolerance for a Type I error. There is another probability known as beta (β), which represents the probability of a Type II error in a hypothesis test. The *power of a test* equals $1 - \beta$. In planning a test, we can specify the acceptable α level in advance, and accept or reject the null hypothesis when we compare the *P*-value to α. This is partly because the P-value depends on a single hypothesized value of the mean, as expressed in the null hypothesis.

The situation is not as simple with β. A Type II error occurs when we don't reject the null hypothesis, but we should have done so. That is to say, the alternative hypothesis — which is always an inequality — is true, but our test missed it. In our example, where the alternative hypothesis was $\mu < 180$, there are an infinite number of possible means that would satisfy the alternative hypothesis. To estimate β, we need to make an educated guess at an actual value of μ. JMP provides another animation to visualize the effects of various possible μ values.

For this discussion of power, we'll alter our example and consider the two-sided version of the significance test: is the mean different from 180 minutes?

1. Click the red triangle next to **Test Mean** and select **Power animation**. This opens an interactive graph (Figure 11.5 below) that illustrates this concept of power.

Figure 11.5: The Power Animation

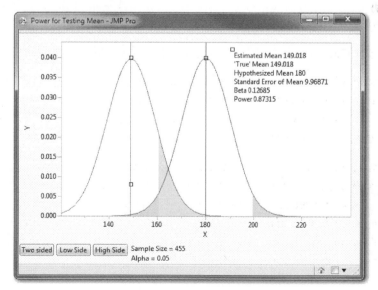

2. Initially, the annotations on the graph (estimated mean and so on) are far in the upper left corner. Grab the small square at the upper left and slide all of text to the right as shown in Figure 11.5 so that you can see it clearly.

This graph is more complex than the earlier one. We initially see two distributions. The red one (on the right), centered at 180, shows the sampling distribution based on the assumption expressed in the null hypothesis. The blue curve on the left shows an alternative sampling distribution (a parallel universe, if you like) based on the possibility that μ actually is equal to 149.018, the value of the sample mean.

So this graph lays out one pre-sampling scenario of two competing visions. The red vision depicts the position of the null hypothesis: the population mean is truly 180, and samples vary around it. The blue vision says "Here's another possible reality: the population mean just so happens to be 149.018." We'll be able to relocate the blue curve, and thereby see how the power of our test would change if the population mean were to be truly any value we choose.

The graph initially displays a two-sided test. In the red curve, the pink-shaded areas now represent not the *P*-value but rather α = .05. That is, each tail is shaded to have 2.5% with 95% of the curve between the shaded areas. If we take a sample and find the sample mean lying in either pink region, we'll reject the null hypothesis. In advance of sampling, we decided to make the size of those regions 5%. This is the rejection rule for this particular test.

The blue-shaded area shows β. As it is depicted, with the population mean really at 149.018, the null hypothesis *is* false and should be rejected. Among the possible samples from the blue universe (which is where the samples would really come from) about 13% of them are outside of the rejection region of the test, and in those 13% of possible samples, we'd erroneously fail to reject the null hypothesis. The power of this test is 0.87, which is to say that if the true mean were really 149.018 minutes, this test would correctly reject the null hypothesis in 87% of possible samples. Not bad. Now let's interact with the animator.

3. First, let's change **Alpha** from **0.05** to 0.01. Click **0.05** and edit it.

Notice what happens in the graph. Because alpha is smaller, the pink shaded area is smaller and the light blue shading gets larger. There is a trade-off between α and β: if we tighten our tolerance for Type I error, we expose ourselves to a greater chance of Type II error and also diminish the power of a test.

4. Restore **Alpha** to 0.05, and now double the **Sample Size** to 910. You'll need to rescale the vertical axis (grab the upper portion with the hand tool) to see the full curves after doing this.

Doubling the size of the sample radically diminishes beta and substantially increases power from 0.87 to 0.99. So there's a lesson here: another good reason to take large samples is that they make tests more powerful in detecting a true alternative hypothesis.

5. Finally, grab the small square at the top of the blue curve, and slide it to the right until the true mean is at approximately 170 minutes.

This action reveals yet another vertical line. It is green and it represents our sample mean. More to the point, however, see what has happened to the power and the chance of a Type II error. If the mean truly were 170 minutes, the alternative is again true and the null hypothesis is false: the population mean is not 180 as hypothesized, but 10 minutes less.

When the discrepancy between the hypothesized mean and the reality shrinks, the power of the test falls and the risk of Type II error increases rapidly. This illustrates a thorny property of power and one that presents a real challenge. Hypothesis testing is better at detecting large discrepancies than small ones. If our null hypothesis is literally false but happens to be close to the truth, there will be a relatively good chance that a hypothesis test will lead us to the incorrect conclusion. Enlarging the sample size would help, if that is feasible and cost-effective.

What if Conditions Aren't Satisfied?

The pipeline data table contains a large number of observations so that we could use the *t*-test despite the indications of a highly skewed population. What are our options if we don't meet the conditions for the *t*-test?

If the population is reasonably symmetric, but normality is questionable, we could use the Wilcoxon Signed Rank test. This is an option presented in the **Test Means** command, as illustrated in Figure 11.2. Wilcoxon's test is identified as a *non-parametric test*, which just indicates that its reliability does not depend on the parent population following any specific probability distribution. Rather than testing the mean, the test statistic computes the difference between each observation and the median, and ranks the differences. This is why symmetry is important: in a skewed data set the gap between mean and median is comparatively large.

We interpret the *P*-values just as we do in any hypothesis test, and further information is available within the JMP Help system. At this point it is useful to know that the *t*-test is not the only viable approach.

In recent years, many statisticians have taken yet another approach known as *bootstrapping*. The idea of bootstrapping is to use software to take a large number of repeated random samples with replacements from the original data table, and to compute the sample mean for each of these samples. This rapidly produces an empirical sampling distribution for the population mean and permits a judgment about the likelihood of the one observed sample mean.

The professional edition of JMP (JMP PRO©) includes a bootstrapping capability. Because it is not included in the standard version of the software, it is not illustrated here.

If the sample is small and strongly skewed, then inference is risky. If possible, one should gather more data. If costs or time constraints prohibit that avenue, and if the costs of indecision exceed those of an inferential error, then the techniques illustrated above might need to inform the eventual decision (however imperfectly) with the understanding that the estimates of error risk are inaccurate.

Using JMP to Estimate a Population Mean

Just as with categorical data, we may want to estimate the value of the population parameter—in this case, the mean of the population. We'll continue examining the number of minutes required to restore an area to safety following a pipeline break. We'll start by looking back at the initial **Distributions** report that we created a few pages back.

You might not have noticed previously, but JMP always reports a confidence interval for the population mean, and it does so in two ways. Box plots include a *Mean Confidence Diamond* (see Figure 11.6) and the **Summary Statistics** panel reports the **Upper 95% Mean** and **Lower 95% Mean**. The mean diamond is a visual representation of the confidence interval, with the left and right (or lower and upper) points of the diamond corresponding to the lower and upper bounds of the confidence interval.

Figure 11.6: Box Plot Detail Showing Mean Confidence Diamond

Now look at the **Summary Statistics** panel, shown earlier in Figure 11.1. If we can assume that the conditions for inference have been satisfied, we can estimate with 95% confidence that the nationwide mean time to secure the area of a pipeline disruption in the U.S. is between 129.4 and 168.6 minutes.

> Notice that JMP calculates the bounds of the interval any time we analyze the distribution of a continuous variable; it is up to us whether the conditions warrant interpreting the interval.

What if we wanted a 90% confidence interval, or any confidence level other than 95%?

1. Click the red triangle next to **STMin**, and select **Confidence Interval ▸ 0.90**.

Figure 11.7: 90% Confidence Intervals

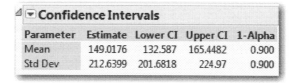

Parameter	Estimate	Lower CI	Upper CI	1-Alpha
Mean	149.0176	132.587	165.4482	0.900
Std Dev	212.6399	201.6818	224.97	0.900

A new panel opens in the results report (see Figure 11.7). The report shows the boundaries of the 90% confidence interval for the population mean. It also shows a confidence interval for the population standard deviation—not only can we estimate the population's mean (μ), but we can similarly estimate the standard deviation, σ.

Matched Pairs: One Variable, Two Measurements

The *t*-distributions underlying most of this chapter were the theoretical contribution of William Gosset, so it is appropriate to conclude this chapter with some data originally collected by Gosset, whose expertise was in agriculture as it related to the production of beer and ale, the industry in which he worked as an employee of the Guinness Brewery.

This particular set of data contains measurements of crop yields from eleven different types of barley (referred to as corn in Gosset's day) in a *split plot* experiment. Seeds of each type of barley were divided into two groups. Half of the seeds were dried in a kiln, and the other half were not. All of the seeds were planted in eleven pairs of farm plots, so that the kiln-dried and regular seeds of one variety were in adjacent plots.

As such, the data table contains 22 measurements of the same variable: crop yield (pounds of barley per acre). However, the measurements are not independent of one another, because adjacent plots used the same variety of seed and were presumably subject to similar conditions. The point of the experiment was to learn if kiln-dried seeds increased the yields in comparison to conventional air-dried seeds. It is important that the analysis compare the results variety by variety, isolating the effect of drying method.

The appropriate technique is equivalent to the one-sample *t*-test, and rather than treating the data as 22 distinct observations, we treat it as 11 *matched pairs* of data. Rather than analyze the individual values, we analyze the *differences* for each pair. Because we have a small sample, the key condition here is that the population differences be approximately normal, just as in the one-sample *t*-test.

1. Open the data table **Gosset's Corn**. The first column shows the yields for the 11 plots with kiln-dried seeds and the second column shows the corresponding yields for the regular corn. Glancing down the columns, we see that the kiln-dried values are sometimes larger and sometimes smaller than the others.

2. Select **Analyze ► Matched Pairs**. Choose both columns and select them as **Y, Paired Response**, and click **OK**. The results are shown in Figure 11.8.

Figure 11.8: The Results of the Matched Pairs Command

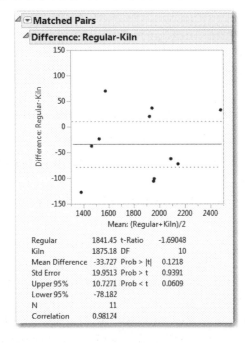

This report provides a confidence interval and *P*-values for the mean of the pairwise differences. First, we see a graph of the eleven differences in comparison to the mean difference of –33.7 pounds per acre. There is variation, but on average the kiln-dried seeds yielded nearly 34 pounds more per acre within the sample (the differences are calculated as regular – kiln). Now we want to ask what we can infer about the entire population based on these results.

Next note the confidence interval estimate. With 95% confidence, we can conclude that the yield difference between regular and kiln-dried seeds is somewhere in the interval from –78.182 to +10.7271. Because this interval includes 0, we cannot confidently say that kiln-drying increases or decreases yield in the population, even though we did find an average increase in this one study.

On the lower right side of the report, we find the results of the significance test. We're interested in the bottom *P*-value, corresponding to the less-than test (that is, our alternative hypothesis was that kiln-dried yields would be greater than regular, so that differences would be negative). Based on this *P*-value, the results fall slightly short of conventional significance; we don't reject the null hypothesis.

Application

Now that you have completed all of the activities in this chapter, use the concepts and techniques that you've learned to respond to these questions.

1. *Scenario:* We'll continue our analysis of the pipeline disruption data. The columns **PRPTY** and **GASPRP** contain the dollar value of property and natural gas lost as a consequence of the disruption.

   a. Do each of these columns satisfy the conditions for statistical inference? Explain.

   b. Report and interpret a 95% confidence interval for the mean property value costs of a pipeline disruption.

   c. Now construct and interpret a 90% confidence interval for the same parameter.

   d. In a sentence or two, explain what happens when we lower the confidence level.

   e. Construct and interpret a 99% confidence for the dollar cost of natural gas loss in connection with a pipeline disruption.

    f.   Conduct a test using an alternative hypothesis that, on average, a pipeline disruption loses less than $20,000 worth of gas. Report on your findings.

    g.   Use the *P*-value animation tool to alter the null hypothesis just enough to change your conclusion. In other words, at what point would the data lead us to the opposite conclusion?

    h.   Use the power animator and report on the power of this test.

2.   *Scenario:* Recall the data table called **Michelson 1879**. These are Michelson's recorded measurements of the speed of light. Because the measurements were taken in different groups, we'll focus only on his last set of measurements. Before doing any further analysis, take the following steps.

- Open the data table.

- Select **Rows ▶ Data Filter**. Highlight **Trial#** and click **Add**.

- Select the **Show** and **Include** check boxes, and select **fifth** from the list. Now your analysis will use only the 20 measurements from his fifth trial.

    a.   Assume this is an SRS. Does the **Velocity** column satisfy our conditions for inference? Explain.

    b.   Construct and interpret a 90% confidence interval for the speed of light, based on Michelson's measurements.

    c.   Michelson was a pioneer in measuring the speed of light, and the available instruments were not as accurate as modern equipment. Today, we generally use the figure 300,000 kilometers per second (kps) as the known constant speed. Put yourself in Michelson's shoes, and conduct a suitable hypothesis test to decide whether 300,000 kps would have been a credible value based on the measurements he had just taken.

    d.   Use the *P*-value animation tool to alter the null hypothesis so that Michelson might have believed the null hypothesis. In other words, at what point would his data no longer be statistically distinguishable from the null hypothesis value?

3.   *Scenario:* Recall the data in the **NHANES** data table. Specifically, remember that we have two measurements for the height of very young respondents: **BMXRECUM** and **BMXHT**, the recumbent (reclining) and standing heights.

    a.   In a sentence or two, explain why you might expect these two measurements of the same individuals to be equal or different.

b. Perform a matched pairs test to decide if the two ways of measuring height are equivalent. Interpret the reported 95% confidence interval, and explain the test results. What do you conclude?

4. *Scenario:* The data table **Airline Delays** contains a sample of flights for two airlines destined for four busy airports. Using the data filter as described in Scenario 2 above, select just the sample of American Airlines flights at Chicago's O'Hare Airport (ORD) for analysis. Our goal is to analyze the delay times, expressed in minutes.

    a. Assume this is an SRS of many flights by the selected carrier over a long period of time. Have we satisfied the conditions for inference about the population mean delay? Explain.

    b. Report and interpret the 95% confidence interval for the mean flight delay for American Airlines flights bound for Chicago.

    c. Let L stand for the lower bound of the interval and U stand for the upper bound. Does this interval tell you that there is a 95% chance that your flight to Chicago will be delayed between L and U minutes? What *does* the interval tell you?

    d. Using a significance level of 0.05, test the null hypothesis that the population mean delay is twelve minutes against a one-sided alternative that the mean delay is less than twelve minutes.

    e. Now suppose that the true mean of the delay variable is actually ten minutes. Use the power animation to estimate the power of this test under that condition. Explain what you have found.

5. *Scenario:* We have additional flight data in the data table **Airline delays 3**, which contains data for every United Airlines flight departing Chicago's O'Hare International Airport from March 8 through March 14, 2009. Canceled flights have been dropped from the table. Our goal is to analyze the departure delay times, expressed in minutes.

    a. Assume this is an SRS. Have we satisfied the conditions for inference about the population of United Flights?

    b. "Departure Delay" refers to the time difference between the scheduled departure and the time when the aircraft's door is closed and the plane pulls away from the gate. Develop a 95% confidence interval for the mean departure delay **DepDelay**, and interpret the interval.

c. Sometimes there can be an additional delay between departure from the gate and the start of the flight. "Wheels off time" refers to the moment when the aircraft leaves the ground. Develop a 95% confidence interval for the mean time between scheduled departure and wheels-off (**SchedDeptoWheelsOff**). Report and interpret this interval.

d. Now use a matched-pairs approach to estimate (with 95% confidence) the mean difference between departure delay and wheels off delay. Explain what your interval tells you.

6. *Scenario:* In this exercise, we return to the **FAA Bird Strikes** data table. Recall that this set of data is a record of events in which commercial aircraft collided with birds in the United States.

a. Do the columns **speed** and **cost of repairs** satisfy the conditions for statistical inference? Explain.

b. Report and interpret a 95% confidence interval for the mean speed of flights.

c. Now construct and interpret a 99% confidence interval for the same parameter.

d. In a sentence, explain what happens when we increase the confidence level.

e. Conduct a test using an alternative hypothesis that, on average, flight repairs caused by bird strikes cost less than $170,000. Report on your findings.

f. Use the *P*-value animation tool to alter the null hypothesis just enough to change your conclusion. In other words, at what point would the data lead us to the opposite conclusion?

g. Use the power animator and report on the power of this test.

7. *Scenario:* In 2004, the state of North Carolina released to the public a large data set containing information on births recorded in this state. Our data table **NC Births** is a random sample of 1,000 cases from this data set. In this scenario, we will focus on the birth weights of the infants (**weight**) and the amount of weight gained by mothers during their pregnancies (**gained**). All weights are measured in pounds.

   a. Do the data columns labeled weight and gained satisfy the conditions for inference? Explain.

   b. Test the hypothesis that the mean amount of weight gained by all N.C. mothers was more than 30 pounds in 2004. Explain what you did and what you decided. Use a significance level of 0.05.

   c. Construct and interpret a 95% confidence interval for the mean birth weight of NC infants in 2004.

Chi-Square Tests

Overview

This chapter takes us further into inferential analysis of categorical data. Two prior chapters have concentrated on categorical variables. In Chapter 6, we learned to compute relative frequencies, which we then used to illustrate concepts and rules of probability. In Chapter 10, we studied dichotomous categorical variables to learn about confidence intervals and hypothesis tests for a single population proportion.

In this chapter, we'll consider two basic issues. First, we'll learn to draw inferences about a categorical variable that can take on more than two possible values. Second, we'll use sample data to infer whether two categorical variables are independent or related. The thread that unifies most of the discussion in this chapter is another family of continuous density functions called *chi-square* distributions. The test statistics that we'll encounter in this chapter follow the shape and proportions of a chi-square (χ^2) distribution, enabling us to calibrate the likelihood of sampling error.

Chi-Square Goodness-of-Fit Test

In Chapter 10, we worked with the **Pipeline Safety** data table to learn about estimating a population proportion. In that chapter, we looked at the **IGNITE** column—a dichotomous variable indicating whether natural gas ignited in connection with a reported disruption. At this point, let's consider a different variable in the data table that indicates the type of disruption. The **LRTYPE_TEXT** column shows four different values: **LEAK**, **RUPTURE**, **OTHER,** and **N/A** (not available) corresponding to the reported type of disruption involved in the pipeline incident. Suppose we naively wondered whether all disruption types are equally likely to occur.

Here's how we'll approach the question. First we'll specify a hypothesis reflecting our question and express the hypothesis as a model describing the causes. There are four categories, and we're provisionally asserting that all four categories happen with equal probability in the United States. In short, our hypothesis becomes this:

$$H0: Pr(\textbf{LEAK}) = Pr(\textbf{RUPTURE}) = Pr(\textbf{OTHER}) = Pr(\textbf{N/A}) = 0.25$$

Next, we'll examine the sample data in comparison to the model. As usual, the software does the heavy work, but the underlying logic is simple. For this variable, our sample size happens to be $n = 447$. Before we look at the distribution of the observed data, we can note that if our null hypothesis is true, then we anticipate observing that approximately $0.25 \times 447 = 111.75$ incidents are classified as each of the four types.

Obviously, we won't have observed 111.75 instances of any type—that's not possible. If the model is credible, though, we should discover somewhere in the vicinity of 112 leaks, 112 ruptures, and so on. As usual in statistical testing, we will need to decide precisely what qualifies as "in the vicinity" of 111.75. If some of the observed frequencies deviate too far from 111.75, we'll conclude that there is a mismatch between the model and the empirical data. If the observed frequencies are all close to 111.75, we'll conclude that the model is credible. Appropriately, the name of the test we'll use is the *goodness-of-fit* test. The goal of this test is to make a decision about how well the model fits with the data.

1. Open the **Pipeline Safety** data table.

2. Select **Analyze ▶ Distribution**. Cast **LRTYPE_TEXT** as **Y** and click **OK**.

3. Click the red triangle next to **LRTYPE_TEXT** and select **Test Probabilities**. You should see a new panel like the one shown in Figure 12.1.

Figure 12.1: Test Probabilities Panel

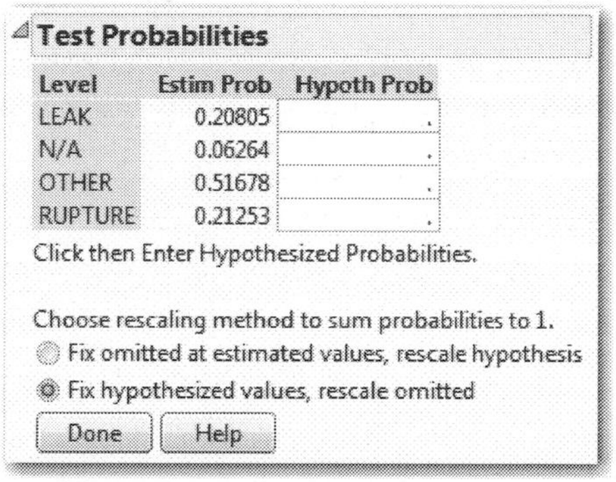

4. Our hypothesized probabilities are all 0.25, so go ahead and enter .25 into each of the four cells in the **Hypoth Prob** column and click **Done**. The new results are in Figure 12.2 on the next page.

> JMP anticipates that we might have hypothesized different probabilities for each category, so we separately enter each probability value. In fact, we could just enter a 1 into each cell and JMP will rescale all values to equal 0.25; in this example, we'll take the trouble to enter the hypothesized probabilities to reinforce the fact that we are testing the observed relative frequencies against these specific values. If we had a model with different null values, the procedure would be the same, and we would enter the appropriate values. The software needs to ensure that the sum of the probabilities is 1, and it needs our instructions about how to correct things if it isn't. By selecting the **Fix hypothesized values, rescale omitted** option, we're telling JMP to lock in the numbers we enter, and change any omitted cell to balance the sum to 1.

JMP reports two test statistics and their associated P-values. Under the null hypothesis with sufficiently large samples, both test statistics asymptotically follow an approximate chi-square distribution with $k - 1$ degrees of freedom (DF), where k refers to the number of categories for the variable in question. In this discussion, we'll consider Pearson's chi-square[1]; both test results are interpreted in the same way.

Figure 12.2: The Results of the Goodness-of-Fit Test

Test Probabilities		
Level	**Estim Prob**	**Hypoth Prob**
LEAK	0.20805	0.25000
N/A	0.06264	0.25000
OTHER	0.51678	0.25000
RUPTURE	0.21253	0.25000

Test	ChiSquare	DF	Prob>Chisq
Likelihood Ratio	192.9603	3	<.0001*
Pearson	195.6756	3	<.0001*

Method: Fix hypothesized values, rescale omitted

The reported test statistic was computed using this formula:

$$\chi^2 = \sum_{i=1}^{k} \frac{(O_i - E_i)^2}{E_i}$$

where O_i is the observed frequency of category i and E_i is the expected frequency of category i assuming that the null hypothesis is true. In this example, the expected frequency for each of the four categories is 111.75, as explained earlier.

Pearson's chi-square statistic aggregates the discrepancies between the observed and expected counts for each category. If the data perfectly corresponded to the model, the numerator of each term in the above sum would be zero, and the test statistic would equal zero. To the extent that the data and the model don't correspond, the statistic grows.

When the null is true, there will be some small aggregated discrepancy due to sampling error and to the fact that observed frequencies are integers. That aggregated discrepancy follows a chi-square distribution, enabling JMP to compute a P-value for the test statistic. When the discrepancy is so large as to cast doubt on the null hypothesis (as reflected in a very tiny P-value), we reject the null. That's the case in this instance. The P-value is reported as being smaller than 0.0001, so we reject the null hypothesis and conclude that our initial model of equal likelihoods does *not* fit with the data.

What Are We Assuming?

This chi-square test has comparatively few assumptions. As with other forms of inference, we can generalize to the population if we can assume that the sample is representative of the population, and hence we need to believe that the observations in the sample are independent of one another. If the sample is randomly selected, we can safely assume independence.

The test can be unreliable if the *expected frequency* of any one category is less than 5. This condition involves both the number of categories as well as the sample size. Note that JMP will compute and report test results whether this condition is satisfied. It is up to the analyst to make the judgment about whether the conditions warrant interpretation of the test.

Inference for Two Categorical Variables

Let's carry the analysis a step further. In Chapter 10, we looked at the likelihood of ignition in connection with a major disruption, and we estimated that approximately 81% of incidents involved ignition. Is the risk of ignition somehow related to the type of disruption? In other words, are **IGNITE** and **LRTYPE_TEXT** *independent*? You might remember that we discussed independence back in Chapter 5 and used the conditional probabilities from a contingency table to make decisions about whether two events were independent. In Chapter, 5 the subject was the behavior of dolphins and we considered two events at a time. With the current example, we'll ask a broader question: are the variables independent of one another?

Contingency Tables Revisited

Before considering the inferential question, let's start with a review of contingency tables by making one for this example. We created several contingency tables in Chapter 5.

1. Select **Analyze ▶ Fit Y by X**. Choose **IGNITE** as **Y** and **LRTYPE_TEXT** as **X**.

The results window shows a mosaic plot, a contingency table, and test results. Let's look initially at the mosaic plot and contingency table, shown in Figures 12.3 and 12.4. As you look at these two figures, remember that they both display the same information.

In the mosaic plot, the vertical dimension represents whether there was ignition and the horizontal represents the various incident types. The visual prevalence of dark blue[2] shows that ignition occurs in the vast majority of incidents. At the far right, we see the marginal breakdown with about 80% of the incidents involving gas that ignited. The left portion of the plot shows that the prevalence of ignition within the sample observations seems to vary a bit. Notably, leaks were less likely to have ignited—among those incidents reported in this particular sample.

Figure 12.3: Mosaic Plot for Pipeline Data

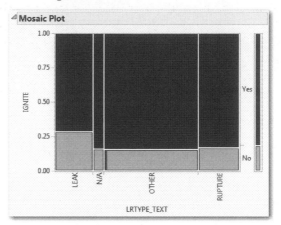

Now let's consider the contingency table and track down the corresponding numerical values. The table provides the specific values that underlie the visual impression conveyed by the graph.

Figure 12.4 Basic Contingency Table for Ignite vs. Type

	Count / Total % / Col % / Row %	No	Yes	
LEAK	Count	25	63	88
	Total %	5.90	14.86	20.75
	Col %	31.65	18.26	
	Row %	28.41	71.59	
N/A	Count	4	21	25
	Total %	0.94	4.95	5.90
	Col %	5.06	6.09	
	Row %	16.00	84.00	
OTHER	Count	34	182	216
	Total %	8.02	42.92	50.94
	Col %	43.04	52.75	
	Row %	15.74	84.26	
RUPTURE	Count	16	79	95
	Total %	3.77	18.63	22.41
	Col %	20.25	22.90	
	Row %	16.84	83.16	
		79	345	424
		18.63	81.37	

As in Chapter 6, we'll look in the margins of the table for the categories and their marginal frequencies, and at the inner cells for joint and conditional frequencies. To find the probability (relative frequency) of any one category, we look in the opposite margin. For example, 79 of 424 incidents did not ignite. The relative frequency was 18.63%, so we find that *Pr(No)* = 0.1863.

Within the interior cells, the upper number is the joint frequency showing the number of incidents with or without ignition for each disruption type. The next number is the *joint relative frequency*. Thus, in the upper-left cell, we see that 25 of the 424 incidents were leaks without ignition for a joint relative frequency of 5.9% of all incidents.

The upper-left cell of the table reports that the next two numbers within each cell are the column percentage and the row percentage. These are *conditional* values. A column percentage is the relative frequency of the row event within the column only. In other words, restricting ourselves to the condition represented by the column, it is the relative frequency of each row. Similarly, the row percentages give the probability of the column events conditioned by each row.

The contingency table provides an informative summary of the sample data. Within the sample, the relative frequency of ignition plainly varies across the different incident types. Based on this evidence, can we conclude a similar pattern of variation occurs in the entire population of incidents?

Chi-Square Test of Independence

The chi-square test of independence uses sample data to test the null hypothesis that two categorical variables are independent of one another. In this example, if ignition is independent of incident type, then gas is just as likely to ignite in one type of incident as in another. Or, to put it another way, knowing the type of disruption would give no information about the chance of ignition.

In this test the null hypothesis is always the same:

> H₀: The two variables are independent.

Naturally, the alternative hypothesis is that they are not independent, but rather related in some way. This test works much the same as the goodness-of-fit test. Our null hypothesis establishes expectations about the joint frequencies we would observe if the null were true. We can then compare the actual observed values to the expected values, aggregate the discrepancies, and gauge the size of the accumulated differences using a chi-square distribution. To understand the test more fully, do this:

1. Press the ALT key and click the red triangle next to **Contingency Table**. This opens a dialog titled **Select Options and click OK** (not shown here). Clear the options **Total %**, **Col %**, and **Row %**.

2. Select **Expected**, **Deviation**, and **Cell Chi Square** and then click **OK**.

Figure 12.5: Modified Contingency Table with Chi-Square Results

Contingency Table

	IGNITE		
Count Expected Deviation Cell Chi^2	No	Yes	
LEAK	25	63	88
	16.3962	71.6038	
	8.60377	-8.6038	
	4.5148	1.0338	
N/A	4	21	25
	4.65802	20.342	
	-0.658	0.65802	
	0.0930	0.0213	
OTHER	34	182	216
	40.2453	175.755	
	-6.2453	6.24528	
	0.9691	0.2219	
RUPTURE	16	79	95
	17.7005	77.2995	
	-1.7005	1.70047	
	0.1634	0.0374	
	79	345	424

(LRTYPE TEXT)

Tests

N	DF	-LogLike	RSquare (U)
424	3	3.2646115	0.0160

Test	ChiSquare	Prob>ChiSq
Likelihood Ratio	6.529	0.0885
Pearson	7.055	0.0702

After modifying the contingency table in this way, your results should match those in Figure 12.5.

Let's look closely at the numbers within the cell in the upper left of the table that correspond to leaks that did not ignite. There are four numbers in the cell, as there are in each cell of the table. The legend again appears in the upper-left corner of the table.

Each cell contains the observed count, the expected count, the deviation $(O - E)$, and the cell chi-square value. We'll consider these one at a time. As in Figure 12.4, we see that 25 of the 424 incidents were, in fact, leaks where gas did not ignite. Next is the expected count of 16.3962, computed in the following way.

Among all observed incidents 79 of 424, or 18.6321%, did not ignite. If the null hypothesis is true—that in the on-going process of pipeline disruptions ignition is independent of disruption type—then we'd expect 18.63% of all incidents to be free from ignition, regardless of incident type. If so, then 18.63% of the 88 leaks would have no ignition. 18.6321% of 88 equals 16.3962.

The deviation value of 8.60377 is just the difference between the observed and expected frequencies: $Oi - Ei$. Finally, the cell Chi^2 is $\dfrac{(O_i - E_i)^2}{E_i}$, or $8.60377^2 / 16.3962$

= 4.5148.

Each cell's deviation makes a contribution to the aggregated total chi-square value, as reported at the bottom of the panel. Pearson's chi-square for the two-dimensional contingency table is 7. 055. In a contingency table, the number of degrees of freedom equals the (number of rows – 1) x (number of columns – 1). Here, we have four rows and two columns, so DF = 3. If the null hypothesis is true, the chance of obtaining a chi-square statistic of 7.055 or larger is about 0.07. Assuming a conventional significance level of $\alpha = 0.05$, we'd conclude that this evidence is insufficient to reject the null. Although the evidence raises doubt about the independence of these two variables, the evidence is not compelling enough to conclude that they are related.

What Are We Assuming?

The assumptions for this test are the same as for the goodness-of-fit test. We can rely on the results if our observations are independent and if the expected count in each cell exceeds 5.

Notice that, unlike with the *t*-tests of earlier chapters, there is no assumption here about the shape or nature of the population distribution. For this reason, chi-square tests are also sometimes classified as "distribution-free" or *nonparametric* methods. The reliability of the test results is not dependent on how the categorical variables are distributed in the entire population.

Application

Now that you have completed all of the activities in this chapter, use the concepts and techniques that you've learned to respond to these questions. For each question, explain which test you've chosen, report the relevant test results, and interpret the results in the context of the scenario.

1. *Scenario:* Before we leave the pipeline data, let's ask two more questions:

   a. Are disruptions equally likely to occur in all regions of the United States?

   b. Are the variables **Explosion** and **LRTYPE_TEXT** independent?

   c. Are the variables **Region** and **Explosion** independent?

2. *Scenario:* The main example in Chapter 5 was based on the data table called **Dolphins**. Please note that this data table contains frequencies rather than raw data; refer to Figure 5.2 for instructions about creating the contingency table.

   a. Are the variables **Activity** and **Period** independent?

   b. Are dolphins equally likely to be observed feeding, socializing, and traveling? (Again, recall that the data table shows frequencies rather than raw data).

3. *Scenario:* We'll continue to examine the World Development Indicators data in **BirthRate 2005**. We'll focus on three categorical variables in the data table.

- **Region**: Global region of the country in question

- **Provider**: Source of maternity leave benefits (public, private, or a combination of both)

- **MatLeave90+**: Mandated length of maternity leave—fewer than 90 days, or 90 days and more.

   a. Are the variables **Provider** and **Region** independent?

   b. Are the variables **Provider** and **MatLeave90+** independent?

   c. Are the variables **MatLeave90+** and **Region** independent?

4. *Scenario:* In these questions, we return to the **NHANES** data table containing survey data from a large number of people in the U.S. in 2005. For this analysis, we'll focus on just the following variables:

- **RIDRETH1**: Respondent's racial or ethnic background
- **DMDMARTL**: Respondent's reported marital status.

   a. Are these two categorical variables independent?

5. *Scenario:* The following questions were inspired by a 2009 article on alcohol consumption and risky behaviors (Hingson *et al.* 2009) based on responses to a large national survey. The data table **BINGE** contains summary frequency data calculated from tables provided in the article. We'll focus on the self-reported frequency of binge drinking (defined as five or more drinks in a two-hour period for males and four or more drinks in a two-hour period for females) and whether the respondents reported having ever been involved in a car accident after drinking.

   a. Are the two variables independent? Explain.

6. *Scenario:* Let us once again look at the **Sleeping Animals** data table.

   a. Are mammalian species equally likely to be classified across the five categories of the exposure index?

   b. Are all mammalian species equally likely to be classified across the five categories of the predation index?

   c. Are the variables **Predation** and **Exposure** independent? Explain.

7. *Scenario:* For these questions we return to the **TimeUse** data table. Use the data filter to confine the analysis to the responses from 2007.

   a. Are employment status and gender independent? Explain.

   b. Are employment status and region independent? Explain.

   c. Are employment status and marital status independent? Explain.

   d. Are marital status and gender independent? Explain.

8. *Scenario:* Open the **NC Births** data table. Recall that this data table contains a simple random sample of 1,000 births in North Carolina in 2004.

   a. Use a chi-square test of goodness of fit to test the hypothesis that a women is as likely to give birth to a girl as to a boy. Explain what you did and what you concluded.

    b.   Are baby gender and whether or not the mother smoked independent? Explain.

    c.   Are low birth weight and whether or not the mother smoked independent? Explain.

    d.   Are low birth weight and mature mother independent? Explain.

9. *Scenario:* Open the **FAA Bird Strikes CA** data table, which contains data from three airports in California: Los Angeles, Sacramento, and San Francisco. For each question below, we want to know if the attributes in question varied by location (that is, is the response variable independent of airport?). Conduct a test of independence and report on your conclusions, explaining the reasoning for your conclusion. For all tests, use a significance level of 0.05.

    a.   Wildlife strikes occur during different phases of flights. Does the frequency of strikes in each phase vary by airport?

    b.   Does the airline involved in the incident vary by airport?

    c.   **Birds Struck** is an ordinal variable describing the number of birds struck during the flight. Is this variable independent of the airport?

    d.   Does the size of the bird(s) struck (**Size**) vary by airport?

---

[1] There is more than one method of computing a chi-square statistic. Pearson's is the statistic commonly presented in introductory statistics texts.

[2] On your screen, JMP is using the default mosaic plot color scheme of blue for Yes and red for No; because some readers of this book will see a grayscale version of this image, Figure 12.3 uses dark blue (dark gray) for Yes and light blue (light gray) for No.

Two-Sample Inference for a Continuous Variable

Overview

At the end of the prior chapter, we studied the matched pairs design to make inferences about the mean of differences within a population. In this chapter, we take up the comparison of samples drawn independently from two different populations, and learn to estimate the differences between population means and variances. The procedures discussed here assume that samples are randomized (and therefore representative) and are a comparatively small subsets of their respective populations.

Conditions for Inference

We'll first focus on the centers of the two populations, and assume that measurements are independent both within and across samples. As we did in the last chapter, we begin with cases in which the populations are either approximately normal or in which the Central Limit Theorem applies. Then, we'll see a nonparametric (also known as *distribution-free*) approach for non-normal data. Finally, we'll learn to compare the spread of two distributions as measured by the variance.

Using JMP to Compare Two Means

We've used the **NHANES** data to illustrate many statistical concepts, and we'll return to it again now. Within this large data table there are several measurements typically taken during a routine physical exam. For some of these measurements, we expect to find differences between male and female subjects. Men tend to be taller and heavier than women, for example. But are their pulse rates different? In our first illustration, we'll estimate the difference (if any) between the resting pulse rates of adult women and men.

Assuming Normal Distributions or CLT

1. Open the **NHANES** data table.

The subjects in the **NHANES** data table range in age from infants up. Our first step is to exclude subjects under the age of 18 years. We'll use the Data Filter to select just the adults.

2. Select **Rows ▶ Data Filter** As shown in the left side of Figure 13.1, choose the **RIDAGEYR** column and click **Add**.

Figure 13.1: Using the Data Filter to Restrict Analysis to Adults

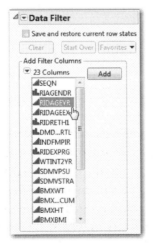
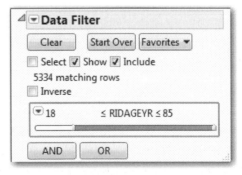

3.   In the filter, check the boxes as shown and edit the minimum age to equal 18.

Before performing any inferential analysis, let's look at the pulse rate data, separating the observations for females and males.

4.   Select **Analyze ▸ Distribution**. Cast **BPXPLS** (pulse rate) into **Y** and **RIAGENDR** into **By**.

5.   In the **Distribution** report, click the red triangle next to **Distributions**, and select **Stack** to create the report shown in Figure 13.2.

Figure 13.2: Distribution of Pulse Rate by Gender

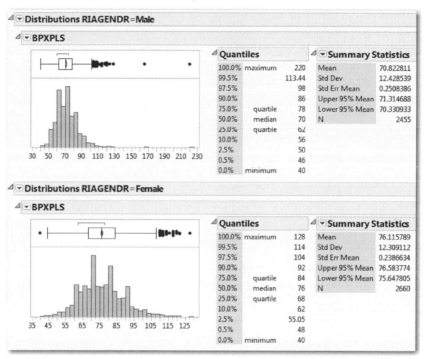

A quick glance suggests that each distribution is reasonably symmetric with a central peak and similar measures of dispersion. Given that fact as well as the large size of the samples, we can conclude that we're on solid ground using a *t*-test. As in the one-sample case, the *t*-test is reliable when the underlying population is approximately normal or when the samples are large enough for the Central Limit Theorem to apply. We have very large samples, so will rely on the CLT to proceed with the analysis.

6. Select **Analyze ▶ Fit Y by X**. Select **BPSXPLS** as **Y, Response** and **RIAGENDR** as **X, Factor,** and click **OK**.

7. Move your cursor over the graph and right-click to open a menu. Choose **Marker Size ▶ 1, Small**. With more than 2,000 observations for each gender, the default dot size makes it difficult to see the individual points.

8. Click the red triangle next to **Oneway Analysis**, and select **Display Options ▶ Points Jittered**. This randomly spaces out the points horizontally for each group, adding further resolution to the image.

9. Click the same red triangle again, and select **Display Options ▶ Box Plots**. When you've done this, your report looks like Figure 13.3.

Figure 13.3: Bivariate Comparison of Pulse Rates

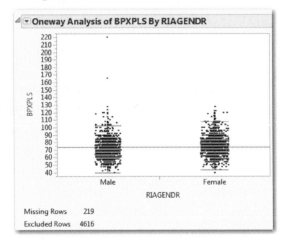

In this graph, we see side-by-side box plots of the observations. The box plots illustrate the symmetry of the two distributions, and suggest that they have similar centers and dispersion. Though this graph re-displays some of the information from Figure 13.2, here we get a clearer sense of just how the points are distributed.

The **Fit Y by X** platform allows us to test whether there is a statistically significant difference between the mean pulse rates of men and women based on these two large subsamples, and also to estimate the magnitude of the difference (if any).

10. Click the red triangle once more, and select **t Test**. This produces the results shown in Figure 13.4.

You may have noticed that, unlike the one-sample *t*-test, this command gives us no option to specify a hypothesized difference of means. Usually the two-sample *t*-test has as its default the null hypothesis of no difference. In other words, the default setting is that the difference in means equals 0. This version of the two sample *t*-test assumes that the variances of the two populations are unequal; we'll discuss that assumption shortly.

Figure 13.4: Results of a Bivariate *t*-Test

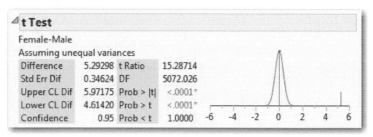

In the results panel, we see that there was a difference of 5.29 beats per minute between the females and males in the sample. We also find the 95% confidence interval for the difference in population means. Based on these two samples of women and men, we estimate that in the population at large, women's pulse rates are 4.61 to 5.98 beats per minute higher than men's.

We also see a *t*-distribution centered at 0 and a red line marking the observed difference within the sample. The observed difference of nearly 5.3 beats is more than 15.287 standard errors from the center—something that is virtually impossible by chance alone. Whether or not you expected to find a difference, these samples lead us to conclude that there is one.

Using Sampling Weights (optional section)

The **NHANES** data comes from a complex sample, and the data table includes sampling weights. To conduct inferential analysis properly with surveys like **NHANES**, we should apply the post-stratification weights. The weights do alter the estimated means and standard errors, and therefore lead to slightly different conclusions. We interpret weighted results exactly as described above, and the software makes it very easy to apply the weights.

1. Select **Analyze ▸ Distribution**. Select **BPXPLS** (pulse rate) as **Y** and **RIAGENDR** as **By**, and also select **WTINT2YR** as **Weight**.

All other instructions and interpretations are as described above. Similarly, you can use the **Weight** option in the **Fit Y by X** platform.

Equal vs. Unequal Variances

The *t*-test procedure we just performed presumed that the variance in pulse rates for women and men are unequal, and estimated the variance and standard deviation separately based on the two samples. That is a conservative and reasonable thing to do.

If we can assume that the populations not only share a common normal shape (or that the Central Limit Theorem applies because both samples are large) but also share a common degree of spread, we can combine the data from the two samples to estimate that shared variance. With smaller samples, there is some gain in efficiency if we can combine, or *pool*, the data to estimate the unknown population variances. Later in this chapter, we'll see a way to test whether the variances are equal or not. At this point, having noted that the samples seem to have similar spread, let's see how the pooled-variance version of this test operates.

1. Click the red triangle, and select **Means/Anova/Pooled t**. This produces the results shown in Figure 13.5.

Figure 13.5: The Pooled Variance *t*-Test

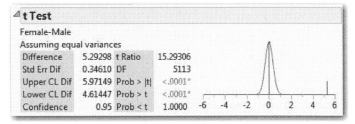

This command produces several new items in our report, all within a panel labeled **Oneway Anova**. We'll devote all of Chapter 14 to this technique, and won't say more about it here. Locate the heading **t Test**, and confirm that the report includes the phrase **Assuming equal variances**. We read and interpret these results exactly as we did in the unequal variance test.

Because the two sample variances are very similar and the samples are so large, these results are extremely close to the earlier illustration. The estimated pooled standard error (**Std Err Dif**) is just slightly smaller, so the resulting confidence interval is a bit narrower in this procedure.

Dealing with Non-Normal Distributions

Let's look again at the pipeline safety data, confining our attention to just two regions of the United States—the South and the Southwest. We'll compare the time required to restore the area to safety after a pipeline disruption in these two regions, assuming that disruptions and their repairs are independent in the two regions over time.

1. Open the **Pipeline Safety SouthSW** data table.

2. Select **Analyze ▸ Distribution**. Select **STMin** as **Y, IREGION** as **By,** and click **OK.** This creates two histograms, one for each region.

According to the reported quantiles and summary statistics, the observed incidents in the Southern region required a median time of 60 minutes and a mean of 176.8 minutes to secure. In contrast, the median was 86 minutes and the mean was 157 minutes in the Southwest. Is there a significant difference between the two regions? Before running a *t*-test as before, we should check for normality in these two small samples. The histograms are strongly skewed to the right, but let's also look at a normal quantile plot.

3. Hold the CTRL key, click the red triangle next to **Distributions**, and check **Normal Quantile Plot**; the Southern results are shown in Figure 13.6.

Figure 13.6: Normal Quantile Plot for a Strongly Skewed Distribution

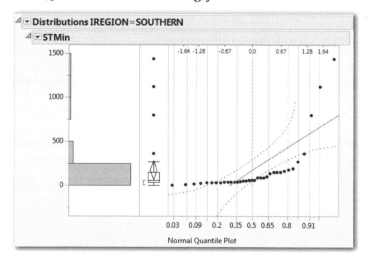

We have two skewed distributions with relatively small samples. We cannot invoke the Central Limit Theorem as needed to trust the *t*-test, but we do have nonparametric options available. It is beyond the scope of most introductory courses to treat nonparametric alternatives in depth, but we'll illustrate one of them here because it is commonly used. The null hypothesis of the test is that the two distributions of **STMin** are centered at the same number of minutes.

1. Select **Analyze ▸ Fit Y by X**. The **Y, Response** column is **STMin** and the **X, Factor** column is **IREGION**. Click **OK**.
2. Click the red triangle next to **Oneway Analysis**, and select **Display Options ▸ Box Plots.** This provides an easy visual comparison of the two distributions.
3. Click the red triangle again, and select **Nonparametric ▸ Wilcoxon Test**. This generates the report shown in Figure 13.7.

Figure 13.7: Results of a Two-Sample Wilcoxon Test

Wilcoxon / Kruskal-Wallis Tests (Rank Sums)

Level	Count	Score Sum	Expected Score	Score Mean	(Mean-Mean0)/Std0
SOUTHERN	35	1295.50	1365.00	37.0143	-0.706
SOUTHWEST	42	1707.50	1638.00	40.6548	0.706

2-Sample Test, Normal Approximation

S	Z	Prob>|Z|
1295.5	-0.70645	0.4799

1-way Test, ChiSquare Approximation

ChiSquare	DF	Prob>ChiSq
0.5063	1	0.4767

When we have two samples, as we do here, JMP conducts a Wilcoxon test, which is the equivalent of another test known as the Mann-Whitney U Test. With more than two samples, JMP performs and reports a Kruskal-Wallis Test (as we'll see in Chapter 14). The common approach in these tests is that they combine all observations and rank them. They then sum up the rankings for each group (region, in this case) and find the mean ranking. If the two groups are centered at the same value, then the mean ranks should be very close. If one group were centered at a very different point, the mean ranks would differ by quite a lot.

Naturally, the key question is, "how much of a difference is considered significant?" Suffice it to say that JMP uses two conventional methods to estimate a *P*-value for the observed difference in mean ranks. The two *P*-values appear in the lower portion of Figure 13.7, and both lead to the same conclusion with this sample. Based on the available sample data, we cannot conclude that there is any statistically significant difference in the time required to secure a gas line disruption in these two regions because the observed difference is unsurprising; it is consistent with sampling error.

Using JMP to Compare Two Variances

In our earlier discussion of the *t*-test, we noted that there are two versions of the test, corresponding to the conditions in which the samples share the same variance or have different variances. Both in the *t*-tests and in techniques we'll see later, we may need to decide whether two or more groups have equal or unequal variances before we choose a statistical procedure. The **Fit Y by X** platform in JMP includes a command that facilitates the judgment.

Let's shift back to the **NHANES** data, and consider the Body Mass Index (BMI) for the adult men and women in the samples. BMI is a measure of body fat as a percentage of total weight. A lean person has lower BMI than an overweight person. We'll test the hypothesis that adult women and men in the US have the same mean BMI, and before choosing a test we'll compare the variances of the two populations.

1. Return to the **NHANES** data table; make sure that you're still showing and including respondents 18 years of age and older.

2. Select **Analyze ▸ Fit Y by X.** Our **Y** column is **BMXBMI** and our **X** column is **RIAGENDR**[1]. Click **OK**.

3. Click the red triangle next to **Oneway Analysis**, and select **UnEqual Variances**. This opens the panel shown in Figure 13.8 (next page).

We find a simple graph comparing the sample standard deviations (*s*) of the two groups, with the numerical summaries below it. For the women, $s = 7.32$, and for the men, $s = 6.12$. Because the variance is the square of the standard deviation, we can determine that the sample variances are approximately 53.6 and 37.5, respectively. Based on these sample values, should we conclude that the population variances are the same or different?

JMP performs five different tests for the equality, or *homogeneity,* of variances. Although they all lead to the same conclusion in this particular example, that's not always the case. For two-sample comparisons the most commonly used tests are probably Levene's test and the F test two-sided. The null hypothesis is that the variance of all groups is equal, and the alternative hypothesis is that variances are unequal.

In this example, the *P*-values fall well below the conventional significance level of 0.05. That is, there is sufficient evidence to reject the null assumption of equal variances, so we can conclude that this set of data should be analyzed with an unequal variance *t*-test. You can perform the test as one of the Application exercises at the end of the chapter.

Figure 13.8: Testing for Equal Variances

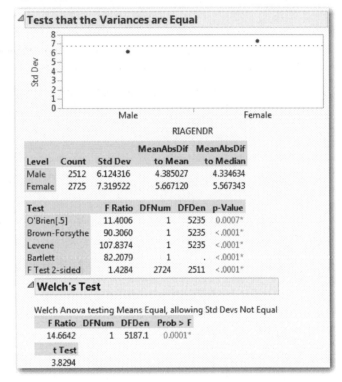

Tests that the Variances are Equal

RIAGENDR

Level	Count	Std Dev	MeanAbsDif to Mean	MeanAbsDif to Median
Male	2512	6.124316	4.385027	4.334634
Female	2725	7.319522	5.667120	5.567343

Test	F Ratio	DFNum	DFDen	p-Value
O'Brien[.5]	11.4006	1	5235	0.0007*
Brown-Forsythe	90.3060	1	5235	<.0001*
Levene	107.8374	1	5235	<.0001*
Bartlett	82.2079	1	.	<.0001*
F Test 2-sided	1.4284	2724	2511	<.0001*

Welch's Test

Welch Anova testing Means Equal, allowing Std Devs Not Equal

F Ratio	DFNum	DFDen	Prob > F
14.6642	1	5187.1	0.0001*

t Test
3.8294

Application

Now that you have completed all of the activities in this chapter, use the concepts and techniques that you've learned to respond to these questions. Be sure to examine the data to check the conditions for inference, and explain which technique you used, why you chose it, and what you conclude.

1. *Scenario:* Let's continue analyzing the data for adults in the **NHANES** data table. For each of the following columns, estimate the difference between the male and female respondents using a 95% confidence level.

    a. What is the estimated difference in Body Mass Index (BMI)?

    b. What is the estimated difference in systolic blood pressure?

    c. What is the estimated difference in diastolic blood pressure?

    d. What is the estimated difference in waist circumference?

2. *Scenario:* We'll continue our analysis of the pipeline disruption data in the Southern and Southwestern regions. The columns **PRPTY** and **GASPRP** contain the dollar value of property and natural gas lost as a consequence of the disruption.

    a. Report and interpret a 95% confidence interval for the difference in mean property value costs of a pipeline disruption.

    b. Test the hypothesis that there is no difference in the mean natural gas loss costs between the two regions.

3. *Scenario:* Go back to the full **Pipeline Safety** data table. Exclude and hide all incidents *except* those involving **LEAKS** and **RUPTURES** of pipelines.

   a. Does it take longer, on average, to secure the area after a rupture than after a leak?

   b. Estimate the difference in property damage costs associated with leak versus ruptures.

   c. Test the hypothesis that the variances are equal in comparing gas loss costs for leaks versus ruptures. Explain your conclusion.

4. *Scenario:* Parkinson's disease (PD) is a neurological disorder affecting millions of older adults around the world. Among the common symptoms are tremors, loss of balance, and difficulty in initiating movement. Other symptoms include vocal impairment. We have a data table, **Parkinsons** (Little, McSharry, Hunter, & Ramig, 2008), containing a number of vocal measurements from a sample of 32 patients, 24 of whom have PD.

   Vocal measurement usually involves having the patient read a standard paragraph or sustain a vowel sound or do both. Two common measurements of a patient's voice refer to the wave form of sound: technicians note variation in the amplitude (volume) and frequency (pitch) of the voice. In this scenario, we will focus on just three measurements.

   • **MDVP:F0(Hz)** –vocal fundamental frequency (baseline measure of pitch)

   • **MDVP:Jitter(Abs)** – jitter refers to variation in pitch

   • **MDVP:Shimmer(db)** – shimmer refers to variation in volume

   a. Examine the distributions of these three columns, distinguishing between patients with and without PD. Do these data meet the conditions for running and interpreting a two-sample *t*-test? Explain your thinking.

   b. Use an appropriate technique to decide whether PD and non-PD patients have significantly different baseline pitch measurements.

   c. Use an appropriate technique to decide whether PD and non-PD patients have significantly different jitter measurements.

   d. Use an appropriate technique to decide whether PD and non-PD patients have significantly different shimmer measurements.

5. *Scenario:* The data table **Airline Delays** contains a sample of flights for two airlines destined for four busy airports. Our goal is to analyze the delay times, expressed in minutes, and compare the two airlines.

   a. Assume this is an SRS of many flights by the selected carrier over a long period of time. Is the variance of delay the same for both airlines?

   b. What can you conclude about the difference in mean delays for these two airlines?

6. *Scenario:* Let's look again at the **TimeUse** data. For each of the following questions, evaluate the extent to which conditions for inference have been met, and then report on what you find when you use an appropriate technique.

   a. Estimate the mean difference in the amount of time men and women spent sleeping in 2003.

   b. Estimate the mean difference in time that individuals devoted to e-mail in 2003 versus 2007.

   c. What can you conclude about the difference in the mean amount of time spent by women and men socializing on a given day?

7. *Scenario:* This question looks at the **NC Births** data. For each of the following questions, evaluate the extent to which conditions for inference have been met, and then report on what you find when you use an appropriate technique.

    a. Do babies born to mothers who smoke weigh less than those born to non-smokers?

    b. Estimate the mean weight gain among pregnant smokers and non-smokers.

    c. Do smokers and non-smokers deliver their babies after the same number of weeks, on average?

---

[1] If you want to use sampling weights in the analysis, use the weight column **WTINT2YR**.

Analysis of Variance

Overview

Previous chapters introduced the concept of bivariate inference and the idea of a response and factor variables. In Chapter 12, we worked with two categorical variables and in Chapter 13, we worked with a continuous response and a dichotomous factor (to represent two independent samples). This chapter introduces several techniques that we can use when we have a *continuous response variable* and one or more categorical *factors* that may have several categorical values.

What Are We Assuming?

As a first example, consider an experiment in which subjects are asked to complete a simple crossword puzzle, and our response variable is the amount of time required to complete the puzzle. We randomly divide the subjects into three groups. One group completes the puzzle while rock music plays loudly in the room. The second group completes the puzzle with the recorded sound of a large truck engine idling outside the window. The third group completes the puzzle in a relatively quiet room. We refer to the different sound conditions as *treatments*.

Our question might be this: does the mean completion time depend on the treatment for the group? In other words, are the means are the same for the three groups or not?

We might model the completion time for each individual (Y_{ik}) as consisting of three separate elements: an overall population mean, plus or minus a *treatment effect*, plus or minus *individual variation*. We might anticipate that the control group's times might enable us to estimate a baseline value for completion times. Further, we might hypothesize that rock music helps some people work more quickly and efficiently, and therefore that individuals in the rock music condition might complete the puzzle faster, on average, than those in the control group. Finally, regardless of conditions, we also anticipate that individuals are simply variable for numerous unaccountable (or unmeasured) reasons.

Symbolically, we express this as follows:

$$Y_{ik} = \mu + \tau_k + \varepsilon_{ik}$$

where:

μ represents the underlying but unknown population mean,

τ_k is the unknown deviation from the population mean (that is, the treatment effect) for treatment level k, and

ε_{ik} is unique individual variation, or error.

Our null hypothesis is that all groups share a single mean, which is to say that the treatment effect for all factor levels (τ_k) is zero. We can express this null hypothesis in this way:

H_0: $\tau_1 = \tau_2 = \cdots = \tau_k = 0$ versus the complementary alterative hypothesis that H_0 is not true.

The first technique we'll study is called *one-way analysis of variance* (ANOVA), which is a reliable method of inference when our study design satisfies these three assumptions:

- All observations are independent of one another.
- The individual error terms are normally distributed[1].
- The variance of the individual errors is the same across treatment groups.

These three conditions form the foundation for drawing inferences from ANOVA. In practice, it is helpful, though not necessary to have samples of roughly equal size. With varying sample sizes, the technique is particularly sensitive to the equal variance assumption, which is to say that the statistical results can be misleading if our sample data deviate too far from approximately equal variances. Therefore, an important step in conducting an analysis of variance is the evaluation of the extent to which our data set challenges the validity of these assumptions.

One-Way ANOVA

Our first example comes from the United Nation's birth rate data, a source we've used before. In this instance, we'll look at the variation in the birth rate around the world. Specifically, birth rate is defined as the number of births annually per 1,000 population. We'll initially ask if differences in birth rates are associated with one particular matter of public policy: whether maternity leave is provided by the state, by the private sector, or by a combination of public and private sectors.

Note that this is observational data, so that we have not established the different factor levels experimentally or by design. Open the **BirthRate 2005** data table.

We'll start simply by examining the distribution of the response and factor variables, in part to decide whether we've satisfied the conditions needed for ANOVA, although we'll evaluate the assumptions more rigorously below. Even before looking at the data, we can conclude that the observations are independent: it is hard to think of reasons that either the birth rates or type of provider in one country would be affected by those in another.

1. Select **Analyze ▶ Distribution**. Cast the columns **BirthRate** and **Provider** into the **Y, Columns** role, and click **OK**.

When we look at the **BirthRate** histogram, we find that the variable is positively skewed in shape; this could signal an issue with the normality assumption, but we have large enough subsamples that we can rely on the Central Limit Theorem to proceed with the analysis.

2. Click on any of the bars in the **Provider** graph to select the rows that share one provider type. When you do so, note the corresponding values in the **BirthRate** distribution. Does there seem to be any consistent pattern in birth rates that corresponds to how maternity leave benefits are provided? ANOVA will help to clarify the situation.

   To begin the ANOVA, we return to the **Fit Y by X** analysis platform.

3. Select **Analyze ▶ Fit Y by X**. Cast **BirthRate** into the **Y** role and **Provider** into the **X** role, and click **OK**. The default output is a simple scatterplot showing the response values across the different treatment groups; Figure 14.1 has been modified to jitter the points for visual clarity:

Figure 14.1: One-Way ANOVA Initial Graph (with points jittered)

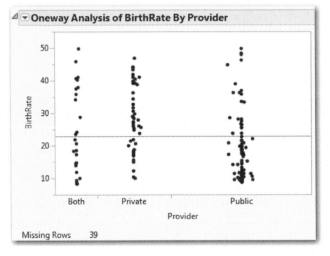

This graph reveals at least two things of interest. The dispersion of points within each group appears comparable, which suggests constant variance and some indication of a treatment effect—that countries with publicly provided maternity-leave benefits might have somewhat lower birth rates on average than those with privately provided benefits. In the graph, notice the concentration of points above the overall mean line in the **Private** group and the large number of points below the line for the **Public** group. Let's look more closely.

4. Click the red triangle next to **Oneway Analysis of BirthRate by Provider** and select **Means/Anova**.

As shown in Figure 14.2, you'll see a green *means diamond* superimposed on each set of points. In this case, the diamonds have different widths that correspond to differences in sample size.

Figure 14.2: Graph Including Means Diamonds

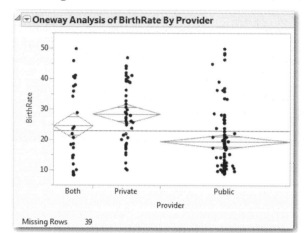

The center line of the diamond is the sample mean, and the upper and lower points of the diamond are the bounds of a 95% confidence interval. The horizontal lines near the top and bottom of the diamonds are *overlap marks*; these are visual boundaries that we can use to compare the means for the groups. Diamonds with overlapping areas between the overlap marks indicate insignificant differences in means. In this picture, the means diamond for the **Public** group does not overlap the **Private** group, suggesting that we do have significant differences here. Before we go too much further with the analysis of variance, we need to check our assumptions.

Does the Sample Satisfy the Assumptions?

The independence assumption is a logical matter; no software can tell us whether the observations were collected independently of each other. Independence is a function of the study and sampling design; and, as noted above, we can conclude that we have satisfied the independence condition.

1. We can test for the equality of variances by again clicking the red triangle and choosing **UnEqual Variances**, which generates the result in Figure 14.3.

Direct your attention to the lower table titled **Test**. There are several standard tests used to decide whether we've satisfied the equal variance condition. In all cases, the null hypothesis is that the variance of all groups is equal, and the alternative hypothesis is that variances are unequal. Levene's test is probably the most commonly used, so you should use it here.

Figure 14.3: Tests for Equality of Variances

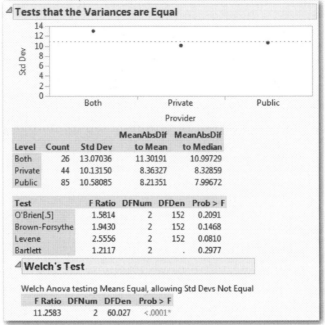

In this example, the *P*-values for all four tests are relatively large by conventional standards. That is, there is insufficient evidence to reject the null assumption of equal variances, so we can feel safe in continuing to assume that the sample satisfies the equal variance assumption.

When we test for equality of variances, JMP also reports the results of Welch's test. This test adjusts for the inequality of variance to provide an alternative to the standard one-way analysis of variance. In other words, Welch's test is a fallback option in case the variances are unequal; we'll say more about this in a few pages.

Although the Central Limit Theorem applies in this case, it is still prudent to check whether the individual error terms might be normally distributed. We have no way to measure the individual errors directly, but we can come close by examining the deviations between each individual measurement and the mean of its respective group.

We call these deviations *residuals*, and can examine their distribution as an approximation of the individual random error terms.

To analyze the residuals, we proceed as follows:

2.   Click the red triangle next to **Oneway**, and select **Save ▸ Save Residuals**. This creates a new column in the data table.

3.   In the data table, select **Birthrate centered by Provider**. Right-click, select **Column Info**, and change the column name to **Residuals**.

4.   With the data table highlighted, select **Analyze ▸ Distribution**. Choose the new residuals column as **Y, Columns**, and click **OK**.

5.   Click the hotspot next to **Residuals** and select **Normal Quantile Plot**. We can use the normal quantile plot (Figure 14.4) to judge whether these residuals are decidedly skewed or depart excessively from normality.

Figure 14.4: Normality Check for Residuals

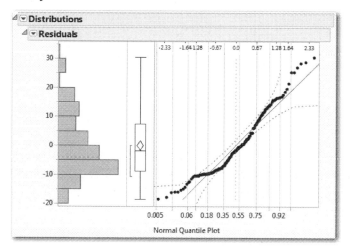

The reliability of ANOVA is not dramatically affected by data that is non-normal. We say that the technique is *robust* with respect to normality. As long as the residuals are generally unimodal and symmetric, we can proceed.

Figure 14.4 shows the distribution of residuals in this case, and based on what we see here we can conclude that the normality assumption is satisfied. Thus, we've met all of the conditions for a one-way analysis of variance, and can proceed.

Factorial Analysis for Main Effects

In the scatterplot with means diamonds, we found visual indications that the mean of the **Public** provision group might be lower than the other two groups. We find more definitive information below the graph. Having validated the assumptions, it is time to look back at the **Oneway Anova** report. We find three sets of statistics, shown below in Figure 14.5. At this point in our studies, we'll focus on just a few of the key results.

Figure 14.5: One-Way ANOVA Results

◢ **Oneway Anova**

◢ **Summary of Fit**

Rsquare	0.119247
Adj Rsquare	0.107658
Root Mean Square Error	10.90897
Mean of Response	22.82065
Observations (or Sum Wgts)	155

◢ **Analysis of Variance**

Source	DF	Sum of Squares	Mean Square	F Ratio	Prob > F
Provider	2	2449.084	1224.54	10.2898	<.0001*
Error	152	18088.861	119.01		
C. Total	154	20537.945			

◢ **Means for Oneway Anova**

Level	Number	Mean	Std Error	Lower 95%	Upper 95%
Both	26	24.5018	2.1394	20.275	28.729
Private	44	28.4277	1.6446	25.178	31.677
Public	85	19.4039	1.1832	17.066	21.742

Std Error uses a pooled estimate of error variance

The **Summary of Fit** table provides an indication of the extent to which our ANOVA model accounts for the total variation in all 155 national birth rates. The key value here is Rsquare[2], which is always between 0 and 1, and describes the proportion of total variation in the response variable that is associated with the factor. Here, we find that only 12% of all variation in birth rates is associated with the variation in provision of maternity benefits; that's not very impressive. Other factors must account for the other 88% of the variation.

In the **Analysis of Variance** table (similar to the tables described in your main text), we find our test statistic (10.2897) and its *P*-value (< .0001). Here, the *P*-value is very small indicating statistical significance. We reject the null hypothesis of equal treatment effects, and conclude that at least one treatment has a significant effect.

Looking at the last table, **Means for Oneway Anova**, we see that the confidence interval for the mean of the private group is higher than the mean for the public group, and that the interval for the "both" group overlaps the other two. For a more rigorous review of the differences in group means, we can also select a statistical test to compare the means of the groups.

We have several common methods available to us for such a comparison. With two factor levels, it makes sense to use the simple two-sample *t*-test comparison, but with more than two levels the *t*-tests will overstate differences, and therefore we should use a different approach here.

In this particular study, we do not have a control group as we might in a designed experiment. Because there isn't a control, we'll use *Tukey's HSD* (Honestly Significant Difference), the most conservative of the approaches listed. Had one of the factor levels referred to a control group, we'd use *Dunnett's method* and compare each of the other group means to the mean of the control group.

1. Click the **Oneway** hotspot once again, and choose **Compare Means ▸ All Pairs, Tukey's HSD.**

Figure 14.6: Graphical Representation of Tukey's HSD

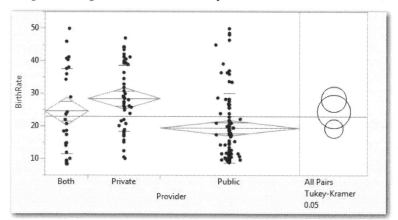

In the upper panel (shown above in Figure 14.6), we now see a set of three overlapping rings representing the three groups. Note that the upper ring (**Private** group) overlaps the middle **Both** ring, that the middle and lower rings overlap, but that the **Private** and **Public** rings do not overlap.

Click the uppermost (**Private**) ring; notice that it turns red, as does this middle (**Both**) ring, but the lower **Public** ring becomes blue. This indicates a significant difference between **Private** and **Public**, but no significant difference between **Private** and **Both**.

We'll scroll down to find the numerical results of Tukey's test, shown here in Figure 14.7.

Figure 14.7: Tukey HSD Results

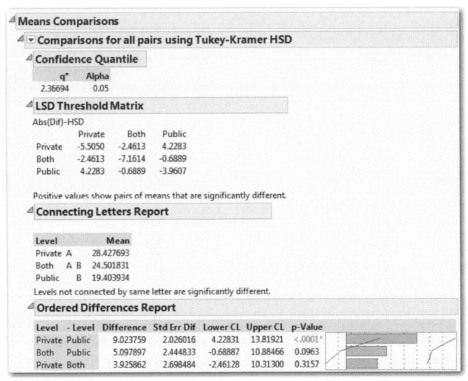

How do we read these results? The bottom portion of the output displays the significance levels for estimated differences between each group. We find one significant *P*-value, which is associated with the difference between the **Private** and the **Public** provision of benefits. We can confidently conclude that mean birth rates are different (apparently higher) in countries with private provision of benefits than in those with public provision of benefits, but cannot be confident about any other differences based on this sample.

You might also want to look at the connecting letters report just above the bottom portion. Here the idea is simple: each level is tagged with one or two letters. If a mean has a unique letter tag, we conclude that it is significantly different from the others. If a level has two tags, it could be the same as either group. In this analysis, **Private** is A and **Public** is B, but **Both** is tagged as A B. So, we conclude that the means for public and private are different from one another, but can't make a definitive conclusion about the **Both** category.

What if Conditions Are Not Satisfied?

In this example, our data were consistent with the assumptions of normality and equal variances. What happens when the sample data set does not meet the necessary conditions? In most introductory courses, this question probably doesn't arise or if it does, it receives very brief treatment. As such, we will deal with it in a very limited way. Thankfully, JMP makes it quite easy to accommodate "uncooperative" data.

Earlier we noted that the assumption of equal variances is a critical condition for reliable results in an ANOVA, and we saw how to test for the equality of variances. If we find that there are significant differences in variance across the groups, we should not rely on the ANOVA table to draw statistical conclusions. Instead, JMP provides us with the results of **Welch's ANOVA**, a procedure that takes the differences in subgroup variation into account.

1. Look back in your output for the **Test that the Variances are Equal** (which appears in Figure 14.3 above).

At the bottom of that output, you'll find the results of Welch's test. In our example, that test finds $F = 11.2583$ with a P-value < 0.0001. We would decide to reject the null hypothesis and conclude that the group means are not all equal. With Welch's test, there are no procedures comparable to Tukey's HSD to determine where the differences lie.

If the residuals are not even close to looking normal—multimodal, very strongly skewed, or the like—we can use a *nonparametric* equivalent test. Nonparametric tests make no assumptions about the underlying distributions of our data (some statisticians refer to them as "distribution-free" methods).

2. Return to the hotspot, and select **Nonparametric ▸ Wilcoxon Test**.

Now look at the results, labeled **Wilcoxon/Kruskal-Wallis Tests**. First a note on terminology: strictly speaking, the Wilcoxon test is the nonparametric equivalent of the two-sample *t*-test, while the Kruskal-Wallis test is the equivalent of the one-way ANOVA.

These tests rely on chi-squared, rather than the F, distribution to evaluate the test statistic. Without laying out the theory of these tests, suffice it to say that the null hypothesis is this: the distributions of the groups are centered[3] in the same place. A *P*-value less than 0.05 conventionally means that we'll decide to reject the null.

Including a Second Factor with Two-Way ANOVA

Earlier we noted that our model accounted for relatively little of the variation in national birth rates, and that other factors might also play a role. To investigate the importance of a second categorical variable, we can extend our Analysis of Variance to include a second factor. The appropriate technique is called *Two-Way ANOVA*.

In some countries new mothers typically have lengthy maternity leaves from their jobs, while in others the usual practice is to offer much briefer time off from work. Let's use the variation in this aspect of prevailing national policies as an indication of a country's commitment to the well-being of women who give birth. We'll focus on the factor called **MatLeave90+,** a dichotomous variable indicating whether a country provides 90 days or more of maternity leave for new mothers.

1. As we did before, use **Analyze ▶ Distribution**. This time, select **BirthRate, Provider**, and **MatLeave90+** as the **Y, Columns.**

2. Click on different bars in the three graphs, looking for any obvious visual differences in birth rates in countries with different insurance and maternity leave practices. Think about what you see in this part of the analysis.

Before performing the two-way analysis, we need to consider one potential complication. It is possible that the duration of the maternity leave and the provider of benefits each have separate and independent associations with birth rate, and it is also possible that the impact of one variable is different in the presence or absence of the other. In other words, perhaps the duration of the leave is relevant where it is privately provided, but not where it's publicly provided.

Because of this possibility, in a two-way ANOVA, we investigate both the *main effects* of each factor separately as well as any possible *interaction effect* of the two categorical variables taken together. When we specify the two-way model, we'll identify the main factors and their possible interactions.

3. Now let's perform the two-way ANOVA. Select **Analyze ▶ Fit Model**. As shown in Figure 14.8, cast **Birthrate** as the **Y** variable, and then highlight both **Provider** and **MatLeave90+** and click **Add**.

4. Now highlight the same two categorical columns again in **Select Columns** and click **Cross**. Crossing the two categorical variables establishes an *interaction term* with six groupings (three provider groups times two leave duration groups).

5. Now click **Run** to perform the analysis.

Figure 14.8: Fit Model Dialog Box for Two-Way Model

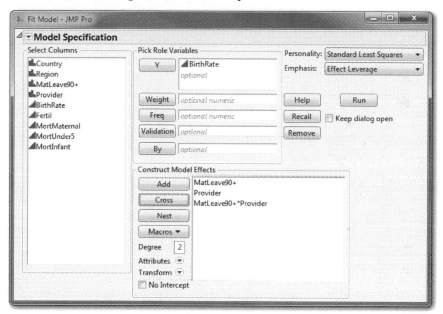

This command generates extensive results, and we'll now focus on a few critical elements that will be illustrated in the next several pages. In a two-way analysis, it is good practice to review the results in a particular sequence. We see in the Analysis of Variance table that we have a significant test (F = 5.5602, *P*-value = 0.0001), so we might have found some statistically interesting results. First we evaluate assumptions (which are the same as for one-way ANOVA), look at the significance of the entire model, then check to determine whether there are interaction effects, and finally, evaluate main effects.

> The **Fit Model** results are quite dense, and contain elements that go far beyond the content of most introductory courses in statistics. We'll look very selectively at portions of the output that correspond to topics that you are likely to see in an introductory course. If your interests lead you to want or need to interpret portions that we bypass, you should use the context-sensitive **Help** tool (?) and just point at the results you want to explore.

Evaluating Assumptions

As before, we can reason that the observations are independent of one another, and should examine the residuals for indications of non-normality or unequal variances.

1. Scroll to the bottom of your output, and find the **Residual by Predicted Plot** (Figure 14.9).

This graph displays the computed residuals in six vertical groupings, corresponding to the six possible combinations of the levels of both factors. We use this graph to make a judgment as to whether each of the groups appears to share a common variance; in this case, the spread of points within each group appears similar, so we can conclude that we've met the constant variance assumption. If the spreads were grossly different, we should be reluctant to go on to evaluate the results any further.

Figure 14.9: Plot of Residuals vs. Predicted Means

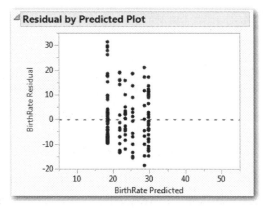

Let's also check the residuals to see how reasonable it is to assume that the individual error terms are normally distributed.

2. We can again save the residuals in the data table by clicking on the hotspot at the top of the analysis and selecting **Save Columns ▶ Residuals**.

3. As you did earlier, use **Analyze ▶ Distribution** to create a histogram and normal probability plot of the residuals.

You should see that the residuals are single-peaked and somewhat skewed. While not perfectly normal, there's nothing to worry about here.

Interaction and Main Effects

Next we should evaluate the results of the test for the interaction term. We do so because a significant interaction can distort the interpretation of the main effects, so before thinking about the main effects, we want to examine the interaction. An interaction exists when the relationship between one factor and the response is *different* across the groups of the second factor.

1. Below the ANOVA table, locate the **Effect Tests** table, as shown in Figure 14.10.

This table lists the results of three separate F-tests, namely the test for equality of group means corresponding to different provider arrangements, different lengths of maternity leave, and the interaction of the two factors. The null hypothesis in each test is that the group means are identical.

Figure 14.10. Tests of Main and Interaction Effects

Effect Tests					
Source	Nparm	DF	Sum of Squares	F Ratio	Prob > F
MatLeave90+	1	1	749.9762	6.4562	0.0121*
Provider	2	2	1037.8525	4.4672	0.0131*
MatLeave90+*Provider	2	2	28.2710	0.1217	0.8855

Each row of the table reports the number of parameters (**Nparm**), corresponding to one less than the number of factor levels for each main factor, and the product of the nparms for the interaction. We also find degrees of freedom, the sum of squares, and the F-ratio and *P*-value for each of the three tests.

We first look at the bottom row (the interaction) and see that the F statistic is quite small with a very high *P*-value. Therefore, we conclude that we cannot reject the null for this test, which is to say that there is no significant interaction effect. Whatever difference (if any) in birth rates are associated with the length of maternity leave, it is the same across the different provider conditions. Ordinarily, having found no evidence of an interaction, an investigator would probably re-run the model without the interaction term. However, to illustrate additional visualizations of main and interaction effects, we'll leave the term in the model for now.

Finally, we consider the main effects where we find that both *P*-values are significant. This indicates that both of the factors have main effects, so the natural next step is to understand what those effects are.

For a visual display of the differences in group means we can look at *factor profiles*.

2. Click the red triangle at the very top of the report (next to **Response Birthrate**), and select **Factor Profiling ▶ Profiler.**

JMP generates two graph panels that display confidence intervals for the mean birth rates for each main factor level grouping.

3. Use the Grabber (Hand) tool to stretch the vertical axes to magnify the scale of the two graphs, or enlarge the graphs by dragging the lower-right corner diagonally. The result will look like something like Figure 14.11.

Figure 14.11: Prediction Profiler for Main Effects

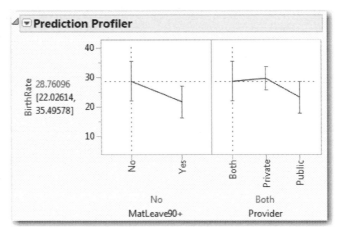

The profiler graphs are easy to interpret. In the left panel, we see confidence intervals for mean birth rate in the countries with maternity leaves less than 90 days and those with 90 days or more. Similarly, the right panel compares the intervals for the different provider arrangements. As with other JMP graphs, the panels here are linked and interactive.

The red dashed lines indicate the level settings for the profiler. Initially, we are estimating the birth rate for countries with shorter maternity leaves that are provided jointly by the public and private sectors. On the far left, we see the confidence interval of approximately 22.03 to 35.50.

4. To see how the mean birth rate prediction changes, for example, if we look at countries with shorter maternity leaves that are publicly provided, just use the arrow tool and click on the blue interval corresponding to **Public**.

The prediction falls to approximately 18.09 to 28.73.

5. Experiment by clicking on the different interval bars until you begin to see how the **BirthRate** estimates change with different selections in the Profiler.

We can also construct a profiler for interaction effects. In this example, the interaction is insignificant, but we can still learn something from the profiler graph.

6. Click the red triangle again and select **Factor Profiling ▶ Interaction Plots**. Enlarge the plot area or use the Grabber to adjust the vertical axes so that you can see the plotted lines clearly as in Figure 14.12.

Figure 14.12: Interaction Plots for the Two-Way Analysis of Variance

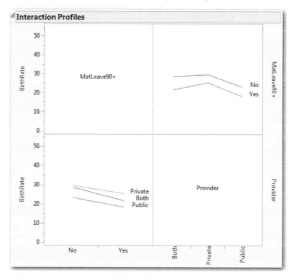

The interaction profiles show the mean birth rates corresponding to both categorical variables. In the lower-left graph, the different provider arrangements are represented by three color-coded lines, and we see that regardless of provider, the mean **BirthRate** is lower for those countries with longer maternity leaves (**MatLeave90+ = Yes**). The plot in the upper right swaps the arrangement. In this graph, we have two color-coded lines, corresponding to the two different levels of maternity leave durations. The lines display the different birth rate mean values for the three different provider systems.

When there is no interaction, we'll find lines that are approximately parallel. If there is a significant interaction, we'll see lines that cross or intersect. The intersection shows that the pattern of different means for one categorical factor varies depending on the value of the other factor.

If we want more precision than a small graph offers, we can once again look at multiple comparisons just as we did for the one-way ANOVA. Because this command performed three tests (two main effects and one interaction), we'll need to ask for three sets of multiple comparisons.

7. Return to the top of the output, scroll to the right slightly to locate the **Leverage Plots**[4], and click the red triangle next to **MatLeave90+**. Because this categorical variable is dichotomous, we can compare the means using a two-sample *t*-test. Choose **LSMeans Student's t** as shown in Figure 14.13.

Figure 14.13: Requesting Multiple Comparisons for Two-Way ANOVA

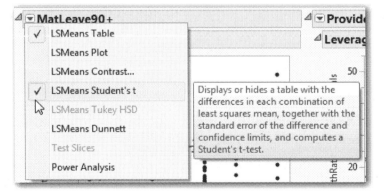

8. Click the red triangle next to **Provider** and select **LSMeans Tukey HSD.**

Now look below each of the Leverage Plots and locate the connecting letters reports (shown here in Figure 14.14; scroll down a bit to adjust the report in this fashion).

Figure 14.14: Connecting Letters Report for the Two Factors

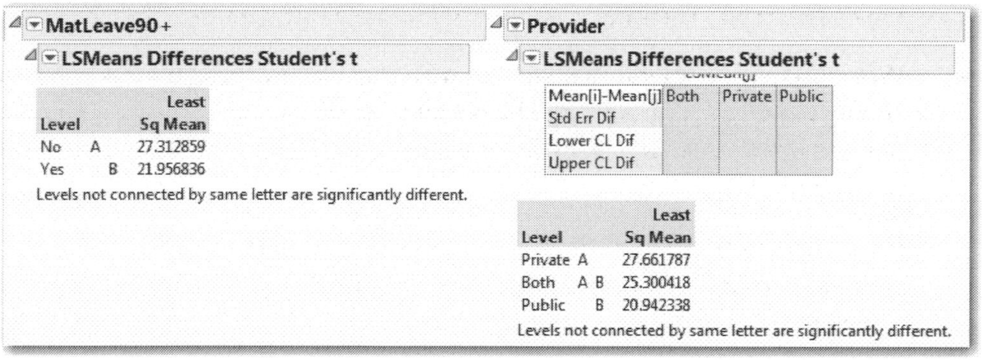

Here we see the significant contrasts: there is a significant difference in means between countries with longer and shorter maternity leaves, and also between those that provide the benefits privately and publicly.

Application

Now that you have completed all of the activities in this chapter, use the techniques that you've learned to respond to these questions. In all problems, be sure to evaluate the reasonableness of the ANOVA assumptions and comment on your conclusions.

1. *Scenario:* We'll continue to examine the World Development Indicators data in the data table **BirthRate 2005**. We'll broaden our analysis to work with other columns in that table:

   - **Region**: country's geographic region
   - **Fertil**: number of births per woman
   - **Mortmaternal**: maternal mortality, estimated deaths per 100,000 births
   - **MortUnder5**: deaths, children under 5 years per 1,000
   - **MortInfant**: deaths, infants per 1,000 live births

   a. Perform a one-way ANOVA to decide whether birth rates are consistent across regions of the world. If you conclude that the rates differ, determine where the differences are.

b. Perform a one-way ANOVA to decide whether fertility rates are consistent across regions of the world. If you conclude that the rates differ, determine where the differences are.

c. Perform a two-way ANOVA to see how maternal mortality varies by length of maternity leaves and provider. If you conclude that the rates differ, determine where the differences are, testing for both main and interaction effects.

d. Perform a two-way ANOVA to see how mortality under five years of age varies by length of maternity leaves and provider. If you conclude that the rates differ, determine where the differences are, testing for both main and interaction effects.

2. *Scenario:* A company that manufactures insulating materials wants to evaluate the effectiveness of three different additives to a particular material. The company performs an industrial experiment to evaluate the additives. For the sake of this example, these additives are called **Regular, Extra**, and **Super**. The company now uses the regular additive, so that serves as the control group.

The experimenter has a temperature-controlled apparatus with two adjoining chambers. He can insert the experimental insulating material in the wall that separates the two chambers. Initially, he sets the temperature of chamber A at 70°F and the temperature of the other chamber (B) at 30°F. After a pre-determined amount of time, he measures the temperature in chamber A, recording the change in temperature as the response variable. A good insulating material prevents the temperature in chamber A from falling far.

Following good experimental design protocols (see Chapter 18) he repeats the process in randomized sequence thirty times, testing each insulating material a total of 10 times.

We have the data from this experiment in a file called **Insulation.JMP**.

a. Perform a one-way ANOVA to evaluate the comparative effectiveness of the three additives.

b. Which multiple comparison method is appropriate in this experiment, and why?

c. Should the company continue to use the regular additive, or change to one of the others? If they should change, which additive should they use?

3. *Scenario:* How do prices of used cars differ, if at all, in different areas of the United States? Our data table **Used Cars** contains observational data about the listed prices of three popular compact car models in three different metropolitan areas in the U.S. The cities are Phoenix, AZ; Portland, OR; and Raleigh-Durham-Chapel Hill, NC. The cars models are the Chrysler PT Cruiser Touring Edition, the Honda Civic EX, and the Toyota Corolla LE. All of the cars are two years old.

   a. Perform a two-way ANOVA to see how prices vary by city and model. If you find differences, perform an appropriate multiple comparison and explain where the differences are.

   b. Perform a two-way ANOVA to see how mileage of these used cars varies by city and model. If you find differences, perform an appropriate multiple comparison and explain where the differences are.

4. *Scenario:* Does a person's employment situation influence the amount that he or she sleeps? Our **TimeUse** data table can provide a foundation for analysis of this question.

   a. Perform a one-way ANOVA to see how hours of sleep vary by employment status. If you find differences, perform an appropriate multiple comparison and explain where the differences are.

   b. Perform a two-way ANOVA to see how hours of sleep vary by employment status and gender. If you find differences, perform an appropriate multiple comparison and explain where the differences are.

5. *Scenario:* In this scenario, we return to the **Pipeline Safety** data table to examine the consequences of pipeline disruptions.

   a. Perform a one-way ANOVA to see how the dollar amount of property damage caused by a disruption varies by **Region**. If you find differences, perform an appropriate multiple comparison and explain where the differences are.

   b. Perform a two-way ANOVA to see how the dollar amount of property damage caused by a disruption varies by **Region** and **Ignite**. If you find differences, perform an appropriate multiple comparison and explain where the differences are.

   c. Perform a one-way ANOVA to see how the time required to make the area safe caused by a disruption varies by type of disruption. If you find differences, perform an appropriate multiple comparison and explain where the differences are.

   d. Perform a two-way ANOVA to see how the dollar amount of property damage caused by a disruption varies by type of disruption and **Region**. If you find differences, perform an appropriate multiple comparison and explain where the differences are.

6. *Scenario:* A company manufactures catheters, blood transfusion tubes, and other tubing for medical applications. In this particular instance, we have diameter measurements from a newly designed process to produce tubing that is designed to be 4.4 millimeters in diameter. These measurements are stored in the **Diameter** data table.

Each row of the table is a single measurement of tubing diameter. There are six measurements per day, and each day one of four different operators was responsible for production using one of three different machines. The first 20 days of production were considered Phase 1 of this study; following Phase 1 some adjustments were made to the machinery, which then ran an additional 20 days in Phase 2. Use the data filter to analyze just the Phase 1 data.

a. Perform a one-way ANOVA to see how the tubing diameters vary by **Operator**. If you find differences, perform an appropriate multiple comparison and explain where the differences are.

b. Perform a one-way ANOVA to see how the tubing diameters vary by **Machine**. If you find differences, perform an appropriate multiple comparison and explain where the differences are.

c. Perform a two-way ANOVA to see how the tubing diameters vary by **Operator** and **Machine**. If you find differences, perform an appropriate multiple comparison and explain where the differences are.

d. Explain in your own words what the interaction plots tell you in this case. Couch your explanation in terms of the specific machines and operators.

7. *Scenario:* JMP ships illustrative data tables with the software. One of these is the **Popcorn** data table. It is one of the few realistic but artificial examples we'll use in this book, but because it is based on the work of three highly influential statisticians[5] it appears here.

a. Perform a series of three one-way ANOVAs to decide if **Yield** varies by **popcorn** type, by the amount of oil used (**oil amt**), or size of **batch**. Verify assumptions; if you find differences, perform an appropriate multiple comparison and explain where the differences are.

b. Perform a series of two-way ANOVAs to see if there are any interactions among factors, and report on your findings.

[1] Alternatively, we might express this assumption as saying that the individual responses are normally distributed within each group.

[2] We've seen Rsquare before in Chapter 4, and will work with it again in Chapter 15 and 16 when we study regression analysis.

[3] Different nonparametric procedures take different approaches to the concept of a distribution's "center." None use the mean *per se* as the One-way ANOVA does.

[4] Leverage plots are discussed more fully in Chapter 15.

[5] The Popcorn data table is inspired by experimental data from Box, Hunter, Hunter (1978).

Simple Linear Regression Inference

15

Overview

In Chapter 4, we learned to summarize two continuous variables at a time using scatterplot, correlations, and line fitting. In this chapter, we'll return to that subject, this time with the object of generalizing from the patterns in sample data in order to draw conclusions about an entire population. The main statistical tool that we'll use is known as *linear regression analysis*. We'll devote this chapter and the three later chapters to the subject of regression.

Because Chapter 4 is now many pages back, we'll begin by reviewing some basic concepts of bivariate data and line fitting. Then, we'll discuss the fundamental model used in simple linear regression. After that, we'll discuss the crucial conditions necessary for inference, and finally, we'll see how to interpret the results of a regression analysis.

Fitting a Line to Bivariate Continuous Data

We introduced regression in Chapter 4 using the data table **Birthrate 2005**. This data table contains several columns related to the variation in the birth rate and the risks related to childbirth around the world as of 2005. In this data table, the United Nations reports figures for 194 countries. Let's briefly revisit that data now to review some basic concepts, focusing on two measures of the frequency of births in different nations.

1. Open the **Birthrate 2005** data table now.

As we did in Chapter 4, let's look at the columns labeled **BirthRate** and **Fertil**. A country's annual birth rate is defined as the number of live births per 1,000 people in the country. The fertility rate is the mean number of children that would be born to a woman during her lifetime. We plotted these two variables in Chapter 4; let us do that again now.

2. Select **Analyze ▶ Fit Y by X**. Cast **Fertil** as **Y** and **BirthRate** as **X**, and click **OK**.

Your results will look like those shown in Figure 15.1.

Figure 15.1: Relationship Between Birth Rate and Fertility Rate

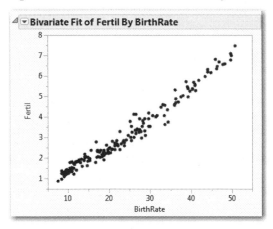

This is the same graph that we saw in Figure 4.10. Again, we note that general pattern is upward from left to right: fertility rates increase as the birth rate increases, although there are some countries that depart from the pattern. The pattern can be described as linear, although there is a mild curvature at the lower left. We also see that a large number of countries are concentrated in the lower left, with low birth rates and relatively low maternal mortality.

In Chapter 4, we illustrated the technique of line-fitting using these two columns. Because these two columns really represent two ways of thinking about a single construct ("how many babies?"), let us turn to a different example to expand our study of simple linear regression analysis.

We'll return to a subset of the NHANES data[1], and look at two body measurement variables. Because adult body proportions are different from children and because males and females differ, we'll restrict the first illustrative analysis to male respondents ages 18 and up. Our subset is a simple random sample of 465 observations drawn from the full NHANES data table, representing approximately 5% of the original data.

3.   Open the data table called **NHANES SRS**. This table contains young and female respondents in addition to the males. To use only the males 18 years and older in our analysis, we'll use the Data Filter.

4.   Select **Rows ▸ Data Filter.**

5.   While pressing the CTRL key, highlight **RIAGENDR** and **RIDAGEYR**, and click **Add**.

6.   In the **Data Filter** (see Figure 15.2 after step 6 below), select the **Show** and **Include** options (the **Select** option is already selected).

7.   Then click **Male** under **RIAGENDR** to include just the male subjects.

8.   Finally, click the number 0 to the left **RIDAGEYR** and replace it with 18. This sets the lower bound for **RIDAGEYR** to be just 18 years. We want to select any respondent who is a male age 18 or older.

Figure 15.2: Selection Criteria for Males Age 18 and Older

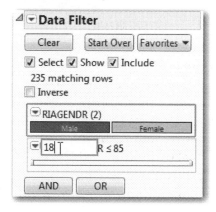

We've restricted the analysis to male respondents who are 18 years of age and older. Now we can begin the regression analysis. We'll examine the relationship between waist circumference and body mass index, or BMI, which is the ratio of a person's weight to the square of height. In the data table, waist measurements are in centimeters, and BMI is kilograms per square meter. In this analysis, we'll see if there is a predictable relationship between men's waist measurements and their BMIs.

We begin the analysis as we have done so often, using the **Fit Y by X** platform.

1. Select **Fit Y by X**. Cast **BMXBMI** as **Y** and **BMXWAIST** as **X** and click **OK**.

This graph (see Figure 15.3) illustrates the first thing that we want to look for when planning to conduct a linear regression analysis—we see a general linear trend in the data. Think of stretching an elliptical elastic band around the cloud of points; that would result in a long and narrow ellipse lying at a slant, which would contain most, if not all, of the points. In fact, we can use JMP to overlay such an ellipse on the graph.

2. Click the red triangle next to **Bivariate Fit** and select **Density Ellipse ▸ 0.95**.

The resulting ellipse appears incomplete because of the default axis settings on our graph. We can customize the axes to show the entire ellipse using the grabber to shift the axes.

Figure 15.3: A Linear Pattern of BMI vs. Waist

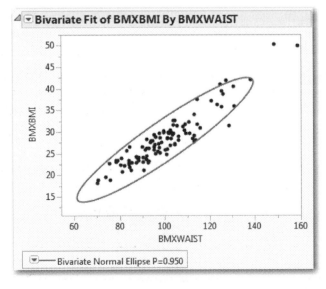

3. Move the grabber tool near the origin on the vertical axis and slide upward until you see a hash mark below 15 appear on the Y axis. Do the same on the horizontal axis until the waist value of 60 cm appears on the X axis.

This graph is a fairly typical candidate for linear regression analysis. Nearly all of the points lie all along the same sloped axis in the same pattern, with consistent scatter. Before running the regression, let's step back for a moment and consider the fundamental regression model.

The Simple Regression Model

When we fit a line to a set of points, we do so with a model in mind and with a provisional idea about how we came to observe the particular points in our sample. The reasoning goes like this. We speculate or hypothesize that there is a linear relationship between Y and X such that whenever X increases by one unit (centimeters of waist circumference, in this case), then Y changes, on average, by a constant amount. For any specific individual, the observed value of Y could deviate from the general pattern.

Algebraically, the model looks like this:

$$Y_i = \beta_0 + \beta_1 X_i + \varepsilon_i$$

where Y_i and X_i are the observed values for one respondent, β_0 and β_1 are the intercept and slope of the underlying (but unknown) relationship, and ε_i is the amount by which an individual's BMI departs from the usual pattern. Generally speaking, we envision *εi* as purely random noise. In short, we can express each observed value of Y_i as partially reflecting the underlying linear pattern, and partially reflecting a random deviation from the pattern. Look again at Figure 15.3. Can you visualize each point as lying in the vicinity of a line? Let's use JMP to estimate the location of such a line.

1. Click the red triangle next to **Bivariate Fit** and select **Fit Line**.

Now your results will look like Figure 15.4 on the next page. We see a green *fitted line* that approximates the upward pattern of the points.

Below the graph, we find the equation of that line:

BMXBMI = –5.888872 + 0.3410071 * BMXWAIST

The slope of this line describes how these two variables co-vary. If we imagine two groups of men whose waist circumferences differ by 1 centimeter, the group with the larger waists would average BMIs that are 0.34 kg/m² higher. As we learned in Chapter 4, this equation summarizes the relationship among the points in this sample. Before learning about the inferences that we might draw from this, let's refine our understanding of the two chunks of the model: the linear relationship and the random deviations.

Figure 15.4: Estimated Line of Best Fit

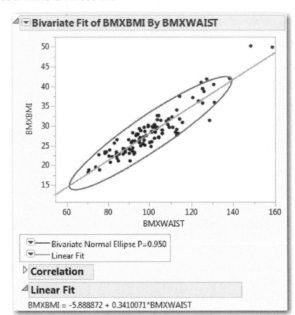

Thinking About Linearity

If two variables have a linear relationship, their scatterplot forms a line or at least suggests a linear pattern. In this example, our variables have a *positive* relationship: as *X* increases, *Y* increases. In another case, the relationship might be negative, with *Y* decreasing as *X* increases. But what does it mean to say that two variables have a *linear* relationship? What kind of underlying dynamic generates a linear pattern of dots?

As noted earlier, linearity involves a constant change in *Y* each time *X* changes by one unit. *Y* might rise or fall, but the key feature of a linear relationship is that the shifts in *Y* do not accelerate or diminish at different levels of *X*. If we plan to generalize from our sample, it is important to ask if it is reasonable to expect *Y* to vary in this particular way as we move through the domain of realistically possible *X* values.

Random Error

The regression model also posits that empirical observations tend to deviate from the linear pattern, and that the deviations are themselves a random variable. We'll have considerably more to say about the random deviations in Chapter 16, but it is very useful at the outset to understand this aspect of the regression model.

Linear regression analysis doesn't demand that all points line up perfectly, or that the two continuous variables have a very close (or "strong") association. On the other hand, if groups of observations systematically depart from the general linear pattern, we should ask if the deviations are truly random, or if there is some other factor to consider as we untangle the relationship between Y and X.

What Are We Assuming?

The preceding discussion outlines the conditions under which we can generalize using regression analysis. First, we need a logical or theoretical reason to anticipate that Y and X have a linear relationship. Second, the default method[2] that we use to estimate the line of best fit works reliably. We know that the method works reliably when the random errors, ε_i, satisfy four conditions:

- They are normally distributed.
- They have a mean value of 0.
- They have a constant variance, σ^2, regardless of the value of X.
- They are independent across observations.

At this early stage in the presentation of this technique, it might be difficult to grasp all of the implications of these conditions. Start by understanding that the following might be red flags to look for in a scatter plot with a fitted line:

- The points seem to bend or oscillate predictably around the line.
- There are a small number of outliers that stand well apart from the mass of the points.
- The points seem snugly concentrated near one end of the line, but fan out toward the other end.
- There seem to be greater concentrations of points distant from the line, but not so many points concentrated near the line.

In this example, none of these trouble signs is present. In the next chapter, we'll learn more about looking for problems with the important conditions for inference. For now, let's proceed assuming that the sample satisfies all of the conditions.

Interpreting Regression Results

There are four major sections in the results panel for the linear fit (see Figure 15.5), three of which are fully disclosed by default. We've already seen the equation of the line of best fit and discussed its meaning. In this part of the chapter, we'll discuss the three other sections in order.

Figure 15.5: Regression Results

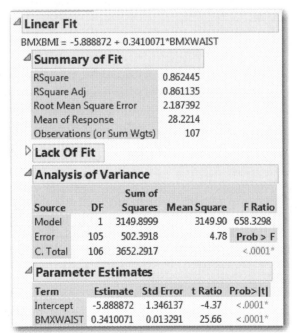

Linear Fit

BMXBMI = -5.888872 + 0.3410071*BMXWAIST

Summary of Fit

RSquare	0.862445
RSquare Adj	0.861135
Root Mean Square Error	2.187392
Mean of Response	28.2214
Observations (or Sum Wgts)	107

Lack Of Fit

Analysis of Variance

Source	DF	Sum of Squares	Mean Square	F Ratio
Model	1	3149.8999	3149.90	658.3298
Error	105	502.3918	4.78	Prob > F
C. Total	106	3652.2917		<.0001*

Parameter Estimates

Term	Estimate	Std Error	t Ratio	Prob>|t|
Intercept	-5.888872	1.346137	-4.37	<.0001*
BMXWAIST	0.3410071	0.013291	25.66	<.0001*

Summary of Fit

Under the heading **Summary of Fit**, we find five statistics that describe the fit between the data and the model.

- **RSquare** and **RSquare Adj** both summarize the strength of the linear relationship between the two continuous variables. The RSquare statistics range between 0.0 and 1.0, where 1.0 is a perfect linear fit. Just as in Chapter 4, think of

RSquare as the proportion of variation in Y that is associated with X. Here, both statistics are approximately 0.86, suggesting that a man's waist measurement could be a very good predictor of his BMI.

- **Root Mean Square Error** (RMSE) is a measure of the dispersion of the points from the estimated line. Think of it as the sample standard deviation of the random noise term, ε. When points are tightly clustered near the line, this statistic is relatively small. When points are widely scattered from the line, the statistic is relatively large. Comparing the RMSE to the mean of the response variable (next statistic) is one way to assess its relative magnitude.

- **Mean of Response** is just the sample mean value of Y.

- **Observations** is the sample size. In this table, we have complete waist and BMI data for 107 men.

Lack of Fit

The next heading is **Lack of Fit**, but this panel is initially minimized in this case. Lack of fit tests typically are considered topics for more advanced statistics courses, so we only mention them here without further comment.

Analysis of Variance

These ANOVA results should look familiar if you've just completed Chapter 14. In the context of regression, ANOVA gives us an overall test of significance for the regression model. In a one-way ANOVA, we hypothesized that the mean of a response variable was the same across several categories. In regression, we hypothesize that the mean of the response variable is the same regardless of X—that is to say that Y does not vary in tandem with X.

We read the table just as we did in the previous chapter, focusing on the F-ratio and the corresponding P-value. Here F is over 658 and the P-value is smaller than 0.0001. This probability is so small that it is highly unlikely that the computed F-ratio came about through sampling error. We reject the null hypothesis that waist circumference and BMI are unrelated, and conclude that we've found a statistically significant relationship.

Not only can we say that the pattern describes the sample, we can say with confidence that the relationship generalizes to the entire population of men over age 17 in the United States.

Parameter Estimates and *t*-tests

The final panel in the results provides the estimated intercept and slope of the regression line, and the individual *t*-tests for each. The slope and intercept are sometimes called the *coefficients* in the regression equation, and we treat them as the *parameters* of the linear regression model.

In Figure 15.6, we reproduce the parameter estimate panel, which contains five columns. The first two columns—**Term** and **Estimate**—are the estimated intercept and slope that we saw earlier in the equation of the regression line.

Figure 15.6: Parameter Estimates

◁ **Parameter Estimates**

Term	Estimate	Std Error	t Ratio	Prob>|t|
Intercept	-5.888872	1.346137	-4.37	<.0001*
BMXWAIST	0.3410071	0.013291	25.66	<.0001*

Because we're using a sample of the full population, our estimates are subject to sampling error. The **Std Error** column estimates the variability attributable to sampling. The **t Ratio** and **Prob>|t|** columns show the results of a two-sided test of the null hypothesis that a parameter is truly equal to 0.

Why do we test the hypotheses that the intercept and slope equal zero? The reason relates to the slope and what a zero slope represents. If X and Y are genuinely independent and unrelated, then changes in the value of X have no influence or bearing on the values of Y. In other words, the slope of a line of best fit for two such variables should be zero. For this reason, we always want to look closely at the significance test for the slope. Depending on the study and the meaning of the data, the test for the intercept may or may not have practical importance to us.

In a simple linear regression, the ANOVA and *t*-test results for the slope will always lead to the same conclusion[3] about the hypothesized independence of the response and factor variables. Here, we find that our estimated slope of 0.341 kg/m^2 change in BMI per 1 cm. increase in waist circumference is very convincingly different from 0: in fact, it's more than 25 standard errors away from 0. It's inconceivable that such an observed difference is the coincidental result of random sampling.

Testing for a Slope Other Than Zero

In some investigations, we might begin with a theoretical model that specifies a value for the slope or the intercept. In that case, we come to the analysis with hypothesized values of either β_0 or β_1 or both, and we want to test those values. The **Fit Y by X** platform does not accommodate such significance tests, but the **Fit Model** platform does. We used **Fit Model** in the prior chapter to perform a two-way ANOVA. In this example, we'll use it to test for a specific slope value other than 0.

We'll illustrate with an example from the field of classical music, drawn from an article by Prof. Jesper Rydén of Uppsala University in Sweden (Rydén 2007). The article focuses on piano sonatas by Franz Joseph Haydn (1732–1809) and Wolfgang Amadeus Mozart (1756–1791) and investigates the idea that these two composers incorporated the *golden mean* within their compositions. A sonata is a form of instrumental music consisting of two parts. In the first part, the composer introduces a melody—the basic tune of the piece—known formally as the exposition. After the exposition comes a second portion that elaborates upon the basic melody, developing it more fully, offering some variations, and then recapitulating or repeating the melody. Some music scholars believe that Haydn and Mozart strove for an aesthetically pleasing but asymmetric balance in the lengths of the exposition and development or recapitulation sections. More specifically, they might have divided their sonatas (deliberately or not) so that the relative lengths of the shorter and longer portions approximated the golden mean.

The golden mean (sometimes called the golden ratio), characterized and studied in the West at least since the ancient Greeks, refers to the division of a line into a shorter segment *a*, and a longer segment *b*, such that the ratio of *a:b* equals the ratio of *b:(a+b)*. Equivalently,

$$\frac{a}{b} = \frac{b}{(a+b)} = \phi \approx 0.61803.$$

We have a data table called **Mozart** containing the lengths, in musical measures, of the shorter and longer portions of 29 Mozart sonatas. If, in fact, Mozart was aiming for the golden ratio in these compositions, then we should find a linear trend in the data. Moreover, it should be characterized by this line:

a = 0 + 0.61803(b)

So, we'll want to test the hypothesis that $\beta_1 = 0.61803$ rather than 0.

1. Open the data table called **Mozart**.

2. Select **Analzye ▸ Fit Model**. Select **Parta** as **Y**, then add **Partb** as the only model effect, and run the model.

Both the graph and the **Summary of Fit** indicate a strong linear relationship between the two parts of these sonatas. Figure 15.7 shows the parameter estimates panel from the results.

Figure 15.7: Estimates for Mozart Data

Term	Estimate	Std Error	t Ratio	Prob>|t|
Intercept	1.3596328	2.882715	0.47	0.6410
Partb	0.6259842	0.030851	20.29	<.0001*

Parameter Estimates

Rounding the estimates slightly we can write an estimated line as *Parta = 1.3596 + 0.626(Partb)*. On its face, this does not seem to match the proposed equation above. However, let's look at the *t*-tests. The estimated intercept is not significantly different from 0, so we cannot conclude that the intercept is other than 0. The hypothesized intercept of 0 is still credible.

Now look at the results for the slope. The estimated slope is 0.6259842, and its standard error is 0.030851. The reported *t*-ratio of about 20 standard errors implicitly compares the estimated slope to a hypothesized value of 0. To compare it to a different hypothesized value, we'll want to compute the following ratio:

$$\frac{estimate - hypothesized}{std.error} = \frac{0.6259842 - 0.61803}{0.030851} = 0.2578$$

We can have JMP compute this ratio and its corresponding *p*-value as follows:

3. Click the red triangle next to **Response Parta** and select **Estimates ▸ Custom Test**. Scroll to the bottom of the results report where you will see a panel like the one shown in Figure 15.8.

4. The upper white rectangle is an editable field for adding a title; type Golden **Mean** in the box.

5. In the box next to **Partb**, change the 0 to a 1 to indicate that we want to test the coefficient of **Partb**.

6. Finally, enter the hypothesized value of the golden mean, .61803 in the box next to **=**, and click the **Done** button.

Figure 15.8: Specifying the Column and Hypothesized Value

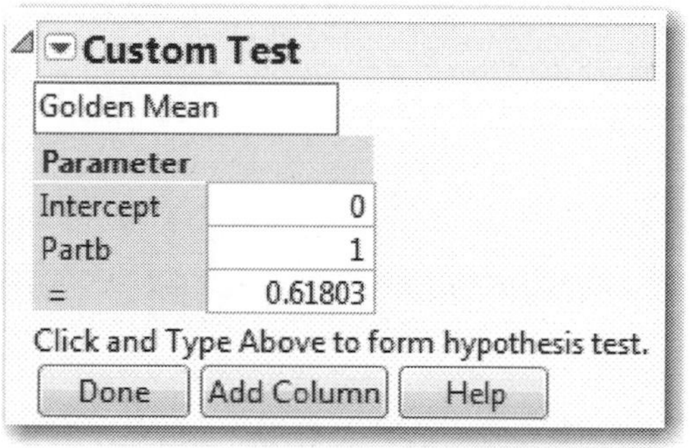

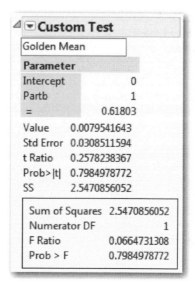

The **Custom Test** panel now becomes a results panel, presenting both a *t*-test and an *F* test, as shown in Figure 15.9. As our earlier calculation showed, the estimated slope is less than 0.26 standard errors from the hypothesized value, which is very close. Based on the large *p*-value of 0.798, we fail to reject the null hypothesis that the slope equals the golden mean.

Figure 15.9: Custom Test Results

In other words, the golden mean theory is credible. As always, we cannot prove a null hypothesis, so this analysis does not definitively establish that Mozart's sonatas conform to the golden mean. This is an important distinction in the logic of statistical testing--our tests are able to discredit a null hypothesis with a high degree of confidence, but we cannot confirm a null hypothesis. What we can say is that we have put a hypothesis to the test, and it is still plausible.

Application

Now that you have completed all of the activities in this chapter, use the concepts and techniques that you've learned to respond to these questions.

1. *Scenario:* Return to the **NHANES SRS** data table.

   a. Exclude and hide respondents under age 18 and all males, leaving only adult females. Perform a regression analysis for BMI and waist circumference for adult women, and report your findings and conclusions.

   b. Is waist measurement a better predictor (in other words, a better fit) of BMI for men or for women?

   c. Perform one additional regression analysis, this time looking only at respondents under the age of 17. Summarize your findings.

2. *Scenario:* High blood pressure continues to be a leading health problem in the United States. In this problem, continue to use the **NHANES SRS** data table. For this analysis, we'll focus on just the following variables:

- **RIAGENDR**: respondent's gender

- **RIDAGEYR**: respondent's age in years

- **BMXWT**: respondent's weight in kilograms

- **BPXPLS**: respondent's resting pulse rate

- **BPXSY1**: respondent's systolic blood pressure ("top" number in BP reading)

- **BPXD1**: respondent's diastolic blood pressure ("bottom" number in BP reading)

   a. Investigate a possible linear relationship of systolic blood pressure versus age. What, specifically, tends to happen to blood pressure as people age? Would you say there is a strong linear relationship?

   b. Perform a regression analysis of systolic and diastolic blood pressure. Explain fully what you have found.

    c. Create a scatterplot of systolic blood pressure and pulse rate. One might suspect that higher pulse rate is associated with higher blood pressure. Does the analysis bear out this suspicion?

3. *Scenario:* We'll continue to examine the World Development Indicators data in **BirthRate 2005**. We'll broaden our analysis to work with other variables in that file:

- **MortUnder5**: deaths, children under 5 years per 1,000 live births

- **MortInfant**: deaths, infants per 1,000 live births

    a. Create a scatterplot for **MortUnder5** and **MortInfant**. Report the equation of the fitted line and the Rsquare value, and explain what you have found.

4. *Scenario:* How do the prices of used cars vary according to the mileage of the cars? Our data table **Used Cars** contains observational data about the listed prices of three popular compact car models in three different metropolitan areas in the U.S. All of the cars are two years old.

    a. Create a scatterplot of price versus mileage. Report the equation of the fitted line and the Rsquare value, and explain what you have found.

5. *Scenario*: Stock market analysts are always on the lookout for profitable opportunities and for signs of weakness in publicly traded stocks. Market analysts make extensive use of regression models in their work, and one of the simplest ones is known as the *random* (or *drunkard's*) *walk* model. Simply put, the model hypothesizes that over a relatively short period of time the price of a particular share of stock is a random deviation from its price on the prior day. If Y_t represents the price at time t, then $Y_t = Y_{t-1} + \varepsilon$. In this problem, you'll fit a random walk model to daily closing prices for McDonald's Corporation for the first six months of 2009 and decide how well the random walk model fits. The data table is called **MCD**.

a. Create a scatterplot with the daily closing price on the vertical axis and the prior day's closing price on the horizontal. Comment on what you see in this graph.

b. Fit a line to the scatterplot, and test the credibility of the random walk model. Report on your findings.

6. *Scenario*: Franz Joseph Haydn was a successful and well-established composer when the young Mozart burst upon the cultural scene. Haydn wrote more than twice as many piano sonatas as Mozart. Use the data table **Haydn** to perform a parallel analysis to the one we did for Mozart.

a. Report fully on your findings from a regression analysis of **Parta** versus **Partb**.

b. How does the fit of this model compare to the fit using the data from Mozart?

7. *Scenario*: Throughout the animal kingdom, animals require sleep and there is extensive variation in the number of hours in a day that different animals sleep. The data table called **Sleeping Animals** contains information for more than 60 mammalian species, including the average number of hours per day of total sleep. This will be the response column in this problem.

a. Estimate a linear regression model using gestation as the factor. Gestation is the mean number of days that females of these species carry their young before giving birth. Report on your results and comment on the extent to which gestational period is a good predictor of sleep hours.

b. Now perform a similar analysis using brain weight as the factor. Report fully on your results and comment on the potential usefulness of this model.

8. *Scenario*: For many years, it has been understood that tobacco use leads to health problems related to the heart and lungs. The **Tobacco Use** data table contains recent data about the prevalence of tobacco use and of certain diseases around the world.

a. Using cancer mortality (**CancerMort**) as the response variable and the prevalence of tobacco use in both sexes (**TobaccoUse**), run a regression analysis to decide whether total tobacco use in a country is a predictor of the number of deaths from cancer annually in that country.

b. Using cardiovascular mortality (**CVMort**) as the response variable and the prevalence of tobacco use in both sexes (**TobaccoUse**), run a regression analysis to decide whether total tobacco use in a country is a predictor of the number of deaths from cardiovascular disease annually in that country.

c. Review your findings in the earlier two parts. In this example, we're using aggregated data from entire nations rather than individual data about individual patients. Can you think of any ways in which this fact could explain the somewhat surprising results?

9. *Scenario*: In Chapter 2, our first illustration of experimental data involved a study of the compressive strength of concrete. In this scenario, we look at a set of observations all taken at 28 days (4 weeks) after the concrete was initially formulated. The data table is **Concrete28**. The response variable is the **Compressive Strength** column, and we'll examine the relationship between that variable and two candidate factor variables.

a. Use **Cement** as the factor and run a regression. Report on your findings in detail. Explain what this slope tells you about the impact of adding more cement to a concrete mixture.

b. Use **Water** as the factor and run a regression. Report on your findings in detail. Explain what this slope tells you about the impact of adding more water to a concrete mixture.

10. *Scenario*: Prof. Frank Anscombe of Yale University created an artificial data set to illustrate the hazards of applying linear regression analysis without looking at a scatterplot (Anscombe 1973). His work has been very influential, and JMP includes his illustration among the sample data tables packaged with the software. You'll find **Anscombe** both in this book's data tables and in the JMP sample data tables. Open it now.

    a. In the upper-left panel of the data table, you'll see a red triangle next to the words **The Quartet**. Click the triangle, and select **Run Script**. This produces four regression analyses corresponding to four pairs of response and predictor variables. Examine the results closely, and write a brief response comparing the regressions. What do you conclude about this quartet of models?

    b. Now return to the results, and click the red triangle next to **Bivariate Fit of Y1 By X1**; select **Show Points** and re-interpret this regression in the context of the revised scatterplot.

    c. Now reveal the points in the other three graphs. Is the linear model equally appropriate in all four cases?

11. *Scenario*: Many cities in the U.S. have active used car markets. Typically, the asking price for a used car varies by model, age, mileage, and features. The data table called **Used Cars** contains asking prices (**Price**) and mileage (**Miles**) for three popular budget models; all cars were two years old at the time the data were gathered, and we have data from three U.S. metropolitan areas. All prices are in dollars. In this analysis, **Price** is the response and **Miles** is the factor.

a. Because the car model is an important consideration, we'll begin by analyzing the data for one model: the Civic EX. Use the **Data Filter** to isolate the Civic EX data for analysis. Run a regression; how much does the asking price decline, on average, per mile driven? What would be a mean asking price for a two-year old Civic EX that had never been driven? Comment on the statistical significance and goodness-of-fit of this model.

b. Repeat the previous step using the Corolla LE data.

c. Repeat one more time using the PT Cruiser data.

d. Finally, compare the three models. For which set of data does the model fit best? Explain your thinking. For which car model are you most confident about the estimated slope?

12. *Scenario:* We'll return to the World Development Indicators data in **WDI**. In this scenario, we'll investigate the relationship between access to improved sanitation (the percent of the population with access to sewers and the like) and life expectancy. The response column is **life_exp** and the factor is **sani_acc**.

a. Use the **Data Filter** to **Show** and **Include** only the observations for the **Year** 2010, and the **Latin America & Caribbean Region** nations. Describe the relationship you observe between access to improved sanitation and life expectancy.

b. Repeat the analysis for East Asia & Pacific countries in 2010.

c. Now do the same one additional time for the countries located in Sub-**Saharan Africa**.

d. How do the three regression models compare? What might explain the differences in the models?

13. *Scenario:* The data table called **USA Counties** contains a wide variety of measures for every county in the United States.

a. Run a regression casting **sales_per_capita** (retail sales dollars per person, 2007) as **Y** and **per_capita_income** as **X**. Write a short paragraph explaining why county-wide retail sales might vary with per capita income, and report on the strengths and weaknesses of this regression model.

---

[1] Why a subsample? Some of the key concepts in this chapter deal with the way individual points scatter around a line. With a smaller number of observations, we'll be able to better visualize these concepts.

[2] Like all statistical software, JMP uses a default method to line-fitting that is known as *ordinary least squares estimation*, or *OLS*. A full discussion of OLS is well beyond the scope of this book, but it's worth noting that these assumptions refer to OLS in particular, not to regression in general.

[3] They will have identical *P*-values and the F-ratio will be the square of the *t* ratio.

Residuals Analysis and Estimation

Overview

Regression analysis is a powerful and flexible technique that we can reliably apply under certain conditions. Given a data table with continuous X-Y pairs, we can always use the ordinary least squares (OLS) method to fit a line and generate results like those we saw in Chapter 15. However, if we want to draw general conclusions or use an estimated regression equation to estimate, predict, or project Y values, it is critical that we decide whether our sample data satisfy the conditions underlying OLS estimation. This chapter explains how to verify whether we are on solid ground for inference with a particular regression analysis, then covers three types of inference using a simple linear regression. The techniques of this chapter are also relevant in later chapters.

Conditions for Least Squares Estimation

Recall that the linear regression model is

$$Y_i = \beta_0 + \beta_1 X_i + \varepsilon_i$$

where Y_i and X_i, are the observed values for one case, β_0 and β_1 are the intercept and slope of the underlying relationship, and ε_i is the amount by which an individual response deviates from the line.

OLS will generate the equation of a line approximating $\beta_0 + \beta_1 X_i$ so that the first condition is that the relationship between X and Y really is linear. Additionally, the model treats ε as a normal random variable whose mean is zero, whose variance is the same all along the line and whose observations are independent of each other. Thus, we have four other conditions to verify before using our OLS results. Also, the logic of the simple linear model is that X is the only variable that accounts for a substantial amount of the observed variation in Y.

In this chapter, we'll first learn techniques to evaluate the degree to which a sample satisfies the OLS conditions, and then learn to put a regression line to work when the conditions permit. We'll consider all of the conditions in sequence and briefly explain criteria for judging whether they are satisfied in a data table, as well as pointing out how to deal with samples that do not satisfy the conditions.

Residuals Analysis

Several of the conditions just cited refer directly to the random error term in the model, ε_i. Because we cannot know the precise location of the theoretical regression line, we cannot know the precise values of the εs. The random error term values are the difference between the observed values of Y and the position of the theoretical line. Although we don't know where the theoretical line is, we do have a fitted line and we'll use the differences between each observed Y_i and the fitted line to approximate the ε values.

These differences are called *residuals*. Again, a residual is the difference between an observed value of Y and the fitted value of Y corresponding to the observed value of X. In a scatterplot, it is the vertical distance between an observed point and the fitted line. If the point lies above the line, the residual is positive, and if the point lies below the line, the residual is negative. By examining the residuals after fitting a line, we can judge how well the sample satisfies the conditions. To illustrate the steps of a residual analysis, we'll first look at the **NHANES SRS** table again and repeat the regression that we ran in Chapter 15.

As you did in the previous chapter, take the following steps now to fit a line to estimate the relationship between body mass index (BMI) and waist circumference for our random sample of adult males.

1. Open the data table called **NHANES SRS**.

As you may recall, we want to use data just from the males 18 years and older in our analysis. We'll use the Data Filter to restrict the analysis; by now, you know how to do this. To refresh your memory, return to the previous chapter.

2. After you have set the filter to **Show** and **Include** only males 18 and older, select **Analyze ▶ Fit Y by X**. Cast **BMXBMI** as **Y**, and **BMXWAIST** as **X** and click **OK**.

3. Click the red triangle next to **Bivariate Fit** and choose **Fit Line**.

This reproduces the regression analysis from the prior chapter. Now we can evaluate the conditions and see whether it would be wise to use this estimated model for practical purposes.

4. Click the red triangle below the graph, next to **Linear Fit,** and select **Plot Residuals**.

5. Click the same red triangle again and click **Save Residuals**.

The **Plot Residuals** option creates five graphs at the bottom of the report. These graphs help us judge the validity of the OLS conditions for inference. We'll start with the linearity condition.

Linearity

In the prior chapter we discussed the concept of linearity, noting that if a relationship is linear then *Y* changes by a constant each time *X* changes by one unit. Look at the **Bivariate Fit** report with the fitted line. The scatterplot and fitted line in this example do suggest a linear pattern.

Sometimes mild curvature in the points is difficult to detect in the initial bivariate fit scatterplot, and we need a way to magnify the image to see potential non-linear patterns. One simple way to do this is to look at a plot of residuals versus fitted values;

scroll down in the report window until you find the **Diagnotics Plots** heading. The first graph (shown in Figure 16.1) is the plot we want.

Figure 16.1: Residuals versus Predicted Values Plot

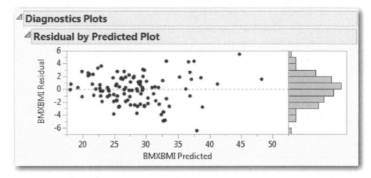

In this graph the fitted (predicted) values of BMXBMI are on the horizontal and the residuals are on the vertical axis. For the most part in this case, we see that the points are scattered evenly above and below the horizontal line at 0, without any sign of curvature. If the relationship were non-linear, we'd see the points bending upward or downward. For example, think back to the first scatterplot we made in Chapter 15 (Refer back to Figure 15.1.) The plot of Fertility Rate vs. Birthrate from the **Birthrate 2005** data showed a very slight upward curve. Figure 16.2 is the residual plot from that regression and it shows such a curvilinear pattern. Notice that on the left side of the graph, corresponding to predicted fertility values from 0 to 2, the large majority of residuals are positive. Between 2 and 5, most of the residuals are negative, and then for predicted fertility rates greater than 5, the residuals are positive. Taken together, the overall trend is a U-shaped curve. As it happens, this plot also shows non-constant variance and we'll discuss that issue in a later section.

Figure 16.2: Residual by Predicted Plot Showing Non-linearity

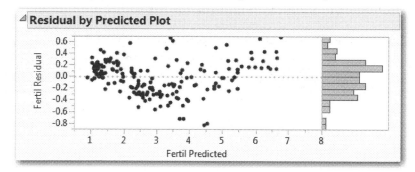

Curvature

What do we do when we find a curved, rather than a straight line, pattern? We have several options, which are discussed at length in Chapter 19. For the moment, the most important thing to recognize is that we should not interpret or apply the results of a simple linear regression if the residual plot indicates curvature. Without a linear pattern, the results will be of very limited use and could lead to very poor decisions.

Influential Observations

Another issue related to linearity is that of influential observations. These are points that are outliers in either the X or the Y directions. Such points tend to bias the estimate of the intercept, the slope, or both.

For example, look back at the scatter plot with the fitted line on your screen (reproduced here as Figure 16.3). Notice the point farthest to the upper-right corner of the graph, indicated with an arrow. This point represents a man whose waist circumference was 158.6 cm, or 62.4 inches, which was an outlier in the sample. His BMI was 49.8 kg/m², the second highest in the sample—lies very close to the line. The OLS method is sensitive to such extreme X values, and as a consequence the line nearly passes through the point.

Figure 16.3: Some Influential Observations

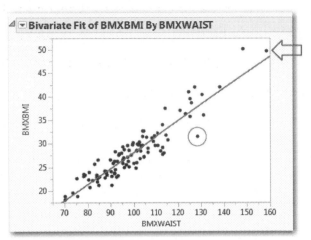

As another example, find the point that is circled in Figure 16.3. This represents a man whose waist measurement and BMI are both within the third quartile. However, this point is a bivariate outlier because it departs from the overall linear pattern of the points. His BMI is low considering his waistline.

With outliers such as these, we can take one or more of the following steps, even in the context of an introductory statistics course.

- If possible, verify the accuracy of the data. Are the measurements correct? Is there a typographical error or obvious mistake in a respondent's answer?

- Verify that the observation really comes from the target population. For example, in a survey of a firm's regular customers, is this respondent a regular customer?

- Temporarily exclude the observation from the analysis and see how much the slope, intercept, or goodness-of-fit measures change. We can do this as follows:

1. Click the red triangle next to **Bivariate Fit of BMXBMI by BMXWAIST** and select **Script ▶ Automatic Recalc.**

2. Now click the scatterplot and click once on the point that is shown circled in Figure 16.3 above. You'll see that it is the observation in row 368, and it appears darkened. Clicking on the point selects it.

3. Now right-click and choose **Row Exclude**[1]. This excludes the point from the regression computations, but leaves it appearing in the graph.

When we omit this point from consideration, the estimated regression line is

$$BMXBMI = -6.533056 + 0.3480755*BMXWAIST$$

as compared to this line when we use all 107 observations:

$$BMXBMI = -5.888872 + 0.3410071*BMXWAIST$$

When we eliminate that particular outlier, the intercept decreases by about 0.7 kg/m^2, but the slope barely changes. A point near the mass of the data with a large residual has little effect on the slope (or on conclusions about whether the two variables are related) but can bias the estimate of the intercept. Omitting that observation also improves the goodness of fit measure R^2 from 0.862 to 0.874. This makes sense because we are dropping a point that doesn't fit the general pattern.

What about that point in the far upper-right corner, highlighted with an arrow (Row 81)? When we exclude it, the estimated equation becomes

$$BMXBMI - 5.504107 + 0.33699*BMXWAIST$$

This changes the equation very little, although R^2 falls a bit farther to 0.84. In this regression, none of the outliers is particularly influential.

Normality

The set of standard residual plots includes two ways to ask whether the residuals follow an approximate normal distribution. The first is a histogram appearing in the margin of the first graph of residuals by predicted values. As shown in Figure 16.1, these residuals form a mound-shaped, symmetric pattern.

The last graph in the **Bivariate Fit** report is a Normal Quantile Plot (see Figure 16.4). Though the residuals do not track perfectly along the red diagonal line, this graph provides no warning signs of non-normality. Here we are looking for any red flag that would suggest that the residuals are grossly non-normal. We want to be able to continue to interpret the results, and would stop for corrective measures if the residuals were heavily skewed or multimodal.

Figure 16.4: Normal Quantile Plot for the BMI Residuals

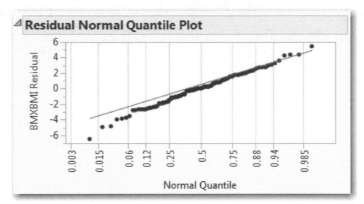

What happens when residuals do *not* follow a normal distribution? What kinds of corrective measures can we take? Generally speaking, non-normal residuals might indicate that there are other factors in addition to X that should be considered. In other words, there could be a lurking factor that accounts for the non-random residuals. Sometimes it is helpful to transform the response variable or the factor using logarithms. In Chapter 18, we'll see how we can account for additional factors and use transformations.

Constant Variance

Look back at Figures 16.1 and 16.2. In an ideal illustration, we would see the residuals scattered randomly above and below the horizontal zero line. Figure 16.1, even though it's not a perfect illustration, comes closer to the ideal than Figure 16.2. In Figure 16.2 not only do the residuals curve, but they increase in variability as you view the graph from left to right. On the left side, they are tightly packed together, and as we slide to the right, they drift farther and farther from the center line. The variability of the residuals is not constant, but rather variable in itself.

The constant variance assumption goes by a few other names as well, such as *homogeneity of variance* or *homoskedasticity*. Valid inference from a regression analysis depends quite strongly on this particular assumption. When we have non-constant variance (or, you guessed it, *heterogeneity of variance* or *heteroskedasticity*) we should refrain from drawing inferences or using the regression model to estimate values of Y. Essentially, this is because *heteroskedasticity* makes the reported standard errors unreliable, which in turn biases the *t*- and *F*-ratios.

What to do? Sometimes a variable transformation helps. In other cases, multiple regression (Chapter 18) can ameliorate the situation. In other instances, we turn to more robust regression techniques that do not use the standard OLS technique. These techniques go beyond the topics typically covered in an introductory course.

Independence

The final condition for OLS is that the random errors are independent of each other. As a general rule, this is quite a common problem with time series data, but much less so with a cross-sectional sample like NHANES. Much like the other conditions, we check for warning signs of severe non-independence with an inclination to continue the analysis unless there are compelling reasons to stop.

In regression, the assumption of independent errors means that the sign and magnitude of one error is independent of its predecessor. Using the residuals are proxies for the random errors, we will reason that if the residuals are independent of one another, then when we graph them in sequence we should find no regular patterns of consecutive positive or negative residuals.

Within the standard set of **Diagnostics Plots**, there is a **Residual by Row Plot**. This is the graph we want, though if the sample is large, it will be easier to detect consecutive clusters of positive and negative residuals if the points in the graphs are connected in sequence. We can make our own version of the graph to better serve the purpose. This is the reason for saving the residuals earlier. The result of that request is a new data table column called **Residuals BMXBMI**.

To illustrate this method, we'll use both our current example as well as some time series data. In both cases, we want to graph the residuals in sequence of observation, looking for repeated patterns of positive and negative residuals. There are multiple ways to accomplish this; we'll use the **Overlay Plot** here because it enables us easily to connect residuals with lines to clearly visualize their sequence.

1. Select **Graph ▶ Overlay Plot**. As shown in Figure 16.5, select **BMXBMI** as **Y**. Clear the **Sort X** option and click **OK**.

Figure 16.5: The Overlay Plot Dialog to Graph Residuals in Sequence

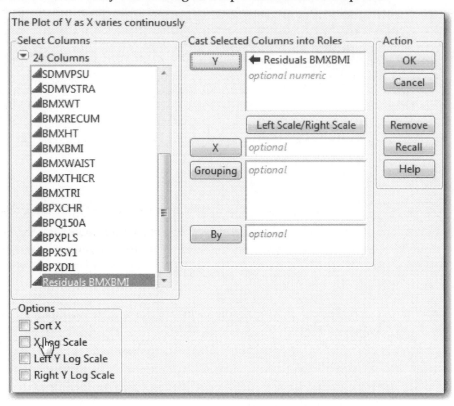

2. Click the red triangle next to **Overlay Plot** and select **Connect Thru Missing** to create the graph shown in the left portion of Figure 16.6.

The graph of the BMI residuals has a jagged irregular pattern of ups and downs, indicating independence. The sign and magnitude of one residual does little to predict the next residual in order.

In contrast, look at the overlay plot on the right in Figure 16.6, based on data from the **Stock Index Weekly** data table. The residuals come from a regression using the weekly closing index value for the Istanbul Stock Exchange as the response variable, and the index from the London Stock Exchange as the factor. Here we see a group of positive values, followed by a group of negatives, then more positives. The line oscillates in a wave, suggesting that there is a non-random pattern to the weekly residual values.

Figure 16.6: Two Sets of Residuals in Sequence

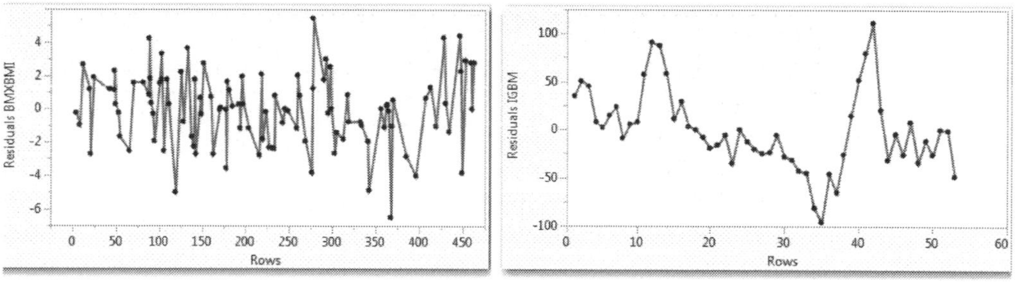

Estimation

Once we have confirmed that our sample satisfies the conditions for inference, we can use the results in three estimation contexts. This section explains the three common ways that regression results can be used and introduces ways to use JMP in all three contexts.

- In some studies, the goal is to estimate the magnitude and direction of the linear relationship between X and Y, or the magnitude of the intercept, or both.

- In some situations, the goal is to estimate the *mean value of Y* corresponding to a particular value of X.

- In some situations, the goal is to estimate an *individual value of Y* corresponding to a particular value of X.

Earlier, we estimated this regression line based on our simple random sample:

$$BMXBMI = -5.888872 + 0.3410071 * BMXWAIST$$

If we had drawn a different random sample, our estimates of the intercept and slope would have been different just because the individuals in the sample would have been different. To put it another way, the specific parameter estimates in the equation are estimated with some degree of sampling error. The sampling error also cascades through any calculations that we make using the estimated line. For example, if we want to estimate BMI for men whose waistlines are 77 cm. (about 30 inches), that calculation is necessarily affected by the slope and intercept estimates, which in turn depend on the particular sample that we drew.

Confidence Intervals for Parameters

Rather than settle for point estimates of β_0 and β_1, we can look at confidence intervals for both. By default JMP reports 95% confidence intervals, and we interpret these as we would any confidence interval. Confidence intervals enable us to more fully express an estimate and the size of potential sampling error.

1. Return to the **Fit Y by X** report, hover the cursor over the table of parameter estimates, and right-click. This brings up a menu, as shown in Figure 16.7.

Figure 16.7: Requesting Confidence Intervals for Parameter Estimates

Remember: In JMP, right-click menus are context sensitive. If you right-click over the title **Parameter Estimates** you'll see a different menu than if you right-click over the table below that title. Be sure that the cursor is over the table before right-clicking.

2. Select **Columns ▶ Lower 95%**.

3. Right-click once again and select **Columns ▶ Upper 95%**.

The revised table of parameter estimates should now look like Figure 16.8 below. On the far right, there are two additional columns representing the upper and lower bounds of 95% confidence intervals for the intercept and slope.

Figure 16.8: Confidence Intervals for Parameters

Parameter Estimates						
Term	Estimate	Std Error	t Ratio	Prob>|t|	Lower 95%	Upper 95%
Intercept	-5.888872	1.346137	-4.37	<.0001*	-8.558013	-3.219732
BMXWAIST	0.3410071	0.013291	25.66	<.0001*	0.3146545	0.3673598

How do we interpret the intervals for the slope and intercept? Consider the slope. Based on this sample of individuals, this regression provides an estimate of the amount by which BMI varies for a 1-cm increase in waist circumference among all adult males. The point estimate is, most likely, inaccurate. That is, we don't really know that the marginal increase in BMI is 0.34 per centimeter of waist circumference. It is more accurate and genuine to say that we are 95% confident that the slope is between approximately 0.31 and 0.37.

Confidence Intervals for $Y|X$

As noted above, we can use the estimated equation to compute one or more *predicted* (or *fitted*) values of Y corresponding to specific values of X. Such an estimate of Y is sometimes called *Y-hat,* denoted $\hat{Y}$. Computing a *Y-hat* value is a simple matter of substituting a specific value for X and then solving the regression equation. If we want to build an interval around a fitted value of Y, we first need to look to the context of the estimate.

A *confidence interval* for $Y|X$ estimates the mean value of Y given the value of X. In our current model of BMI, for example, we might want to estimate the mean BMI among men with 77 cm waists. The point estimate would be

$$\text{BMXBMI} = -5.888872 + 0.3410071 * 77 \approx 20.37 \text{ kg/m}^2$$

This value, which is the location of the estimated line at $X = 77$, becomes the center of the confidence interval, and then JMP computes an appropriate margin of error to form the interval. In the next chapter, we'll discuss this technique in greater depth. As an introductory illustration, we'll see how JMP constructs a *confidence band* along the entire length of the estimated line.

1. In the **Fit Y by X** report, return to the scatter plot with the fitted regression line and enlarge the graph by clicking and dragging the lower-right corner.

2. Click the red triangle next to **Linear Fit** and select **Confid Curves Fit**. Repeat to select **Confid Shaded Fit**. Your graph should now look like Figure 16.9.

Figure 16.9: Fitted Line Plot with Confidence Bands

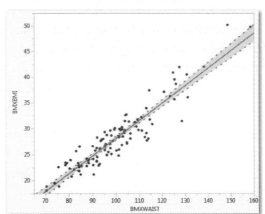

Prediction Intervals for *Y|X*

Finally, we consider the situation in which one might want to estimate or predict a single observation of *Y* given a value of *X*. Conceptually, the procedure is exactly the same as the prior example, but it differs in one critical detail. The magnitude of the sampling error is greater when predicting an individual value than when estimating the mean value of *Y*.

1. Just as before, click the red triangle next to **Linear Fit**.

2. Select **Confid Curves Indiv** and then repeat to select **Confid Shaded Indiv**.

Now look at your scatterplot (not shown here). You should see a new shaded region bounded by *prediction bands,* which are much wider than the confidence bands from Figure 16.9. Both shaded regions are centered along the line of best fit, but the prediction band reflects the substantially greater uncertainty attending the prediction of an individual value rather than the mean of values.

Application

Now that you have completed all of the activities in this chapter, use the concepts and techniques that you've learned to respond to these questions. Notice that these problems are continuations of the corresponding scenarios and questions posed at the end of Chapter 15. Each problem begins with a residual analysis to assess the conditions for inference.

1. *Scenario:* Return to the **NHANES SRS** data table.

   a. Exclude and hide respondents under age 18 and all males, leaving only adult females. Perform a regression analysis for BMI and waist circumference for adult women, evaluate the OLS conditions for inference, and report your findings and conclusions.

   b. If you can safely use this model, provide the 95% confidence interval for the slope and explain what it tells us about BMI for adult women.

   c. If you can safely use this model, use the linear fit graph to read off an approximate confidence interval for the mean BMI for women whose waist measurements are 68 cm.

2. *Scenario:* High blood pressure continues to be a leading health problem in the United States. In this problem, continue to use the **NHANES SRS** table. For this analysis, we'll focus on just the following variables:

   RIAGENDR: respondent's gender

   RIDAGEYR: respondent's age in years

   BMXWT: respondent's weight in kilograms

   BPXPLS: respondent's resting pulse rate

   BPXSY1: respondent's systolic blood pressure ("top" number in BP reading)

   BPXD1: respondent's diastolic blood pressure ("bottom" number in BP reading)

a. Perform a regression analysis with systolic BP as the response and age as the factor. Analyze the residuals and report on your findings.

b. Perform a regression analysis of systolic and diastolic blood pressure and then evaluate the residuals. Explain fully what you have found.

c. Create a scatterplot of systolic blood pressure and pulse rate. One might suspect that higher pulse rate is associated with higher blood pressure. Does the analysis bear out this suspicion?

3. *Scenario:* We'll continue to examine the World Development Indicators data in **BirthRate 2005**. We'll broaden our analysis to work with other variables in that file:

   MortUnder5: deaths, children under 5 years per 1,000 live births

   MortInfant: deaths, infants per 1,000 live births

   a. Create a scatterplot for **MortUnder5** and **MortInfant**. Run the regression and explain what a residual analysis tells you about this sample.

4. *Scenario:* How do the prices of used cars vary according to the mileage of the cars? Our data table **Used Cars** contains observational data about the listed prices of three popular compact car models in three different metropolitan areas in the U.S. All of the cars are two years old.

   a. Create a scatterplot of price versus mileage. Run the regression, analyze the residuals, and report your conclusions.

   b. If it is safe to draw inferences, provide a 95% confidence interval estimate of the amount by which the price falls for each additional mile driven.

   c. You see an advertisement for a used car that has been driven 35,000 miles. Use the model and the scatterplot to provide an approximate 95% interval estimate for an asking price for this individual car.

5. *Scenario*: Stock market analysts are always on the lookout for profitable opportunities and for signs of weakness in publicly traded stocks. Market analysts make extensive use of regression models in their work, and one of the simplest ones is known as the *Random,* or *Drunkard's, Walk* model. Simply put, the model hypothesizes that over a relatively short period of time the price of a particular share of stock will be a random deviation from the prior day. If Y_t represents the price at time *t*, then $Y_t = Y_{t-1} + \varepsilon$. In this problem, you'll fit a random walk model to daily closing prices for McDonald's Corporation for the first six months of 2009 and decide how well the random walk model fits. The data table is called **MCD**.

a. Create a scatterplot with the daily closing price on the vertical axis and the prior day's closing price on the horizontal. Comment on what you see in this graph.

b. Fit a line to the scatterplot and evaluate the residuals. Report on your findings.

c. If it is safe to draw inferences, is it plausible that the slope and intercept are those predicted in the random walk model?

6. *Scenario*: Franz Joseph Haydn was a successful and well-established composer when the young Mozart burst upon the cultural scene. Haydn wrote more than twice as many piano sonatas as Mozart. Use the data table **Haydn** to perform an analysis parallel to the one we did for Mozart.

a. Evaluate the residuals from the regression fit using Parta as the response variable.

b. Compare the Haydn data residuals to the corresponding residual graphs using the Mozart data; explain your findings.

7. *Scenario*: Throughout the animal kingdom, animals require sleep, and there is extensive variation in the number of hours in the day that different animals sleep. The data table called **Sleeping Animals** contains information for more than 60 mammalian species, including the average number of hours per day of total sleep. This will be the response column in this problem.

a. Estimate a linear regression model using gestation as the factor. Gestation is the mean number of days that females of these species carry their young before giving birth. Assess the conditions using residual graphs and report on your conclusions.

8. *Scenario*: For many years, it has been understood that tobacco use leads to health problems related to the heart and lungs. The **Tobacco Use** data table contains recent data about the prevalence of tobacco use and of certain diseases around the world.

   a. Using Cancer Mortality (**CancerMort**) as the response variable and the prevalence of tobacco use in both sexes (**TobaccoUse**), run a regression analysis and examine the residuals. Should we use this model to draw inferences? Explain.

   b. Using Cardiovascular Mortality (**CVMort**) as the response variable and the prevalence of tobacco use in both sexes (**TobaccoUse**), run a regression analysis and examine the residuals. Should we use this model to draw inferences? Explain.

9. *Scenario*: In Chapter 2, our first illustration of experimental data involved a study of the compressive strength of concrete. In this scenario, we look at a set of observations all taken at 28 days (4 weeks) after the concrete was initially formulated; the data table is called **Concrete 28**. The response variable is the **Compressive Strength** column, and we'll examine the relationship between that variable and two candidate factor variables.

   a. Use **Cement** as the factor, run a regression analysis, and evaluate the residuals. Report on your findings in detail.

   b. Use **Water** as the factor, run a regression analysis, and evaluate the residuals. Report on your findings in detail.

10. *Scenario*: Prof. Frank Anscombe of Yale University created an artificial data set to illustrate the hazards of applying linear regression analysis without looking at a scatterplot (Anscombe 1973). His work has been very influential, and JMP includes his illustration among the sample data tables packaged with the software. You'll find **Anscombe** both in this book's data tables and in the JMP sample data tables. Open it now.

    a. In the upper-left panel of the data table, you'll see a red triangle next to the words **The Quartet**. Click the triangle, and select **Run Script**. This produces four regression analyses corresponding to four pairs of response and predictor variables. Evaluate the residuals for all four regressions and report on what you find.

    b. For which of the four sets of data (if any) would you recommend drawing inferences based on the OLS regressions? Explain your thinking.

11. *Scenario*: Many cities in the U.S. have active used car markets. Typically, the asking price for a used car varies by model, age, mileage, and features. The data table called **Used Cars** contains asking prices (**Price**) and mileage (**Miles**) for three popular budget models; all cars were two years old at the time the data were gathered, and we have data from three U.S. metropolitan areas. All prices are in dollars. In this analysis, **Price** is the response and **Miles** is the factor.

    a. As in the prior chapter, run three separate regressions (one for each model). This time, analyze the residuals and report on your findings.

    b. Finally, compare the analysis of residuals for the three models. Which one seems to satisfy the conditions best?

12. *Scenario:* We'll return to the World Development Indicators data in **WDI**. In this scenario, we'll investigate the relationship between access to improved sanitation (the percent of the population with access to sewers and the like) and life expectancy. The response column is **life_exp** and the factor is **sani_acc**.

    a. Use the **Data Filter** to **Show** and **Include** only the observations for the **Year** 2010, and the **Latin America & Caribbean Region** nations. Evaluate the residuals and report on how well the data satisfy the OLS conditions.

    b. Repeat the residuals analysis for **East Asia & Pacific** countries in 2010.

    c. Now do the same one additional time for the countries located in **Sub-Saharan Africa**.

    d. How do the three residuals analyses compare? What might explain the differences?

13. *Scenario:* The data table called **USA Counties** contains a wide variety of measures for every county in the United States.

    a. Run a regression casting sales_per_capita (retail sales dollars per person, 2007) as **Y** and per_capita_income as **X**. Analyze the residuals and report on your findings.

---

[1] You can accomplish the same effect by pressing CTRL-E.

Review of Univariate and Bivariate Inference

Overview

The unifying thread connecting Chapters 10 through 16 has been *statistical inference*, or, how do we assess and manage the risks of drawing conclusions based on limited probabilistic samples? Before moving on to the final portion of the book where we continue with further topics in the practice of inference, we pause again to review the fundamental concepts, assumptions, and techniques of conventional inference methods. As in the earlier review chapters, we'll use the World Development Indicators (WDI) data for our investigations. In particular, we'll revisit some of the descriptive analysis we performed in Chapter 5 to extend the investigation into the realm of inference.

As in the earlier review chapters, the objective is to recapitulate and draw connections between important concepts. The emphasis in this chapter is on the choice of techniques, assessment of assumptions, and interpretation of results. We won't review each previous technique, but focus on a few needed to pursue some specific research questions. We'll also see some "next steps" that anticipate later topics.

Research Context

In Chapter 2, some of our investigations centered on the variability of life expectancy around the world in 2010 and some factors that might account for or be associated with that variation. In this chapter, we'll apply the techniques of inference to the same data and take a closer look at the ways in which differences in national income are associated with changes in life expectancy at birth. To focus the discussion, we'll restrict the investigation to just three variables: life expectancy, income group, and GDP per capita. We'll treat life expectancy as the response variable and both income group and GDP per capita as potential factors.

One Variable at a Time

Our eventual goal is to investigate two bivariate models (a) to determine whether there are statistically significant relationships in either or both models and (b) if so, to decide whether we can use the models to understand how differences in wealth can impact life expectancy. With a quantitative (continuous) response variable and an ordinal variable like income group, we can anticipate that we'll eventually consider applying oneway ANOVA to test their hypothesized relationship. Because GDP per capita is also continuous, simple linear regression is the tool of choice for the second pair. Each of these techniques depends on particular characteristics of the sample data and how it was gathered. Hence we'll start with some basic explorations and estimation for each variable in our study.

1. Open the **WDI** data table.

2. To use only the observations from 2010, use the **Data Filter** to **Show and Include** cases where **Year** = 2010.

3. Use the **Distribution** platform to create summary graphs and statistics for each of the three variables.

For the two quantitative variables, in anticipation of inference, we should almost instinctively look to the histograms and boxplots to get a sense of symmetry, outliers, and peaks. The summary statistics provide further detail about center and dispersion.

As we see in Figure 17.1, both distributions are skewed in opposite directions, with GDP per capita being more strongly skewed with numerous outliers. The life expectancy distribution has multiple peaks, which suggests key sub-group differences.

Figure 17.1: Distributions of Life_Exp and gdp_pc

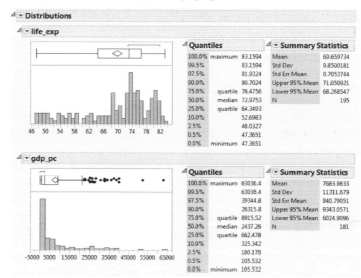

Furthermore, we can say with 95% confidence that in 2010 the mean life expectancy was between 68.3 and 71.1 years. With the same degree of confidence, the mean income per capita (in constant 2000 US dollars) was between $ 6,024.91 and $ 9,343.06.

International agencies define the income groups, so the variation is by design. All five groups are similar in size, with between 31 and 54 nations assigned to each category.

Life Expectancy by Income Group

Checking Assumptions

Both ANOVA and ordinary least squares regression provide reliable inferences under specific conditions. In the case of ANOVA, we want the response variable to be approximately normal within each subgroup of the sample and for variances to be similar for each group. For linear regression, we look for a linear relationship between the two variables, normality in the residuals after first estimating the regression line[1], and constant variance throughout predicted values of Y. We'll consider the linear regression model in the next section. In this part of the review, we'll examine the relationship (if any) between life expectancy and income group, and first consider whether life expectancy can be described as approximately normal.

Earlier, we noted the presence of multiple peaks in the histogram of life expectancy. Might the peaks reflect differences across income groups? Furthermore, might the strongly left-skewed distribution in the aggregate be composed of five separate nearly-normal distributions in the five income groups?

There are several ways we might investigate this. Here is one easy exploratory approach:

1. Open **Graph Builder** and drag **life_exp** into the **Y** drop zone.

2. From the center of the menu bar across the top of the graph, choose the histogram tool.

3. Now drag **Income Group** in the **Group X** drop zone, producing the graph shown below in Figure 17.2.

Figure 17.2: Checking Normality for Life Expectancy by Income Group

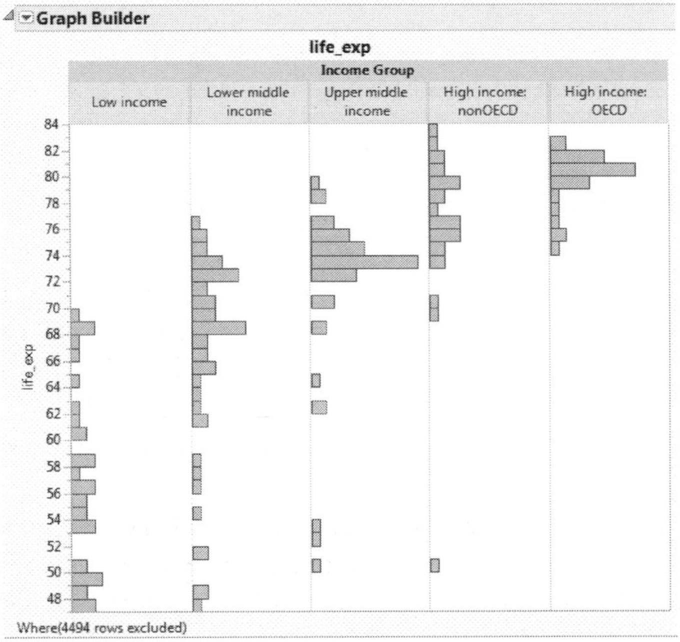

In each income group except for the high income OECD countries, we see substantial departures from normality. We can also see considerable variability in the dispersion of life expectancy values, though it is difficult to judge equality of variances just by looking at these particular graphs. Perhaps an easier way to evaluate the assumptions is by looking at the means and standard deviations in the **Oneway** report from the **Fit Y by X** platform.

4. In the **Fit Y by X** platform, cast **life_exp** as **Y** and **Income Group** as **X.**

5. In the **Oneway** report, click the red arrow and choose **Means and Std Dev** to produce the revised report shown in Figure 17.3.

Figure 17.3: Inspecting Standard Deviations to Check Equal Variance Assumption

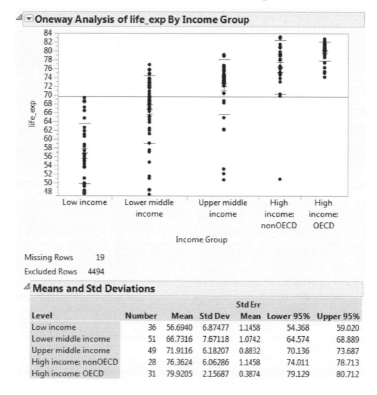

Level	Number	Mean	Std Dev	Std Err Mean	Lower 95%	Upper 95%
Low income	36	56.6940	6.87477	1.1458	54.368	59.020
Lower middle income	51	66.7316	7.67118	1.0742	64.574	68.889
Upper middle income	49	71.9116	6.18207	0.8832	70.136	73.687
High income: nonOECD	28	76.3624	6.06286	1.1458	74.011	78.713
High income: OECD	31	79.9205	2.15687	0.3874	79.129	80.712

The graph shows the individual data values of life expectancy for each income group. In pale blue you should see a short line at the mean value with error bars above and below the mean. Additionally, you'll see longer lines one standard deviation above and below the means. Visually, it is evident that the standard deviation of the high income OECD group is far smaller than the other groups, whose standard deviations are similar. From the table below the graph, we find that the first four groups have standard deviations between approximately 6 and 7.7 years; the last group's standard deviation is only about

2.2. This should signal to us that we might anticipate problems with the assumption of equal variances as well as the normality assumption.

We can formally test for the equality of variances (refer back to Chapter 14 for more details) as follows:

6.   Click the red triangle next to **Oneway** and select **Unequal Variances**.

As we did in Chapter 14, consult the results of Levene's test, and you'll see that there is statistically significant evidence at the 0.0001 level that the variances should not be considered equal. Hence, ANOVA results might be affected by the discrepancies in variance. With unequal variances, Welch's test is a common alternative to ANOVA, and we note that Welch's ANOVA does find a significant difference among the group means.

Conducting an ANOVA

In this case it is probably better to rely on the Welch test than ANOVA, but by way of reviewing the method, let's produce an ANOVA report as well and explore where the differences in group means might occur.

1.   Click the red triangle again and select **Means/Anova**.

2.   Again, this time choose **Compare Means ▶ All Pairs, Tukey HSD**.

Figure 17.4: Visual Summary of the ANOVA and Tukey's HSD

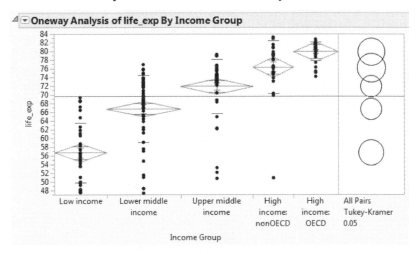

Figures 17.4 and 17.5 are selected from the Oneway report, and will be discussed separately. In Figure 17.4, looking at the means diamonds from left to right as well as the

clicking on the comparison circles on the right side, we see that the means of the two low-income groups are different from each other and from the upper middle income group. The upper middle income group is also different from all others. After that, it becomes more difficult to judge meaningful differences.

Figure 17.5: The Summary of Model Fit and ANOVA Table

⊿ **Oneway Anova**

⊿ **Summary of Fit**

Rsquare	0.598191
Adj Rsquare	0.589731
Root Mean Square Error	6.309155
Mean of Response	69.65973
Observations (or Sum Wgts)	195

⊿ **Analysis of Variance**

Source	DF	Sum of Squares	Mean Square	F Ratio	Prob > F
Income Group	4	11259.402	2814.85	70.7152	<.0001*
Error	190	7563.032	39.81		
C. Total	194	18822.434			

Because of the departures from normality and constant variance, we should be reluctant to draw conclusions from this report, but if conditions had been met we would take note of the significant F-Ratio found in Figure 17.5 (indicating very little likelihood of a Type I error) and the Root Mean Square Error as a goodness-of-fit measure.

Recall from Chapter 14 that when normality cannot be assumed, we should adopt a non-parametric alternative, specifically the Wilcoxon test. Figure 17.6 displays the report from that test. In this instance, the test statistic follows a chi-square distribution (discussed in Chapter 12), and also indicates that we should reject a hypothesis that mean life expectancy is the same in all five income groups.

Figure 17.6: Report of the Wilcoxon/ Kruskal Wallis Test

⊿ **Wilcoxon / Kruskal-Wallis Tests (Rank Sums)**

Level	Count	Score Sum	Expected Score	Score Mean	(Mean-Mean0)/Std0
Low income	36	1101.00	3528.00	30.583	-7.936
Lower middle income	51	3590.00	4998.00	70.392	-4.064
Upper middle income	49	5110.00	4802.00	104.286	0.900
High income: nonOECD	28	4015.00	2744.00	143.393	4.597
High income: OECD	31	5294.00	3038.00	170.774	7.827

⊿ **1-way Test, ChiSquare Approximation**

ChiSquare	DF	Prob>ChiSq
133.8466	4	<.0001*

Life Expectancy by GDP Per Capita

Comparing mean life expectancy across income groups provides a rather crude approach to the relationship between long life and financial resources. We do have a quantitative measure of income, namely, GDP per capita. The variable is measured in inflation-adjusted US dollars, using the year 2000 as the base year. With two quantitative variables and a logical expectation that higher income can provide people with at least some of the preconditions for longer lives, the obvious technique to apply is linear regression.

We will initially use **Graph Builder** to check for a possible linear relationship.

1. Open **Graph Builder**, and place **life_exp** on the **Y** axis and **gdp_pc** on the **X** axis.

The resulting scatterplot is quite obviously *non*-linear. The default smoother rises nearly vertically on the left side of the graph, curves, and then levels off as GDP increases. Such a pattern is typical of *logarithmic growth*, and we can often model such a pattern by taking the logarithm of the X variable; doing so is known as *transforming* the variable.

Within the **Graph Builder**, we can temporarily transform a variable to see how such an adjustment makes a useful difference. Try this:

2. Move the cursor into the **Variables** panel, select **gdp_pc**, and right-click.

3. Choose **Transform ▶ Log**, as shown in Figure 17.7.

Figure 17.7: Making a Temporary Transformed Variable

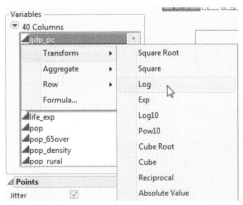

Doing so creates a new temporary variable at the end of the variable list, shown in italics as ***Log[gdp_pc]***.

4. Now drag the temporary variable into the **X** drop zone, placing it just over the axis number values so that it replaces **gdp_pc** as the variable on the horizontal axis.

Figure 17.8: Scatterplot Using life_exp vs. Log (gdp_pc)

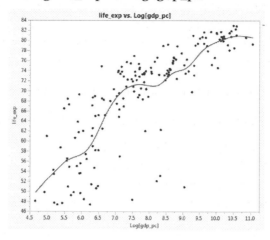

The resulting new graph (Figure 17.8) is much closer to a linear scatterplot than the initial graph. We've surely seen stronger and more distinctly linear patterns, but this is an improvement over the initial graph. To continue with our review, let's return to the Fit *Y* by *X* platform for more extended analysis.

5. Open the **Fit Y by X** dialog and cast **life_exp** into the **Y, Response** role.

6. Hover over the variable name **gdp_pc**, and again apply the log transformation. Be sure to hover over the name, and not the blue triangular continuous variable icon.

7. Cast the new transformed variable as **X, Factor** and click **OK**.

8. In the **Bivariate Fit** report, click the red triangle and choose **Fit Line**.

9. Finally, click the red triangle below the graph, next to **Linear Fit**, and choose **Plot Residuals**.

In Chapter 15, we learned to interpret the series of residual graphs and the regression table, looking for potential violations of the OLS conditions. In these graphs, we find problems with both normality and equal variances (homoskedasticity); as a consequence, the estimates of mean square error, confidence intervals and the *t*-test will be in doubt, but we are still on safe ground to interpret the estimated slope. Consulting the report, we find that the estimated equation is **life_exp = 29.866263 + 5.0379009*Log[gdp_pc]**. To estimate life expectancy for a given country, we would take the natural log of GDP per capita, substitute it into the equation, and find our estimate.

How do we interpret these estimated parameters? In this particular analysis, the intercept has little real-world meaning. If we take its simplest literal meaning, the estimated intercept says that in a nation where the natural log of GDP = 0 (*i.e.,* GDP per capita is $ 1), life expectancy on average would be about 29.9 years, but there is no country that has such a low national income. In our data, the country with the lowest GDP per capita was the Republic of the Congo, with a GDP value of only $ 105.53.

The slope tells us that for each 1 unit increase in the natural log of per capita income, life expectancy increases by slightly more than five years. But, as we noted at the outset, this is not a linear relationship – the five-year increase does not correspond to fixed increases in GDP per capita, but rather to increases in the log of GDP per capita.

Our regression line allows us to estimate that the mean life expectancy among countries with log[gdp_pc] = 6 is approximately 60.09 years. For the log of GDP to be 60.09, GDP per capita would be $ 403.43. The step from log values of 6 to 7 corresponds to per capita income of $ 1,096.63, an increase of more than $ 693. But the next step, from 7 to 8, requires an increase of $1,884.33 in income – hardly a constant rate of increase. Hence, interpretation of this particular regression line involves a more nuanced approach.

Summing Up

Before drawing inferences from sample data, it is important to select a technique appropriate to the type of data at hand, to review the conditions (if any) required by the technique, and to assess whether the data is suitable for drawing inferences. If a hypothesis test is to be done, it is crucial to know what hypothesis is to be tested. If the conditions are met and the data are suitable, we follow well-defined steps to conduct the inferential procedures and interpret the estimates and/or evaluate test results. For all of the statistical tests we've studied (and will study in coming chapters), we'll want to consult the *P*-values and decide if the risk of a Type I error is tolerable.

---

[1] Normality in the residuals can also be thought of as normality in the conditional distribution of Y at each observed value of X. Hence the normality of Y is of some interest.

Multiple Regression 18

Overview

The past two chapters have introduced most of the major concepts important to the practice of regression analysis. At this point, we'll extend the discussion to consider situations in which a response variable Y has simultaneous linear relationships with more than one independent factor. We have been interpreting R^2 as the fraction of the variation in Y associated with variation in X. We've seen examples where a single factor accounts for most of the variation in Y, but we've also seen examples where one X accounts for a small portion of Y's variation. In such cases, we might want to use *multiple regression analysis* to investigate additional factors. Conceptually and in terms of software skills, multiple regression is just an extension of simple regression with one major new concept, known as collinearity. We'll begin by introducing the basic model.

The Multiple Regression Model

Multiple regression substantially expands our ability to build models for variation in Y by enabling us to incorporate the separate effects of more than one factor, and also by enabling us to examine ways in which factors might interact with one another. The multiple regression model accommodates more than one predictive factor:

$$Y_i = \beta_0 + \beta_1 X_{1i} + \beta_2 X_{2i} + ... + \beta_k X_{ki} + \varepsilon_i$$

where Y_i is an observed value of the response variable, each X_{ji} is an observed value for a distinct factor variable, and each of the βs is a parameter of the model. As in the simple model, ε_i is the amount by which an individual response deviates from the model.

As our first illustration, we'll return to a subset of the **Concrete** data. Recall that Prof. Yeh (1998) conducted an experiment to investigate the compressive strength of concrete, measured in a unit call megaPascals (MPa). Compressive strength increases as concrete cures, or ages, after mixture. To simplify our example we'll restrict our attention to a sample of concrete batches that were all measured after curing 28 days.

Concrete is a mixture of water, aggregates (stones, pebbles, sand), and cementitious material. Cementitious materials serve as the binders that hold the mixture together as the water evaporates in curing and are the key source of compressive strength. The most common such binder is called Portland cement, and manufacturers use various cementitious additives to achieve desired properties depending on the structural application.

In our central example, we'll focus on three ingredients: water, Portland cement, and slag—a waste product of iron smelting, which is another cementitious material. We'll build a multiple regression model to predict the compressive strength of concrete after 28 days. Along the way, we'll point out similarities and differences with simple regression analysis. All of the prior conditions and caveats still apply—independent observations, linear relationships, and assumptions about random errors. In addition, we'll add one new concern called *collinearity*, and we'll use a new JMP analysis platform.

Visualizing Multiple Regression

To begin, let's consider the influence of water and Portland cement. In general, the compressive strength of concrete comes from the cementitious ingredients; additional water tends to diminish the compressive strength during curing.

In a simple bivariate regression model, we visualized the relationship between X and Y using a scatterplot. With two independent variables, we could think of the relationship between the response variable and the independent variables in three-dimensional space. JMP makes this easy.

1. Open the data table called **Concrete 28.**

2. Select **Graph ▶ Scatterplot 3D.** Select the columns **Cement, Water,** and **Compressive Strength**, and cast them into the **Y** role (as shown in Figure 18.1). Then click **OK**.

Figure 18.1: Creating a 3D Scatterplot

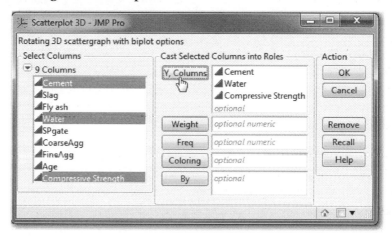

This creates the graph shown below in Figure 18.2. In the graph, you'll notice three axes corresponding to the three columns. Initially, the perspective in this graph is analogous to standing on a balcony looking down into a cube-shaped room from one corner. The points are, as it were, hanging in space; their heights above the floor are set at the observed compressive strength, and their locations reflect the amount of cement and water in the mixture.

JMP enables us to change our perspective interactively.

3. Move the cursor over the grid (into the "room"). You'll note that the cursor shape now becomes 🖑 indicating that we can now rotate the graph.

4. Click and slowly drag the cursor from left to right. The cloud of points will appear to revolve around the vertical axis. Release the mouse button.

5. Now hold the SHIFT key on your keyboard, and click and slowly drag the cursor a short distance from left to right. Release the mouse button, and watch as the graph now spins on its own.

As you watch the spinning graph, notice that the points generally form an *ellipsoid*, which is the three-dimensional version of an ellipse. You might want to imagine a soft rubber ball that has been squashed.

6. To visualize the ellipse, click the red triangle, and select **Normal Contour Ellipsoids**. In the next dialog, just click **OK**.

Figure 18.2: Initial Rendering of 3D Scatterplot with Density Ellipsoid

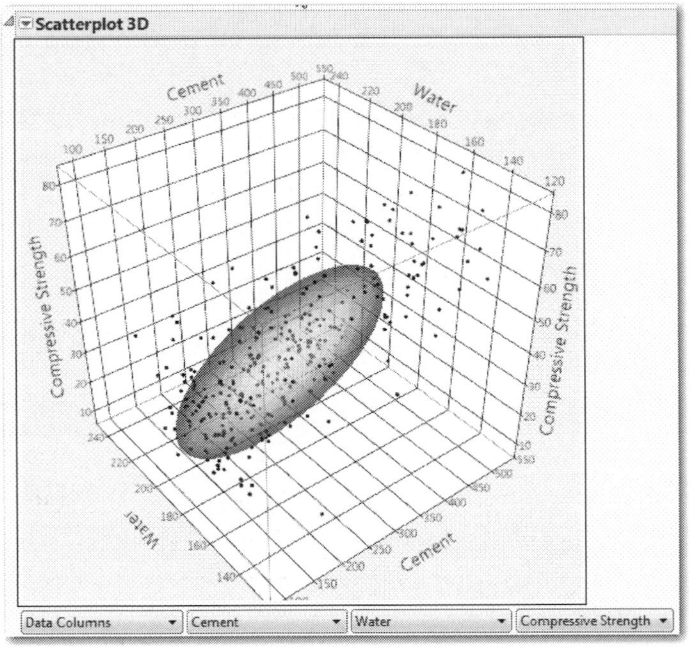

Notice also that from some perspectives, the points may appear quite randomly scattered, but from other perspectives, the ellipsoidal shape of the point cloud is very apparent. This image conveys one very important concept in multiple regression: if Y is related simultaneously to two independent X's, and if we don't look at both X dimensions appropriately, we may fail to detect the nature of the relationship.

In the next section, we will fit a model with three independent (X) variables. Once we move beyond three dimensions, our graphical choices need to change. We'll shortly see how JMP provides graphs that help us see what is going on in the analysis. At this point, though, we're ready to perform a multiple regression analysis, and before we move to three independent variables, let's estimate a model corresponding to Figure 18.2 and see how the visual display corresponds to the multiple regression model.

Fitting a Model

Now let's estimate the model proposed in the previous section.

1. Select **Analyze ▸ Fit Model.**

2. Now select **Compressive Strength** as **Y.**

In our prior regressions, we used both the **Fit Y by X** and the **Fit Model** platforms. The **Fit Y by X** platform is designed for bivariate analysis and cannot be used for multiple regression analyses. We have seen this dialog (Figure 18.3) before, but let's pause now to note how it is organized. As always, available columns appear at the left. In the upper center we can choose the Y variable[1], and in the lower center, we select the factors for the model. The upper right offers some choice of estimation approaches.

Figure 18.3: The Fit Model Dialog Box for this Regression

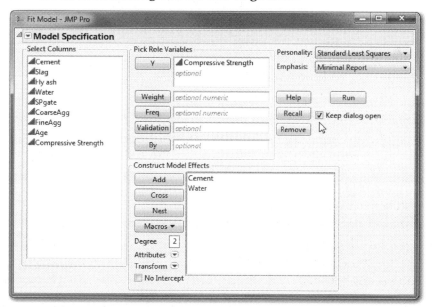

3. In the upper-right corner, select **Minimal Report** as the **Emphasis**.

4. Then, highlight **Cement** and **Water** and click **Add**.

5. Check the box marked **Keep dialog open**, and click **Run**.

Figure 18.4 displays a portion of the results[2]. Although the results window initially looks different from the regressions that we've seen thus far, it actually is very similar. Let's first look at the familiar panels in the output and then learn to interpret the newer parts.

Figure 18.4: Detail from Multiple Regression Report

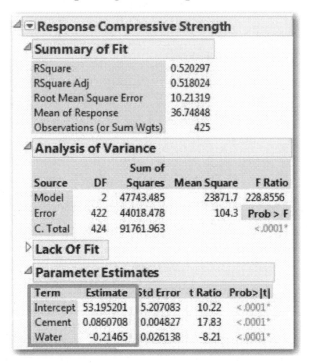

The next section discusses the result reports in detail. Before looking closely at all of the reported statistics, let's consider the coefficients in this model and visualize what their signs tell us about how cement and water together influence compressive strength. Look at the lower portion of Figure 18.4 and locate the estimated slopes. The estimated coefficient of **Cement** is approximately 0.086. The positive coefficient tells us that for each additional kilogram of cement in the mixture, compressive strength increases by 0.086 MPa—assuming that the amount of water remains unchanged. Similarly, the negative coefficient of **Water** indicates that, for a given quantity of cement, additional water weakens compressive strength.

Unlike a simple bivariate regression, this regression model represents a plane in three-dimensional space. The plane slopes upward along the **Cement** dimension and downward along the **Water** dimension. To see this, do the following in the **Fit Model** report:

6. Click the red triangle at the top of the results report. Select **Factor Profiling** ▸ **Surface Profiler**. This opens the panel shown in Figure 18.5 below.

7. Select the **Lock Z Scale** option.

8. Click the disclosure arrow next to **Appearance** and select **Actual**.

Figure 18.5: An Estimated Regression Plane

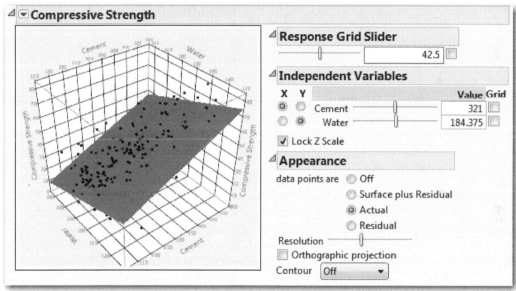

Now you can move the cursor over the 3D grid and rotate the graph, just as you did earlier. You should be able to observe how the actual observed points scatter above and below the estimated regression plane. As you rotate the graph, notice the slope and orientation of the green surface—it rises as the values along the cement axis increase, and it declines along the water axis.

A More Complex Model

We also note that this first model has a modest Rsquare and the points scatter well away from the estimated plane. We can improve the model by adding the third predictor variable: slag. Slag is another "cementitious" component of concrete mixes that serves to bind the mixture together.

1. Return to the **Fit Model** dialog box (still open behind your results) and **Add Slag** to the list of **Model Effects**.

2. Change the **Emphasis** from **Minimal Report** to **Effect Leverage**; we do this to request additional output that helps us decide which factors are influential.

3. **Run** the model.

The first portion of the results is a group of four graphs. Let's first consider the graph on the far left, bypassing the others until a later section. The first graph (Figure 18.6) is essentially a visualization of the goodness of fit of the model. On the horizontal axis, we find the predicted (fitted) values computed by our model; on the vertical axis, we find the actual observed values of compressive strength.

Figure 18.6: Actual by Predicted Plot for the Concrete Strength Model

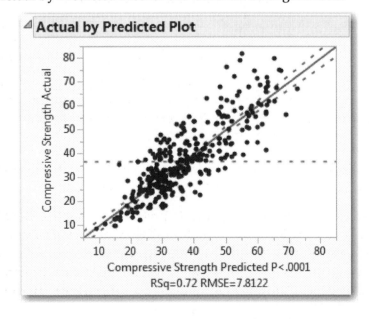

If our model perfectly accounted for all variation in compressive strength, the points would lie exactly on a 45-degree line. To the extent that the model is imperfect and therefore over- or underestimates compressive strength, the points will scatter in the vicinity of the red line. In this graph, we can see that the model performs reasonably well, with predicted values regularly falling close to the actual observed values.

Now look below the group of graphs at three of the next four panels[3]—all of which we've seen before. We interpret these statistics as in the simple case with a few notable exceptions:

- In the **Summary of Fit** panel, we note that now the **RSquare** and **RSquare Adj** statistics are unequal. Because each additional X variable inflates R^2, we prefer the adjusted R^2 because it is a more useful measure when comparing models with different numbers of factors.

- Although the **ANOVA** table is identical in structure to its counterpart in the bivariate context, the unstated null hypothesis has changed. In multiple regression, our null hypothesis is that there is no association between Y and any of the X variables.

- The **Parameter Estimates** panel has expanded, with a row for the intercept and additional rows for each of the X variables in the model.

There are two important changes in interpretation here. First, the coefficients for the slope estimates now represent the estimated change in Y corresponding to a one-unit change in X_j *assuming that all other X values remain unchanged.* So, for example, the estimated coefficient of **Cement** in this model is approximately 0.112. We'll interpret this by saying that the compressive strength of a mixture increases by 0.112 megaPascals for each additional 1 kg. of cement *if* there is no change in the amount of slag or water in the mixture.

Similarly, the negative coefficient of **Water** means that, holding the amount of cement and slag constant, adding water will weaken the mixture.

The second change is the way we interpret the t-statistic and corresponding P-value. As before, a very small P indicates statistical significance. We now note whether a variable makes a significant contribution to the model once all of the other variables have been accounted for. In this example, all of the individual t-tests are significant meaning that each of our three X variables is significant in the presence of the other two.

Residuals Analysis in the Fit Model Platform

The conditions for OLS regression with a multiple regression model are the same as for simple regression. The **Effect Leverage** emphasis automatically generates a plot of residuals versus predicted (Fitted) value, as shown in Figure 18.7. We interpret this graph just as we did in Chapter 16. We see some indication of heteroskedasticity: The points are concentrated closer to the zero line on the left side of the graph, and drift away from zero as we move to the right.

Figure 18.7: Residual vs. Predicted Plot for 3-Variable Compressive Strength Model

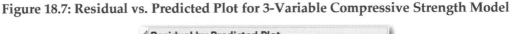
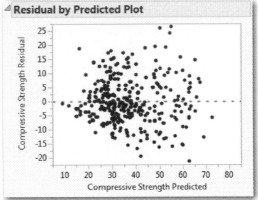

We might be able to come closer to a homoskedastic pattern by including more variables in the model, or by using some approaches illustrated later in Chapter 19. Stay tuned.

> JMP does provide more sophisticated ways to deal with heteroskedasticity, but these go beyond the usual topics covered in an introductory course. Where heteroskedasticity is severe, we should be very reluctant to draw inferences from a least squares regression.

As we have before, we'll need to store and then examine the distribution of the residuals to decide whether the sample satisfies the normality condition.

1.   At the top of the results, click the red triangle next to **Response Compressive Strength** and select **Save Columns ▸ Residuals**.

2.   Select **Analyze ▸ Distribution**. Select **Residual Compressive Strength**.

3. Click the red triangle next to **Residual Compressive Strength** and create a Normal Quantile Plot exactly as we did in Chapter 16.

Although the distribution of the residuals from this model (not shown here) is slightly asymmetric, there are no important departures from normality here. As such, we can conclude that we have the conditions in place to draw general conclusions based on this sample and our least squares regression estimates. That is, we're almost in the clear to go ahead and draw conclusions. Because this is a multiple regression model, there's one new issue to consider: collinearity.

Collinearity

When building a model with multiple X variables, it is important that the X variables vary independently of each other within our sample. If any of the X variables is highly correlated with any subset of X's—a condition called *multicollinearity* or sometimes just *collinearity*—then the estimates of the regression coefficients will likely be distorted and sometimes lead to nonsensical results. Explanatory factors that are highly correlated are essentially redundant. If one is in the model, the next one won't contribute very much additional information. To put it another way: if the experimenters increased the slag content of the mixture every time they increased the cement component, they would never be able to determine how much the strength is affected by each ingredient.

Note two important aspects of collinearity. First, it is a property of a sample rather than the parent population. Whether or not the factors are correlated in the population, we'll have a problem if they are correlated in the sample. Second, it can occur when any subset of X's is correlated with any other subset of X's. This can mean that two factors are correlated or that several X's jointly are correlated with another X. In this section, we'll consider two regression models—one without collinearity and one with.

An Example Free of Collinearity Problems

In a well-designed experiment like this one, the experimenter will have taken care to vary the *X* values independently. After all, in an experiment, the researcher selects and controls the *X* values. In observational or survey studies, collinearity is frequently a problem and we'll look at such an example shortly. First, let's see how we can check for its presence or absence.

JMP provides several ways of detecting collinearity, some of which are quite complex. In this chapter, we'll look at some that are quite easy to use and interpret. First, prior to running a multiple regression, it is advisable to look at possible bivariate correlations between all pairs of *X* columns. We do this as follows:

1. Select **Analyze ▶ Multivariate Methods ▶ Multivariate**. Select the three *X* columns (**Cement, Slag** and **Water**) as **Y** variables in this platform, and click **OK**.

This creates the results shown in Figure 18.8, which displays both scatterplots and a correlation matrix for each pair of the *X* variables. As we can see, none of the relationships are strong[4]—and that is good news. It indicates that no pair of *X*'s is strongly correlated; the strongest correlation is -0.3847, between **Cement** and **Slag**. The scatter plots confirm the absence of strong linear relationships. The ellipses superimposed on the plots are all quite round.

Figure 18.8: Correlations and Scatterplot Matrix for Concrete Data

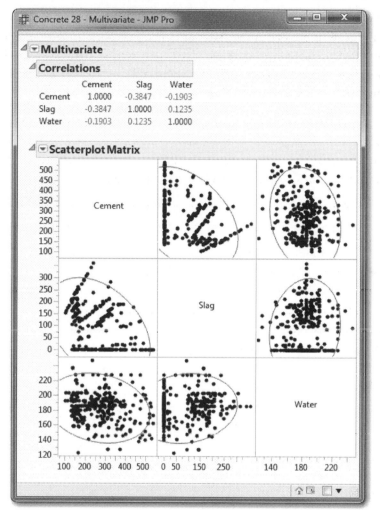

Second, a common indicator of collinearity is that the F test has a small P value indicating a significant model, but the P values for the individual t-tests are all larger. In this example, the individual tests are all quite significant and consistent with the F test.

A third diagnostic tool is the **Leverage Plot** which JMP creates for each factor variable at the top of the regression results. Figure 18.9 shows the leverage plot for the **Cement** factor. The major purpose of a leverage plot is to visually distinguish significant factors from non-significant ones, but these plots can also help show us whether a variable is collinear with other X variables.

Figure 18.9: A Leverage Plot Showing No Collinearity

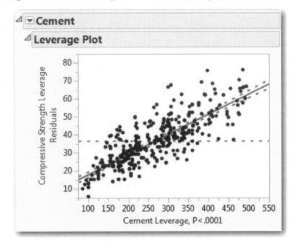

The horizontal dashed blue line represents a reduced regression model excluding the cement variable. The sloped red line with confidence bands represents the full model. Consider a single point in the graph. The vertical distance between the point and the fitted red line is the residual for that observed value of Y. The vertical distance between the point and the horizontal blue line represents the residual in a model that excludes **Cement** from the group of factors. Because the cement variable is significant in this regression, the red line (that is, the full model) "carries" the points better than the blue line. In other words, the model is better when we include cement as a predictor variable.

Figure 18.9 typifies the kind of pattern we're looking for. The red confidence curves exclude the blue horizontal line, except for the small region in the center of the graph where they overlap. If collinearity were a problem in this regression, we'd note that the red dashed curves would largely enclose the blue horizontal line and the points would clump together near the center of the graph.

Another useful approach relies on a diagnostic statistic called the Variance Inflation Factor (VIF). When we have collinearity among X variables, the least squares technique tends to overestimate, or inflate, the standard errors of the estimated regression slopes. As shown in Figure 18.10, JMP computes a VIF for each factor in a model to measure how much the standard errors are inflated by each individual factor.

2. Return to the **Parameter Estimates** panel of the results.

3. Move the cursor over the table of estimates and right-click.

4. Select **Columns ▸ VIF**. The parameter estimates table will include VIFs.

Figure 18.10: Variance Inflation Factors for the 3-Variable Model

Term	Estimate	Std Error	t Ratio	Prob>|t|	VIF
Intercept	42.930012	4.026775	10.66	<.0001*	.
Cement	0.1116149	0.003976	28.08	<.0001*	1.2029719
Water	-0.233898	0.020024	-11.68	<.0001*	1.0407662
Slag	0.0812208	0.004687	17.33	<.0001*	1.1773698

There is no absolute standard for interpreting VIFs. A VIF can never be less than 1 but there is no upper limit. Values close to 1 indicate very little inflation that might be attributed to collinearity. In general practice, values over 10 may signal a problem with collinearity, as we'll see in the next example.

An Example of Collinearity

In this example, we'll use our SRS of the NHANES data to build a model to predict systolic blood pressure for adult men. As we've done before, we'll start by selecting only the observations of adult males:

1. Open the **NHANES SRS** data table.

2. Select **Rows ▸ Data Filter**. As we have done several times, filter the rows to select, show, and include males ages 18 years and older.

Let's initially examine a model that uses a man's age (in months), height, weight, and body mass index (BMI) to predict systolic blood pressure. Note that the BMI is computed directly from height and weight, so it is surely collinear with these two variables. We're deliberately specifying a model with substantial collinearity so that we can learn to detect the problem in the output results. Before running the regression, let's look at the correlations among these predictors.

3. Again select **Analyze ▸ Multivariate Methods ▸ Multivariate**.

4. Select the following four columns as **Y** variables **RIDAGEEX, BMXWT, BMXHT, BMXBMI**. Click **OK**.

The results now look like Figure 18.11. Notice the very strong positive linear association between weight and BMI. This will cause some problems estimating the parameters of a multiple regression model.

Figure 18.11: Correlations in NHANES Data

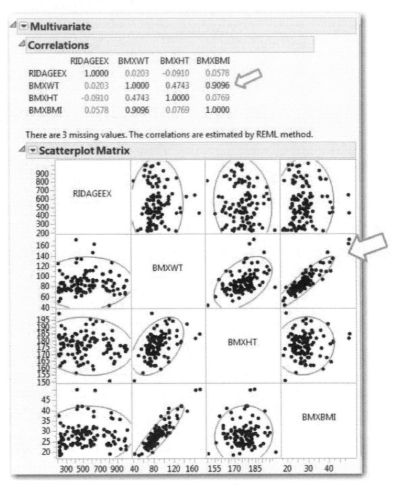

1. Select **Analyze ▸ Fit Model**. The **Y** variable is **BPXSY1**.

2. Now **Add** the columns **RIDAGEEX, BMXWT, BMXHT, BMXBMI**, to the model and click **Run**.

With four variables in the model, we get four leverage plots. In your own results, look first at the leverage plot for age. It looks similar to the plots that we saw earlier; the red confidence curves mostly exclude the blue horizontal line and the points scatter along the length of the red line of fit.

Figure 18.12: Two Leverage Plot for Weight, Showing Collinearity

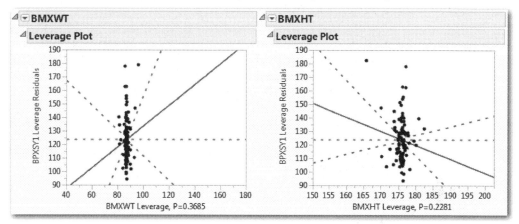

Figure 18.12 shows the plots for weight and height. Notice two key features. First, the red confidence curves bend very sharply to *include* the entire blue horizontal line in both graphs. This indicates that we cannot confidently conclude that the model including the weight and height variable performs any better than the model without the variables. Additionally, all of the points have migrated to the middle of the graph, so that they are all stacked up nearly vertically. In the presence of the other explanatory variables, weight and height do not contribute additional explanatory information.

Once we display the Variance Inflation Factors in the Parameter Estimates table, we see considerable inflation in the estimated standard errors, as shown in Figure 18.13. The three variables which we knew to be related all have very large VIFs.

Figure 18.13: Variance Inflation Factors in a Model with Collinearity

Term	Estimate	Std Error	t Ratio	Prob>|t|	VIF
Intercept	287.37262	152.0513	1.89	0.0618	.
RIDAGEEX	0.0225754	0.007562	2.99	0.0036*	1.0215825
BMXWT	0.770314	0.852579	0.90	0.3685	108.32403
BMXHT	-1.037344	0.85504	-1.21	0.2281	20.399817
BMXBMI	-2.133063	2.702585	-0.79	0.4319	84.221949

Dealing with Collinearity

Once we find it, what do we do about it? There are two basic strategies: we can omit all but one of the collinear variables, or we can *transform* some variables to eliminate the collinearity. We'll discuss transformations in Chapter 19. For now, we'd fit other models using combinations of the remaining variables. In this section, we consider a model without the BMI and weight variables because they had the most inflated variances.

1. At this point, return to the **Fit Model** platform and specify a model for systolic blood pressure using only age and height as explanatory factors.

Figure 18.14 shows the new leverage plot for this simpler, two variable model. This plot shows that the collinearity problem has been ameliorated: the points are now spread along the fitted line. However, the variable still has an insignificant effect. The *P* value is rather high (0.1558) and the red confidence curves include the blue horizontal line.

Figure 18.14: An Improved Leverage Plot

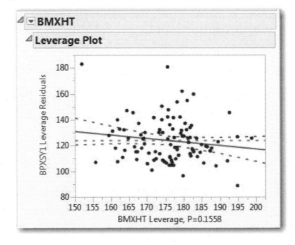

In short, we've dealt with the collinearity issue but still have a deficient model. In general, we don't want to include a variable with no significant predictive value. The height variable does not contribute to the model.

Evaluating Alternative Models

With multiple regression analysis, we often want to compare two or more competing models. In our concrete example, for instance, we earlier estimated a model with three significant factors, reasonably well-behaved residuals, and which accounted for about 72% of the observed variation in compressive strength. Perhaps a four-variable model would do better?

The process of comparing models can easily devolve into a chaotic flurry of trial-and-error attempts. When estimating several models for a single *Y* variable, take these points into consideration:

- Do all of the independent factors make logical sense in the model? Are there solid theoretical reasons to include each factor? If not, why include nonsense factors?

- Does the model under consideration have acceptable residual plots? Does an alternative model more closely meet the conditions for inference? Other things being equal, a model with "good" residuals is preferable.

- Are there signs of collinearity? If so, deal with it. A model with collinearity has limited utility in understanding relationships between factors and a response.

- Do the signs of the estimated coefficients all make sense? Remember to apply the logic of "assuming that the other variable(s) are held constant, what effect do I expect when this independent variable increases by one unit?"

- Are the individual *t*-tests significant for all factors? A variable with a non-significant *t*-test should be removed from the model.

- Among the acceptable models, which one has the smallest root mean square error? In other words, which has the smallest total unexplained variation?

- Which model has the highest adjusted R^2?

To illustrate, let's return to the concrete example and add a variable.

1. Locate the **Fit Least Squares** report for the **Concrete 28** data. Click the red triangle next to **Response Compressive Strength** and select **Model dialog**. We just want to add another variable to the list of independent *X* variables.

Fly ash is yet another cementitious byproduct of coal burning. Those with expertise in concrete formulation would expect that the addition of fly ash will increase compressive strength. Let's estimate the amount by which fly ash increases the strength of concrete. Then we'll work through the bullet points listed above to compare this model to the earlier three-variable model.

2. Highlight **Fly Ash** and click **Add**. Then click **Run**.

Take a moment to review the results of this regression, looking back at the bulleted list of questions above. Your comparison should be similar to that summarized in Table 18.1. Sometimes we'll find that one model is better in one respect, but not in another. In this case, the second model equals or outperforms the first model.

Table 18.1: A Comparison of Two Models

Criterion	First Model (Cement, Water, Slag)	Second Model (Cement, Water, Slag, Fly Ash)	Comments
All factors make sense?	Yes	Yes	
Acceptable Residual Plots?	Yes; mild heteroskedasticity	Yes; mild heteroskedasticity	Both sets of residuals look normal in shape
Evidence of Collinearity?	No	No	Leverage plots show no problem; VIFs small
Estimated signs make sense?	Yes	Yes	Water has negative coefficient; others positive
All *t*-tests significant?	Yes	Yes	
Smallest Root MSE?	RMSE = 7.812	RMSE = 7.236	Second model is better
Largest adjusted RSquare?	AdjR² = .718	AdjR² = .758	Second model is better

It is also worth comparing the estimated coefficients in the two analyses to note another important concept in multiple regression. As you compare the coefficients of cement, slag, and water—the variables included in both models—you'll notice that they are very similar in both regressions, differing only by about ± 0.02. The intercepts of these two models are, however, quite different. In the second model, the intercept has shrunk from 42.93 to about 29.34.

If we omit a variable that should be included in the model—a type of common *specification error*—then the effect of that variable sometimes gets "lumped into" the estimated intercept or the coefficient of another independent variable or both. Once we include the relevant variable, its effect is appropriately assigned to the variable.

Application

Now that you have completed all of the activities in this chapter, use the concepts and techniques that you've learned to respond to these questions. Notice that several of these problems are continuations of the corresponding scenarios and questions posed in the prior two chapters. In each multiple regression analysis, you should examine the residuals for possible violations of regression assumptions and also look for indications of collinearity. When you report your findings, be sure to discuss your conclusions about assumptions and about collinearity.

1. *Scenario:* Return to the **NHANES SRS** data table. Select, include, and show only female respondents ages 18 and older.

    a. Perform a regression analysis for adult women with BMI as Y and use waist circumference and height as the independent variables. Discuss the ways in which this model is or is not an improvement over a model that just uses waist circumference as a predictor.

    b. If you can safely use this model, provide the 95% confidence interval for the slope and explain what it tells us about BMI for adult women.

    c. Perform another regression analysis to predict BMI using waist circumference and wrist circumference as independent variables. Evaluate this model in comparison to the prior model. Report your findings.

    d. Choose another variable that makes sense to you, and perform a third regression analysis. Evaluate this model in comparison to the prior models, and report what you find.

2. *Scenario:* High blood pressure continues to be a leading health problem in the United States. In this problem, continue to use the **NHANES SRS** table. For this analysis, we'll focus on females between the ages of 12 and 19.

    a. Create a 3-D scatterplot of systolic blood pressure (the "top" number in a BP reading), age in months, and weight. Describe what you observe after rotating the scatterplot.

    b. Perform a regression analysis with systolic BP as the response and age (months) and weight as the factors. Report on your findings.

    c.   Perform a regression analysis with systolic BP as *Y* and age (months), weight, and diastolic blood pressure as the independent variables. Explain fully what you have found.

    d.   Review the full list of columns within the data table, and develop another model to predict systolic BP using whichever columns make sense to you. Report on your findings.

3.   *Scenario:* We'll continue to examine the World Development Indicators data in **BirthRate 2005**. Throughout the exercise, we'll use **BirthRate** as the **Y** variable. We'll broaden our analysis to work with other variables in that file:

    MortUnder5: deaths, children under 5 years per 1,000 live births

    MortInfant: deaths, infants per 1,000 live births

    MortMaternal: number of maternal deaths in childbirth per 100,000 live births

    a.   Initially estimate a model using all three of the above-listed variables. Evaluate your model and report on your findings.

    b.   Drop the weakest of the three variables from the model and re-estimate using a reduced model with two independent variables. Report on your findings, comparing these results to those for the prior model.

4.   *Scenario*: Throughout the animal kingdom, animals require sleep and there is extensive variation in the number of hours in the days that different animals sleep. The data table called **Sleeping Animals** contains information for more than 60 mammalian species, including the average number of hours per day of total sleep. Use this as the response column in this problem.

    a.   As in the prior chapters, first estimate a linear regression model using Gestation as the factor. Gestation is the mean number of days that females of these species carry their young before giving birth. Evaluate this regression.

    b.   Now add a second independent variable: brain weight. Re-run the regression and evaluate this model in comparison to the first.

    c.   Finally, estimate a model that also includes body weight as a third explanatory variable. Is this model an improvement over the earlier models? Explain your reasoning.

5.   *Scenario*: Industrial production refers to the manufacture of goods within a country, and is measured in this scenario by India's Index of Industrial Production (IIP).   The IIP is a figure that compares a weighted average of industrial output from different sectors of the economy (for example, mining,

agriculture, and manufacturing) to a reference year in which the index is anchored at a value of 100. As a result, we can interpret a value of the IIP as a percentage of what industrial output was in the base year. So, for example, if the IIP equals 150 we know that production was 50% higher at that point than it was in the base year. The data table **India Industrial Production** contains IIPs for different segments of the Indian economy. All figures are reported monthly from January 2000 through June 2009.

a. First, develop a simple linear regression model to estimate monthly General IIP using one of the sector-specific component indexes. Select the X column by first determining which of the available columns has the highest correlation with the general index. Report on your model.

b. Now add a second X variable, using one of the remaining columns that makes sense to you. Report on your findings.

c. What problem have you encountered in finding an appropriate column to add to your original model? Why should we anticipate encountering this problem?

6. *Scenario*: In the United States, the management of solid waste is a substantial challenge. Landfills have closed in many parts of the nation, and efforts have been made to reduce waste, increase recycling, and otherwise control the burgeoning stream of waste materials. Our data table **Maine SW** contains quarterly measurements of the tonnage of solid waste delivered to an energy recovery facility in Maine. The data come from five municipalities in the state.

a. Bucksport is a lovely coastal community with a population of approximately 5000 people. Use the available data from other Maine cities and towns to develop a multiple regression model for the quarterly tonnage of solid waste in Bucksport. Report on your choice of a model.

b. It seems unlikely that the volume of solid waste in one town can "cause" the generation of solid waste in another community. How can you explain the relationships that your model represents?

---

[1] We actually can choose more than one Y for separate models, but we'll restrict our coverage of the topic to one response variable.

[2] For pedagogical reasons, the **Lack of Fit** panel has been hidden in here. As noted in a prior chapter, the lack of fit statistics are beyond the scope of most introductory courses.

[3] Again, the **Lack of Fit** panel has been hidden in here. As noted in a prior chapter, the lack of fit statistics are beyond the scope of most introductory courses.

[4] In the correlation matrix, very strong positive values appear in dark blue and very strong negative correlations are dark red. The color gradient goes from dark red through gray to dark blue. Weak or insignificant correlations are pale gray.

Categorical, Curvilinear, and Non-Linear Regression Models

Overview

In the past several chapters, we have worked extensively with regression analysis. Two common threads have been the use of continuous data and of linear models. In this chapter, we introduce techniques to accommodate categorical data and to fit several common curvilinear patterns. Throughout this chapter, all of the earlier concepts of inference, residual analysis, and model fitting still hold true. We'll concentrate here on issues of *model specification*, which is to say selecting variables and functional forms (other than simple straight lines) that reasonably and realistically suit the data at hand.

Dichotomous Independent Variables

Regression analysis operates upon numerical data, but for some response variables, a key factor is categorical. We can easily incorporate categorical data into regression models with a *dummy*, or *indicator, variable*—a numeric variable created to represent the different states of a categorical variable. In this chapter, we'll illustrate the use of dummy variables for *dichotomous* variables, which are variables that can take on only two different values.

For our first example, we'll return to the sonatas of Haydn and Mozart and some data that we first analyzed in Chapter 15 (Ryden, 2007). In the sonata form, the composer introduces a melody in the exposition, and then elaborates and recapitulates that melody in a second portion of the piece. The theory that we tested in Chapter 15 was that Haydn and Mozart divided their sonatas so that the relative lengths of the two portions approximated the golden ratio. We hypothesized that our fitted line would come close to matching this line:

$$Parta = 0 + 0.61803(Partb)$$

In this chapter, we'll ask whether the two composers did so with the same degree of consistency. We'll augment our simple regression model of Chapter 13 in two ways. First, we'll just add a dummy variable to indicate which composer wrote which sonata.

1. Open the data table called **Sonatas**.

This data table combines all of the observations from the two tables **Mozart** and **Haydn**. The dummy variable is column 4. Notice that the variable equals 0 for pieces by Haydn and 1 for pieces by Mozart.

2. Select **Analyze ▶ Fit Model**. Cast **Parta** as **Y**, add **Partb** and **Mozart** to the model, then run the model. The parameter estimates appear in Figure 19.1.

Figure 19.1: Estimated Coefficients for the Model with a Dummy Variable

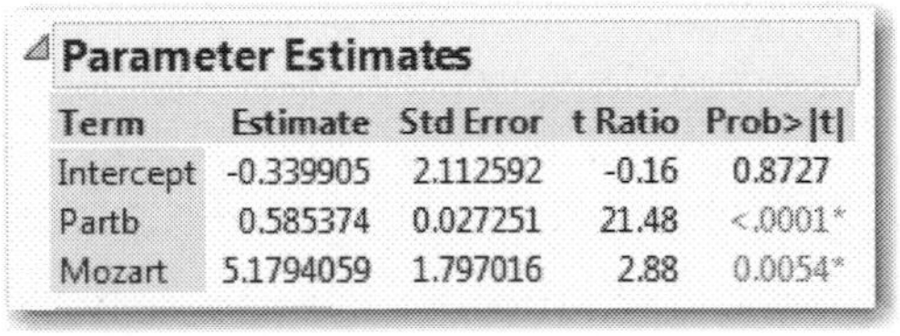

⊿ **Parameter Estimates**

Term	Estimate	Std Error	t Ratio	Prob>|t|
Intercept	-0.339905	2.112592	-0.16	0.8727
Partb	0.585374	0.027251	21.48	<.0001*
Mozart	5.1794059	1.797016	2.88	0.0054*

The fitted equation is:

$$\widehat{Parta} = -0.34 + 0.585(Partb) + 5.179(Mozart)$$

How do we interpret this result? Remember that **Mozart** is a variable that can equal only 0 or 1. Therefore, this equation actually expresses two parallel lines:

Haydn Sonatas: $\widehat{Parta} = -0.34 + 0.585(Partb)$

Mozart Sonatas: $\widehat{Parta} = (5.179 - 0.34) + 0.585(Partb)$

Both of these lines have the same slope but different intercepts. If we want a model that allows for two intersecting lines, we can introduce an *interaction* between the independent variable and the dummy variable, as follows:

3. Return to the **Fit Model** dialog box, which is still open behind the **Least Square** report window.

4. Highlight both **Partb** and **Mozart** in the **Select Columns** box, and click **Cross**. You will now see the term **Partb*Mozart** listed in the **Construct Model Effects** panel.

5. Now click **Run**.

Figure 19.2 shows the parameter estimates for this model, which fits the data slightly better than the previous model. With an interaction, JMP reports the coefficient in terms of adjusted values of the variables, subtracting the sample mean for each variable. We note that all estimates except for the intercept are statistically significant.

Figure 19.2: Results with a Dummy Interaction Term

△ Parameter Estimates				
Term	Estimate	Std Error	t Ratio	Prob>|t|
Intercept	0.6247019	2.109818	0.30	0.7682
Partb	0.5630538	0.028647	19.65	<.0001*
Mozart	5.4173436	1.754171	3.09	0.0030*
(Partb-74.4063)*(Mozart-0.45313)	0.1150727	0.055546	2.07	0.0426*

Just as the first model implicitly represented two intercept values, this model has two intercepts and two slopes: there is a Mozart line and a Haydn line. The default report is a bit tricky to interpret because of the subtraction of means in the interaction line, so let's look graphically at the fitted lines. We'll take advantage of yet another useful visual feature of JMP that permits us to re-express this regression model using the original dichotomous categorical column, **Composer**.[1]

6.   Return again to the **Fit Model** dialog box and highlight **Mozart** and **Partb*Mozart** within the list of model effects. Click the **Remove** button.

7.   Select the **Composer** column, and add it to the model effects.

8.   Highlight **Partb** and **Composer** under **Select Columns** and click **Cross**. The model now lists **Partb**, **Composer**, and **Partb*Composer**. Click **Run**.

Because this regression model has just a single continuous independent variable, JMP automatically plots the regression fit above the scatterplot of fits versus observed values. Your output should now include the graph shown here in Figure 19.3. The red line represents the Haydn sonatas, and the blue represents those by Mozart.

Figure 19.3: Estimated (Fitted) Lines for the Two Composers

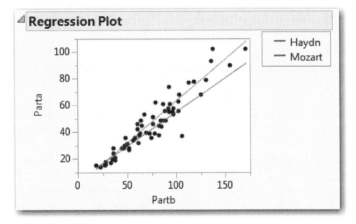

In the table of **Parameter Estimates**, we do find discrepancies between this model and the estimates shown in Figure 19.2. Now, the estimated coefficient of Partb is the same as before, but all of the others are different. The t-ratios and P-values are all the same, but the parameter estimates have changed. This is because JMP automatically codes a dichotomous categorical variable with values of –1 and +1 rather than 0 and 1. Hence, the coefficient of the composer variable is now –2.708672, which is exactly one-half of the earlier estimate. This makes sense, because now the difference between values of the dummy variable is twice what it was in the 0-1 coding scheme.

Dichotomous Dependent Variable

Next, we consider studies in which the dependent variable is dichotomous. In such cases there are at least two important shortcomings of ordinary least squares estimation and a linear model. First, because Y can take on only two values, it is illogical to expect that a line provides a good model. A linear model will produce estimates that reach beyond the 0-1 limits. Second, because all observed values of Y will always be either 0 or 1, it is not possible that the random error terms in the model will have a normal distribution or a constant variance.

With a dichotomous dependent variable, one better approach is to use *logistic regression*[2]. Estimating a logistic model is straightforward in JMP. To illustrate, we'll turn to an example from the medical literature (Little, McSharry et al. 2008) on Parkinson's disease (PD). PD is a neurological disorder affecting millions of adults around the world. The disorder tends to be most prevalent in people over age 60, and as populations age, specialists expect the number of people living with PD to rise. Among the common symptoms are tremors, loss of balance, muscle stiffness or rigidity, and difficulty in initiating movement. There is no cure at this time, but medications can alleviate symptoms, particularly in the early stages of the disease.

People with PD require regular clinical monitoring of the progress of the disease to assess the need for changes in medication or therapy. Given the nature of PD symptoms, though, it is often difficult for many people with PD to see their clinicians as often as might be desirable. A team of researchers collaborating from University of Oxford (UK), the University of Colorado at Boulder, and the National Center for Voice and Speech in Colorado was interested in finding a way that people with PD could obtain a good level of care without the necessity of in-person visits to their physicians. The team noted that approximately 90% of people with PD have "some form of vocal impairment. Vocal impairment might also be one of the earliest indicators for the onset of the illness, and the measurement of voice is noninvasive and simple to administer" (ibid., p. 2). The practical question, then, is whether a clinician could record a patient's voice—even remotely—and by measuring changes in the type and degree of vocal impairment, draw valid conclusions about changes in progress of Parkinson's disease in the patient.

In our example, we'll use the data from the Little, McSharry *et al.* study in a simplified fashion to differentiate between subjects with PD and those without. The researchers actually developed a new composite measure of vocal impairment, but we'll restrict our attention to some of the standard measures of voice quality. In our sample, we have 32 patients; 24 of them have Parkinson's and eight do not.

Vocal measurement usually involves having the patient read a standard paragraph or sustain a vowel sound or both. Two common measurements of a patient's voice refer to the wave form of sound: technicians note variation in the amplitude (volume) and frequency (pitch) of the voice. We'll look at three measurements:

- MDVP:Fo(Hz): vocal fundamental frequency (baseline measure of pitch)[3]
- MDVP:Jitter(Abs): *jitter* refers to variation in pitch
- MDVP:Shimmer(db): *shimmer* refers to variation in volume

In this example, we'll simply estimate a logistic regression model that uses all three of these measures and learn to interpret the output. Interpretation is much like multiple regression, except that we're no longer estimating the absolute size of the change in Y given one-unit changes in each of the independent variables. In a logistic model, we conceive of each X as changing the probability, or more specifically the odds-ratio, of Y. Fundamentally, in logistic regression the goal of the model is to estimate the likelihood that Y is either "yes" or "no" depending on the values of the independent variables.

More specifically, the model estimates $\ln[p/(1-p)]$, where p is the probability that $Y = 1$. Each estimated coefficient represents the change in the estimated log-odds ratio for a unit change in X, holding other variables constant—hardly an intuitive quantity! Without becoming too handcuffed by the specifics, bear in mind at this point that a positive coefficient only indicates a factor that increases the chance that Y equals 1, which in this case means that a patient has Parkinson's.

JMP produces logistic parameter estimates using a *maximum likelihood estimation (MLE[4])* approach rather than the usual OLS (ordinary least squares method). For practical purposes, this has two implications for us. First, rather than relying on F or t-tests to evaluate significance, JMP provides Chi-square test results, which are more appropriate to MLE methods. Second, the residual analysis that we typically perform in a regression analysis is motivated by the assumptions of OLS estimation. With MLE, the key assumption is the independence of observations, but we no longer look for linearity, normality, or equal variance.

1. Open the **Parkinson's** data table.

Each row in the table represents one subject (person) in the study. The **Status** column is a dummy variable where 1 indicates a subject with PD. Hence, this is our dependent variable in this analysis. In the data table, **Status** is numeric but has a nominal modeling type.

2. Select **Analyze ▸ Fit Model**. Select **Status** as **Y** and add **MDVP:Fo(Hz)**, **MDVP:Jitter(Abs)**, and **MDVP:Shimmer(dB)** as model effects. Select with care, as there are several jitter and shimmer columns. Click **Run**.

> Notice that as soon as you specify **Status** as the dependent variable, the personality of the **Fit Model** platform switches from **Standard Least Squares** to **Nominal Logistic**.

The results of the logistic regression appear in Figure 19.4. Next, we review selected portions of several of the panels within the results. For complete details about each portion of the output, consult the JMP Help menu.

Figure 19.4: Logistic Regression Results for the Parkinson's Model

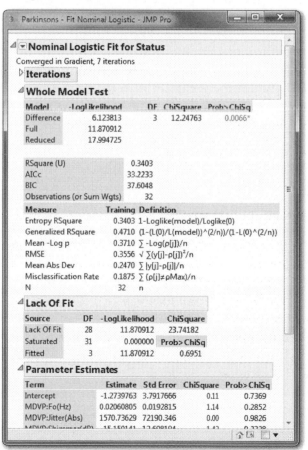

Whole Model Test

This portion of the output compares the performance of our specified model with a naïve default model with no predictor variables. The naïve model essentially posits that, based on the sample ratio of 8/32 or 25% of the patients having no disease, there is a probability of 0.75 that a tested patient has PD. In this panel, the chi-square test is analogous to the customary F test and in this analysis, we find a significant result. In other words, this model is an improvement over simply estimating a 0.75 probability of PD.

We also find **RSquare (U)**, which is analogous to R^2 as a goodness-of-fit statistic. This model provides an unimpressive value of 0.34, indicating a lot of unexplained variation. Such a result is fairly common in logistic modeling due in part to the fact that we're often using continuous data to predict a dichotomous response variable.

Parameter Estimates

As noted earlier, we don't interpret the parameter estimates directly in a logistic model. Because each estimate is the marginal change in the log of the odds-ratio, sometimes we do a bit of algebra and complete the following computation for each estimated β.:

$$e^{\hat{\beta}} - 1$$

The result is the estimated percentage change in the odds corresponding to a one-unit change in the independent variable. Of course, we want to interpret the coefficients only if they are statistically significant. To make judgments about significance, we look at the next panel of output.

Effect Likelihood Ratio Tests

In the bottom-most panel of the results are the analogs to the *t*-tests we usually consult in a regression analysis. We interpret the *P*-values as we do in other tests, although in practice we might tend to use higher alpha levels, given the discontinuous nature of the data. In this example, it appears that there is no significant effect associated with jitter, but shimmer (variation in vocal loudness) does look as if it makes a significant contribution to the model in distinguishing PD patients from non-PD patients. We'd hold off drawing any conclusion about the base frequency (**MDVP:F0(Hz)**) until running an alternative model without the jitter variable.

Curvilinear and Non-Linear Relationships

Earlier in this book, we've seen examples in which there is a clear relationship between X and Y, but the relationship either is a linear combination of powers or roots, or are non-linear (e.g. when the X appears in an exponent). In some disciplines, there are well-established theories to describe a relationship, such as Newton's laws of motion. In other disciplines, non-linear relationships might be less clear. In fact, researchers often use regression analysis to explore such relationships. This section of the chapter introduces two common functions that frequently do a good job of fitting curvilinear data.

Quadratic Models

We begin with a simple descriptive example of a fairly well-known quadratic relationship. A *quadratic* function is one that includes an X^2 term. Quadratic functions belong to the class of *polynomial functions*, which are defined as any functions that contain several terms with X raised to a positive integer power. Let's see how this plays out in our solar system.

1. Open the **Planets** data table. This table contains some basic measurements and facts about the planets. For clarity in our later graphs, the table initially selects all rows and assigns markers that are larger than usual.

2. Select **Analyze ▸ Fit Y by X**. Cast **Sidereal period of orbit (years)** as **Y** and **Mean Distance from the Sun (AU)** as **X** and click **OK**.

The *sidereal period of orbit* is the time, in Earth years, that it takes for a planet to complete its orbit around the sun. *AU* refers to an astronomical unit, which is defined as the distance from the Earth to the sun. So, our response variable is the length of a planet's orbit relative to Earth, and the regressor is the mean distance (as a multiple of the Earth's distance) from the sun. Figure 19.5 shows the scatterplot.

Figure 19.5: Orbital Period vs. Distance from the Sun

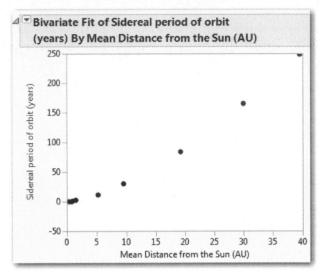

The relationship between these two columns is quite evident, but if we fit a straight line, we'll find that the points actually curve. If we fit a quadratic equation, we get a near-perfect fit.

3. Click the red triangle and select **Fit Line**.

4. Now click the red triangle again and select **Fit Polynomial ▸ 2, quadratic**.

Your analysis report should now look like Figure 19.6. Although the linear model is a good fit, the quadratic model more accurately describes this relationship.

Figure 19.6: Quadratic and Linear Models for Planet Data

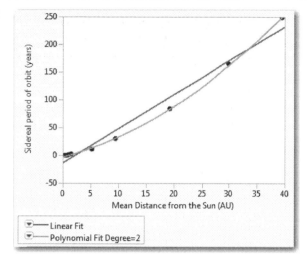

We can use quadratic models in many settings. We introduced regression in Chapter 4 using some data in the data table **Birthrate 2005**. This data table contains several columns related to the variation in the birth rate and the risks related to childbirth around the world as of 2005. In this data table, the United Nations reports figures for 194 countries. You might remember that we saw a distinctly non-linear relationship in that data table. Let's look once more at the relationship between a nation's birth rate and its rate of maternal mortality.

1. Open the **Birthrate 2005** data table.

2. Select **Analyze ▸ Fit Y by X**. Cast **MortMaternal** as **Y** and **BirthRate** as **X** and click **OK**.

We've seen this graph at least twice before. Your results will look like those shown in Figure 19.7, which shows the relationship with both linear and quadratic fits. When we last encountered this graph, we concluded that it would be unwise to fit a linear model, and that we should return to it again in this chapter. Here we are.

Figure 19.7: Relationship Between Maternal Mortality and Birth Rate

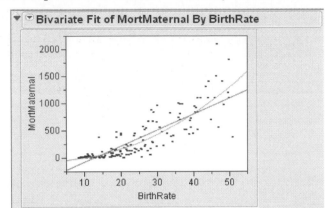

If we were to fit a line, we'd find that a straight-line model is a poor characterization of this pattern. A line will underestimate maternal mortality in countries with very low or very high birth rates, and substantially overstate it in countries with middling birth rates. Given the amount of scatter in the graph, we're not going to find a perfect fit, but let's try the quadratic model as an alternative to a line.

3. Click the red triangle to fit a line.

4. Click the red triangle once more and select **Fit Polynomial ▸ 2, quadratic**.

The goodness-of-fit statistic, R^2, is actually similar for the two models, but the quadratic model captures more of the upward-curving relationship between these two variables. Interpretation of coefficients is far less intuitive than in the linear case, because we now have coefficients for the X and the X^2 terms, and because whenever X increases by one unit, X^2 also increases.

Thus far we've used the **Fit Y by X** analysis platform, which restricts us to a single independent variable. If we want to investigate a model with a combination of quadratic and linear (or even dummy) terms, we can use the **Fit Model** platform.

1. Select **Analyze ▸ Fit Model.** Again, cast **MortMaternal** as Y.

2. Highlight **BirthRate**, but don't click anything yet. Notice the **Macros** button, and the **Degree** box below it. By default, degree is set at 2. Click **Macros**, as shown in Figure 19.8.

Figure 19.8: Building a Polynomial Model

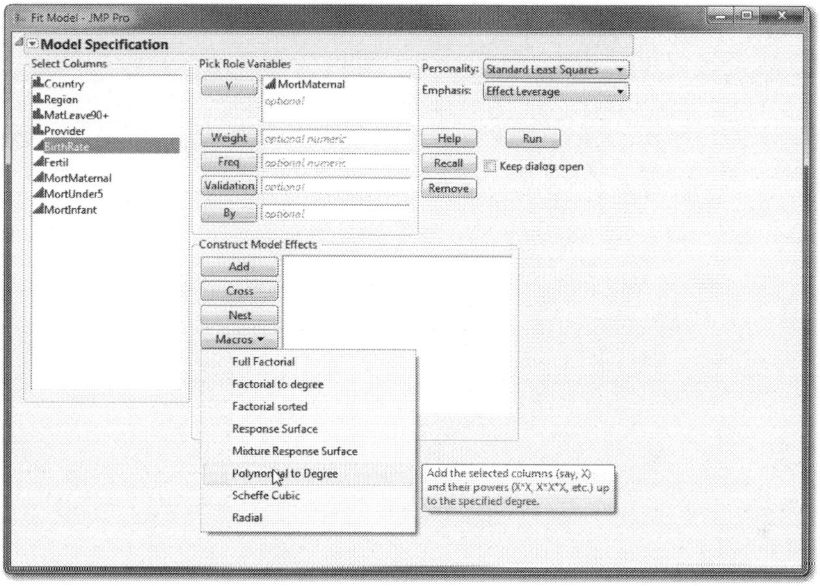

Among the macros, you'll see **Polynomial to Degree**. This choice specifies that you want to add the highlighted variable and its powers up to and including the value in the **Degree** box. In other words, we want to specify a model with X and X^2.

We can use this command to build a polynomial model with any number of factors. If you set **Degree** = 4, the model would include terms for **BirthRate**, **BirthRate**^{2}, **BirthRate**^{3}, and **BirthRate**^{4}.

---

3. Select **Polynomial to Degree** from the drop-down menu. You'll see the model now includes **BirthRate** and **BirthRate*BirthRate**.

4. Click **Run Model**.

The model results (not shown here) are fundamentally identical to the ones that we generated earlier. The **Fit Model** platform provides more diagnostic output than the **Fit Y by X** platform, and we can enlarge our model to include additional independent variables.

Logarithmic Models

Of course, there are many curvilinear functions that might suit a particular study. For example, when a quantity grows by a constant *percentage* rate rather than a constant absolute amount, over time, its growth will fit a logarithmic[5] model. To illustrate, we'll look at the number of cell phone users in Thailand, a nation that was among the earliest to see widespread adoption of cellular technology.

1. Open the data table called **Cell Subscribers**.

This is a table of stacked data, representing 231 countries from 1990 through 2006. We'll first select only those rows containing the Thai data.

2. Select **Rows ▸ Row Selection ▸ Select Where**. Select rows where **Country** equals **Thailand**, and click **OK**.

We want to analyze only the 17 selected rows. We could exclude and hide all others, but let's just subset this data table and create a new 17-row table with only the data from Thailand.

3. Select **Tables ▸ Subset**. You have seen this dialog several times. Click the radio buttons corresponding to **Selected Rows** and also specify that you want **All Columns**.

As we did earlier, we'll learn to create a logarithmic model using both curve-fitting platforms.

4.   Select **Analyze ▸ Fit Y by X**. Cast **Subs per 100 pop** as **Y** and **Year** as **X**. The result looks like Figure 19.9.

Figure 19.9: Growth of Cellular Subscriptions in Thailand

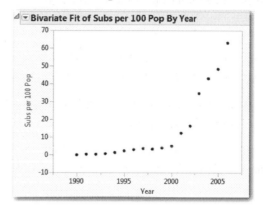

Notice the pattern of these points. For the first several years there is a very flat slope, followed by rapid growth. This distinctive pattern is a common indication of steady *percentage* growth in a series of values.

5.   Click the red triangle and select Fit Special. As shown in Figure 19.10, we'll transform the response variable by taking its natural log and leave *X* alone.

Figure 19.10: Log Transformation of Y

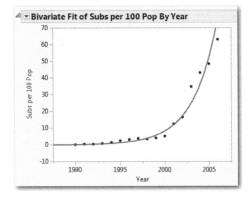

The scatterplot now looks like Figure 19.11. Here we see a smooth fitted curve that maps closely to the points and follows that same characteristic pattern mentioned above. The fitted equation is **Log(Subs per 100 Pop) = –736.7644 + 0.3694427*Year**.

Figure 19.11: A Log-Linear Model—Log(Y) vs. X

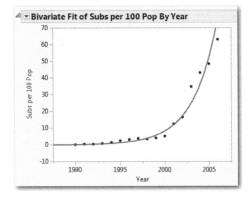

The coefficient of **Year** tells us how the logarithm of subscribers increases each year. How can we use this coefficient to estimate the annual percentage growth rate of cellular phone subscriptions in Thailand from 1990 through 2006? We need to do some algebra starting with a model of steady annual percentage growth. If we let Y_0 represent the initial value of Y and let r equal the unknown annual growth rate, then the number of subscribers per 100 population at any future time, t, equals

$$Y_t = Y_0(1+r)^t$$

If we take the log of both sides of this equation, we get the following:

$$\ln(Y_t) = \ln(Y_0) + t(\ln(1+r))$$

Notice that because Y_0 is a fixed value, this function is linear in form. The intercept of the line is $\ln(Y_0)$, the slope of the line is $\ln(1+r)$, and the independent variable is t. Moreover, we solve for the annual rate of increase, r, as follows:

$$\hat{\beta}_1 = \ln(1+r)$$
$$e^{\hat{\beta}_1} = 1+r$$
$$r = e^{\hat{\beta}_1} - 1$$

So, in this model, the estimated slope is .3694427. If we raise e to that power and subtract 1, we find the rate: $e^{.3694427} - 1 = 1.4469 - 1 = .4469$, or 44.69% per year.

We can also estimate a log-linear model using the **Fit Model** platform. As we did earlier, we'll transform the response variable to equal the log of the number of subscriptions.

1. Select **Analyze ▸ Fit Model**. Again, cast **Subs per 100 Pop** as **Y** and add **Year** to the model.

2. Highlight **Subs per 100 Pop** within the **Y** box.

3. In the center of the dialog box near the very bottom, click the red triangle marked **Transform** and select **Log**, and **Run** the model.

By this point, you should be familiar enough with the results from the **Fit Model** platform that you can recognize that they match those reported from **Fit Y by X**. This command also gives us leverage and residual plots, and permits the addition of other regressors to build more complex multiple regression models.

Application

Now that you have completed all of the activities in this chapter, use the concepts and techniques that you've learned to respond to these questions.

1. *Scenario:* Return to the **NHANES SRS** data table. Exclude and hide respondents under age 18.

   a. Perform a regression analysis for adult respondents with BMI as *Y*, and use waist circumference and gender as the independent variables. Discuss the ways in which this model is or is not an improvement over a model that uses only waist circumference as a predictor.

   b. Perform a regression analysis extending the previous model, and this time, add an interaction term between waistline and gender.

2. *Scenario:* High blood pressure continues to be a leading health problem in the United States. In this problem, continue to use the **NHANES SRS** table. For this analysis, we'll focus on just the following variables, and focus on respondents between the ages of 12 and 19.

   a. Perform a regression analysis with systolic BP as the response and gender, age, and weight as the factors. Report on your findings.

   b. Perform a regression analysis with systolic BP as *Y* and gender, age, weight, and diastolic blood pressure as the independent variables. Explain fully what you have found.

3. *Scenario:* We'll continue to examine the World Development Indicators data in **BirthRate 2005**. Throughout the exercise, we'll use **BirthRate** as the **Y** variable. We'll broaden our analysis to work with other variables in that file:

   a. Earlier, we estimated a quadratic model with maternal mortality as the independent variable. To that model, add the categorical variable **MatLeave90+**, which indicates whether there is a national policy to provide 90 days or more maternity leave. Evaluate your model and report on your findings.

   b. Investigate whether there is any interaction between the maternity leave dummy and the rate of maternal mortality. Report on your findings, comparing these results to those we obtained by using the model without the categorical variable.

4. *Scenario*: The United Nations and other international organizations monitor many forms of technological development around the world. Earlier in the chapter, we examined the growth of mobile phone subscriptions in Thailand. Let's repeat the same analysis for two other countries using the **Cell Subscribers** data table.

    a. Develop a log-linear model for the growth of cell subscriptions in Denmark. Compute the annual growth rate.

    b. Develop a log-linear model for the growth of cell subscriptions in Malaysia. Compute the annual growth rate.

    c. Develop a log-linear model for the growth of cell subscriptions in the United States. Compute the annual growth rate.

    d. We've now estimated four growth rates in four countries. Compare them and comment on what you have found.

5. *Scenario*: Earlier in the chapter, we estimated a logistic model using the Parkinson's Disease (PD) data. The researchers reported on the development of a new composite measure of phonation, which they called **PPE**.

    a. Run a logistic regression using **PPE** as the only regressor, and **Status** as the dependent variable. Report the results.

    b. Compare the results of this regression to the one illustrated earlier in the chapter. Which model seems to fit the data better? Did the researchers succeed in their search for an improved remote indicator for PD?

6. *Scenario*: Occasionally, a historian discovers an unsigned manuscript, musical composition, or work of art. Scholars in these fields have various methods to infer the creator of such an unattributed work, and sometimes turn to statistical methods for assistance. Let's consider a hypothetical example based on the **Sonatas** data.

   a. Run a logistic regression using **Parta** and **Partb** as independent variables and **Composer** as the dependent variable. Report the results.

   b. (Challenge) Suppose that we find a composition never before discovered. Parta is 72 measures long and Partb is 112 measures long. According to our model, which composer is more likely to have written it? Why?

7. *Scenario*: The United Nations and other international organizations monitor many forms of technological development around the world. Earlier in the chapter, we examined the growth of mobile phone subscriptions in Thailand. Let's look at a single year and examine the relationships between adoption levels for several key communication technologies. Open the data table called **MDG Technology 2005**.

   a. Develop linear and quadratic models for the number of cellular subscribers per 100 population, using telephone lines per 100 population as the independent variable. Which model fits the data better?

   b. Develop linear and quadratic models for the number of cellular subscribers per 100 population, using personal computers per 100 population as the independent variable. Which model fits the data better?

8. *Scenario*: The data table called **Global Temperature and CO2** contains estimates and readings for two important environmental measures: the concentration of carbon dioxide (CO_2 parts per million) in the atmosphere and mean global temperature (degrees Fahrenheit).

   a. Using all available years' data, develop linear, quadratic, and log-linear models for the CO_2 concentration. Which model fits the data better?

   b. Use the Data Filter to include and show just the years since 1880, and repeat the previous question using the global temperature variable. Which model fits the data better?

[1] You might wonder why we bother to create the dummy variable if JMP will conduct a regression using nominal character data. Fundamentally, the software is making the 0-1 substitution and saving us the trouble. This is one case where the author believes that there is value in working through the underlying numerical operations. Furthermore, the 0-1 approach initially makes for slightly clearer interpretation.

[2] Logistic regression is one of several techniques that could be applied to the general problem of building a model to classify, or differentiate between, observations into two categories. Logistic models are sometimes called *logit* models.

[3] MDVP stands for the Kay-Pentax Multi-Dimensional Voice Program, which computes these measures based on voice recordings.

[4] For further background, see Freund, R., R. Littell, et al. (2003). *Regression using JMP*. Cary NC, SAS Institute Inc. or Mendenhall, W. and T. Sincich (2003). *A Second Course in Statistics: Regression Analysis*. Upper Saddle River, NJ, Pearson Education.

[5] If one transforms Y by taking the log and leaves X untouched, this is called a log-linear model.

Basic Forecasting Techniques

Overview

For much of this book, we have examined variation across individuals measured simultaneously. This chapter is devoted to variation over time, working with data tables in which columns represent a single variable measured repeatedly and rows represent regularly spaced time intervals. In many instances, *time series* data exhibit common, predictable patterns that we can use to make forecasts about future time periods. As we move through the chapter, we'll learn several common techniques for summarizing variation over time and for projecting future values.

Detecting Patterns Over Time

Typically, we classify patterns in a time series as a combination of the following:

- **Trend**: Long-term movement upward or downward.

- **Cycle**: Multi-year oscillations with regular frequency.

- **Season**: Repeated, predictable oscillations within a year.

- **Irregular**: Random movement up or down, akin to random disturbances in regression modeling.

To illustrate these patterns, we'll first consider industrial production in India. India is one of the largest and fastest-growing economies[1] in the world. Industrial production refers to the manufacture of goods within the country, and is measured in this example by India's *Index of Industrial Production* (IIP). The IIP is a figure that compares a weighted average of industrial output from different sectors of the economy (for example, mining, textiles, agriculture, manufacturing, etc.) to a reference year in which the index is anchored at a value of 100. As a result, we can interpret a value of the IIP as a percentage of what industrial output was in the base year. So, for instance, if the IIP equals 150 we know that production was 50% higher at that point than it was in the base year.

1. Open the table called **India Industrial Production**.

This table contains IIPs for different segments of the Indian economy. All figures are reported monthly from January 2000 through June 2009. In this section, we'll develop some models to forecast one quarter (three months) beyond the available data.

> In this data table, the last three rows are initially hidden and excluded. Standard practice in time series analysis is to hold some observations in reserve while building predictive models, and then test the performance of the models comparing forecasts to later observations.

Let's look at the variability in production of basic goods (processed goods such as paper, lumber, and textiles valued for their use in the manufacture of other products). We'll look at the variation using a *Run Chart*, which simply plots each observed data value in order. We will also look at the patterns (if any) of *autocorrelation* in the series, which is the extent to which monthly values of the index of basic goods are correlated with prior values.

2. Select **Analyze ▶ Modeling ▸ Time Series**. As shown in Figure 20.1, select **BasicGoods** as the **Y, Time Series** to analyze, and identify the observations by **Month**.

3. In the lower left of the dialog box, change the number of **Forecast Periods** to 3.

4. Click **OK.**

Figure 20.1: Launching the Time Series Platform

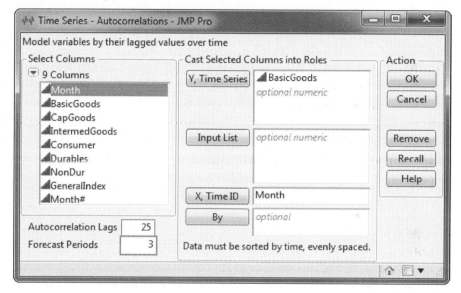

Figure 20.2 Time Series Graph of Production of Basic Goods

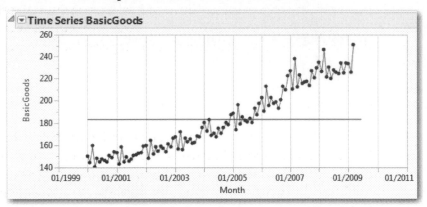

Look at the first simple time series chart (enlarged in Figure 20.2). The horizontal axis shows the months beginning in January 2000 and the vertical axis shows the values of the Index of Industrial Production for basic goods. The upward trend is visible, but it is not a steady linear trend; it has some curve to it as well as some undulation that suggests cyclical movement. Also visible is a sawtooth pattern of ups and downs from month to month, indicating some irregular movement and possibly a seasonal component. In short, this first example exhibits all four of the elements often found in time series data.

Below the graph are two graphs and a table of diagnostic statistics. Although they go beyond the typical content of an introduction to time series techniques, these are a basic feature of the JMP Time Series platform. The two graphs illustrate the extent to which each observation of the time series is correlated with prior observations in recent prior periods. If a time-series is *stationary* (without trend) the *autocorrelation plot* on the left side will have bars of random length and sign. A non-stationary series will have bars that gradually diminish in length. The section about autoregressive models later in the chapter has more to say about using previous values of Y to forecast future values.

In the next two sections of the chapter we'll introduce two common methods for smoothing out the irregular sawtooth movement in a time series, both to better reveal the other components and also to develop short-term forecasts. We'll also introduce some statistics to summarize the performance of the models.

It is valuable to understand at the outset that the techniques of time series analysis share a particular logic. In each of the techniques presented in this chapter, we'll apply the method to a sample of observations to generate retrospective "forecasts" for all or most of the observed data values. We then compare the forecasts to the values that were actually observed, essentially asking "if we had been using this forecasting method all along, how accurate would the forecasts have been?" We'll then select a forecasting method for continued use, based in part on its past performance with a particular set of data.

Smoothing Methods

As a first step in time series analysis it is often helpful to neutralize the effects of the random irregular component and thereby more clearly visualize the behavior of the other three components. Perhaps the simplest method of summarizing a time series is known as a *moving average*. As the name suggests, the method relies on computing means. We begin by computing the average of the first few observations from, say, y_1 through y_m. That average then corresponds either to the chronological center of the first m observations or serves as the "prediction" for y_{m+1}. We then move along in the series, computing the average of observations y_2 through y_{m+1}. This continues until we've passed through the entire sample, continually taking averages of m observations at a time, but always a different group of observations.

Simple Moving Average

Moving averages are commonly used, perhaps chiefly for their simplicity and computability if not for their usefulness as a forecasting method. Simple smoothing methods are most appropriate when the time series is stationary, showing no particularly strong trend or seasonal components. As such it is not well-suited to this particular time series, and we'll see why. Nevertheless, we'll apply it to this series both to demonstrate the technique and to understand its advantages and its shortcomings.

To introduce the concept of a smoothing method, we first look at the Simple Moving average in which each estimate is the mean of a relatively small number of recent observed values. The goal is to smooth out irregularities to make the trend, if any, more visible. To illustrate, we'll compute an average for 3 months at a time. Our first estimate will be the mean of months 1, 2, and 3. The second estimate will be based on months 2, 3, and 4. This is the reason for the name "moving averages": the averaging window moves as we progress through time.

1. Click the red triangle at the top of the time series report window, and choose **Smoothing Model** ▸**Simple Moving Average**.

2. Where noted, enter a 3 (for 3 months) as the smoothing window. Leave the **Centered** box checked.

Conventionally, moving averages are either "centered" or "uncentered." A centered moving average (the default) locates the estimated value at the center of the averaging window. In this case, our first estimated value would correspond to month 2. Otherwise the estimate would correspond to the last month in the averaging window.

Figure 20.3: Specifying a Moving Average Model

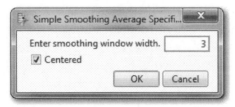

The result is shown in Figure 20.4, which has been enlarged for clarity.

Figure 20.4: A Three-Month Moving Average

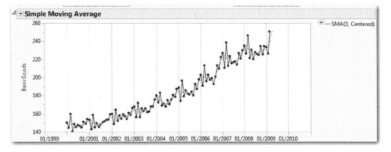

The green **SMA(3)** line (for Simple Moving Average based on 3 observations) begins at Month 2 and is far less erratic than the initial time-series plot. Notice also that the moving average line extends just beyond the final observed month of the time series. This is the forecast value, which is merely equal to the final observed value. Due to the centering, the last estimate will correspond to a month that has already been observed, rather than a future month. More importantly, though, because there is a steady upward trend in the data, a backward looking moving average will predictably underestimate future values.

Simple Exponential Smoothing

Another aspect of simple moving averages is that they rely on a relatively small subset of recent data, and ignore earlier observations. In this case, although we have 111 observations in the data table, each estimate is based on only three values. *Exponential smoothing* methods address this shortcoming. JMP offers six variations of exponential smoothing, and we'll consider three of them.

Exponential smoothing uses a weighted combination of all prior values to compute each successive estimate. The computed values incorporate a weight, or smoothing constant, traditionally selected through an iterative trial-and-error process. In exponential smoothing, each forecast is a weighted average of the prior period's observed and forecast values. Your principal text probably presents the formulas used in various exponential smoothing methods, so we'll move directly into JMP.

The software finds the optimal smoothing constant for us. Let's see how we can use the method with the same set of Indian production data.

1. Click the red triangle next to **Time Series BasicGoods**, and select **Smoothing Model ▸ Simple Exponential Smoothing.**

2. This opens the dialog shown in Figure 20.5. By default JMP reports 95% confidence intervals for the forecast values, and constrains the smoothing constant to be a value between 0 and 1. Just click **Estimate**.

Figure 20.5: Specifying a Simple Exponential Smoothing Model

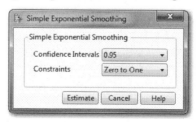

Your results should look like Figure 20.6.

Figure 20.6: Simple Exponential Smoothing Report

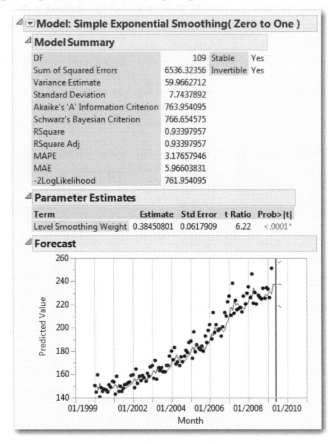

The statistics in the model summary enable us to evaluate the fit and performance of the model, and are directly comparable to subsequent models that we'll build. In this introductory discussion of time series analysis, we'll focus on just four of the summary measures:

Variance Estimate	Sometimes called MSE, a measure of the variation in the irregular component. Models with small variance are preferred.
RSquare	Goodness of fit measure, just as in regression. Values close to 1 are preferred.
MAPE	Mean Absolute Percentage Error. Expresses the average forecast error as a percentage of the observed values. Models with small MAPE are preferred.
MAE	Mean Absolute Error (sometimes called MAD—mean absolute deviation). Average of the absolute value of forecast errors. Models with small MAE are preferred.

Under Parameter Estimates we also see the optimal smoothing constant value of .3845. This value minimizes the model variance and other measures of forecast error.

In the Forecast graph, we see that the model does smooth the series, and to the far right we see the forecasts for the next three periods represented by the short red line bracketed by the blue confidence interval. We can save the forecast values in a new data table as follows:

3. Click the red triangle next to **Model: Simple Exponential Smoothing (Zero to One)** and select **Save Columns.**

This creates an entirely new data table containing the observed and forecast values, as well as residuals, standard errors, and confidence limits. We had requested forecasts for three periods into the future. Because the original data table contained three observations that we hid and excluded from calculations, we actually get forecasts for those three months and then three additional months beyond.

Notice that each of the forecasts is the same value, represented in the graph by the horizontal red line—reflecting the problem with applying this simple approach to forecasting a series with a strong trend. For that, we need a method that captures the trend component. Holt's method does just that.

Linear Exponential Smoothing (Holt's Method)

Linear exponential smoothing, developed by Charles Holt in 1957, expands the model of exponential smoothing by incorporating a linear trend element. Operationally, we proceed very much as we did in generating a simple exponential smoothing model.

4. Return to the Time Series report window and again, click the uppermost red triangle and select **Smoothing Models ▸ Linear Exponential Smoothing.**

5. In the specification panel, accept the default settings and click **Estimate**.

Scroll to the very bottom. In your graph of the linear exponential smoothing model, notice that the forecast for the next three periods does rise slightly, in contrast to the simple model in which the forecasts are flat. This method estimates two optimal smoothing weights, one for the level of the index and another for the trend. We also find the same standard set of summary statistics reported, and near the top of the results window we see a **Model Comparison** panel, as shown in Figure 20.7. The panel on your screen has a horizontal scroll bar at the bottom, and you cannot see the full width unless you enlarge the panel as has been done in in Figure 20.7.

> Each time we fit a different model, JMP adds a row to this table, enabling easy comparison of the different methods.

Figure 20.7: Comparing the Two Exponential Smoothing Models' Performance

Report	Graph	Model	DF	Variance	AIC	SBC	RSquare	-2LogLH	Weights	.2 .4 .6 .8	MAPE	MAE
▾ ☑	☐	Linear (Holt) Exponential Smoothing	107	54.421562	753.39524	758.77794	0.934	749.39524	0.994930		3.236453	6.043387
▾ ☑	☐	Simple Exponential Smoothing(Zero to One)	109	59.966271	763.95409	766.65458	0.934	761.95409	0.005070		3.176579	5.966038

In this case we find that Holt's method has a smaller variance than simple exponential smoothing. On the other hand, the two methods have the same R^2 and the simple model has slightly better MAPE and MAE.

In both of the smoothed graphs, we do see oscillations that might indicate a seasonal component. Neither simple nor linear exponential smoothing incorporates a seasonal component, but Winters' method does offer an explicit consideration of seasons.

Winters' Method

In time series estimation, there are two fundamental approaches to seasonality. One set of models is *additive,* treating each season as routinely deviating by a fixed **amount** above or below the general trend. The other family of models is *multiplicative*, positing that seasonal variation is a fixed **percentage** of the overall trend. In JMP, the implementation of Winters' method, named for Peter Winters who proposed it in 1960, is an additive model. Let's see how it works, and as a starting point we'll assume there are four seasons per year, each consisting of three months.

6. Click the red triangle next to **Time Series BasicGoods** again and select **Smoothing Models ▸ Winters Method.**

7. In the specification panel, change the **Number of Periods per Season** from 12 to 3, accept the other default settings and click **Estimate**.

The resulting parameter estimates and graph are shown in Figure 20.8. The graph is enlarged to more clearly show the forecast for the next three months. Notice in the parameter estimates that there are now three weights, corresponding to the three time series components. In the graph, we see a more refined set of future estimates that continue the upward trend, but show a dip in production in the second month down the road.

Figure 20.8: Results of Winters' Method

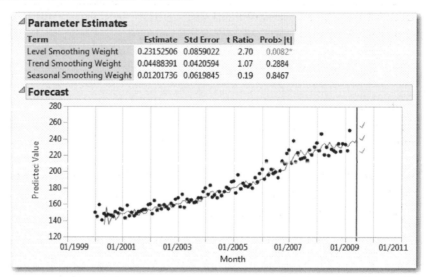

Now look at the comparison of the three models on your screen. Of the three models, the Winters Model has the smallest variance and highest R^2. On the other hand, the simple exponential smoothing model still has the lowest MAPE and MAE.

Trend Analysis

For some time series, we can use the techniques of simple regression to fit a model that uses time as the X variable to describe or predict the general trend of Y. In the current example, we have seen that although there is clearly some irregular variation and sign of season and cycle, the production of basic goods in India has followed a steady upward trend in the years represented in our data table.

For simplicity of interpretation, we'll use the column called **Month#**, which starts at 1 in the first month and runs for 114 months. We can fit the model using the observed 114 months and generate forecasts simply by extrapolating a straight line.

1. Select **Analyze ▸ Fit Y by X**. Cast **BasicGoods** as **Y** and **Month#** as **X**.

2. In the report window, click the red triangle and select **Fit Line**. Your results should look like Figure 20.9.

Figure 20.9: A Linear Trend Analysis

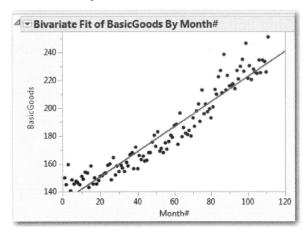

The slope coefficient in this model is approximately 0.9, meaning that with each passing month, the index has increased by 0.9 on average. The goodness-of-fit statistic is quite high (0.92). We can compare these results to the performance of our exponential smoothing models. The Bivariate platform does not report MAPE and MAE, but it does produce a variance estimate, which we find in the ANOVA table as the Mean Square Error value of 75.6. Compared to the R^2 and variances of the exponential smoothing models, the simple linear trend falls short.

Looking back at the scatter plot, we can see that the points bend in a non-linear way. As such, a *linear* model—whether in linear exponential smoothing or linear regression—is not an ideal choice.

For this reason, we may want to try a quadratic or log-linear fit to see whether we can improve the model. To illustrate, let's fit a quadratic model.

3. Click the red triangle next to **Bivariate Fit of BasicGoods by Month** and select **Fit Polynomial ▸ 2, quadratic**. Your results should look like Figure 20.10.

Visually, it is clear that this model fits better than the linear trend, and the model summary measures bear out that impression. RSquare improves slightly from 0.92 to 0.94, and the root Mean Squared Error falls from approximately 8.7 to 7.2.

Figure 20.10: A Quadratic Trend Analysis

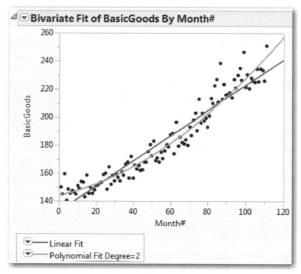

Autoregressive Models

When we use regression analysis to model the trend of a time series, the dependent variable is the observed data Y_t and the independent variable is just the time period, t. An alternative approach is to model Y_t as a function of one or more prior observations of Y. As with exponential smoothing, there are many variations on this basic idea of an autoregressive model. In this section, we'll examine one such model from the general class of AutoRegressive Integrated Moving Average, or ARIMA models. These are also sometimes called Box-Jenkins models, after the two statisticians George Box and Gwilym Jenkins who developed the techniques in the 1960s and early 1970s.

The ARIMA analysis properly is performed on a stationary (without trend) time series. If there is a trend, the Box-Jenkins approach first transforms Y into a stationary series, usually by means of taking differences. That is, if Y were a steadily increasing series, we might find that the series based on first differences (*i.e.*, $Y^* = Y_t - Y_{t-1}$) is a stationary series.

An ARIMA model has three parameters:

Parameter	Role Within the ARIMA Model
p	Autoregressive order: the number of lagged (prior) values of Y to include in the model. A model that only uses one prior period has an AR order=1. If the model includes the two most recent periods, it has AR order=2, and so on.
d	Differencing order: the number of times Y is differenced to generate the stationary series Y*
q	Moving average order: the number of lagged values of error terms (residuals) to include in the model.

The typical notation for an ARIMA model references all three parameters, or at least the nonzero parameters. Thus, an AR(1) model uses one lagged value of the observed values of Y. An AR(1,1) model uses one lagged value of the first differences of Y and so forth.

Our time series plainly is *not* stationary, so we will want to use differencing to "flatten" the curve. However, to illustrate the techniques, we'll first estimate an AR(1) model, and then an AR(1,1) model. We'll compare the results of the two models to each other and then to our exponential smoothing estimates.

1.   Return to the **Time Series** results window.

2.   Once again, click the red triangle at the top of the results window and now select **ARIMA**.

3.   This opens a new window; complete it as shown in Figure 20.11 and click **Estimate**.

Figure 20.11: Specifying an AR(1) Model

Autoregressive models are regression models. The AR(1) model results appear at the bottom of the **Time Series** report window (see Figure 20.12). The model can be expressed as the following equation:

$$Y_t = \beta_0 + \beta_1 Y_{t-1}$$

Figure 20.12: Parameter Estimates for This AR(1) Model

Parameter Estimates

Term	Lag	Estimate	Std Error	t Ratio	Prob>|t|	Constant Estimate
AR1	1	0.95948	0.02796	34.32	<.0001*	7.64489254
Intercept	0	188.67698	19.59393	9.63	<.0001*	

In this case, both the estimated intercept and slope are significant and we see that, at the margin, the estimated index in the next period is nearly 96% of the prior period's index. As with the smoothing methods in the Time Series platform, we can also save the computations from this model by clicking on the red triangle at the top of the **Model: AR(1)** results panel and choosing **Save Columns**.

Before discussing the summary statistics for this model, let's estimate the AR(1,1) model and then compare the results. Recall that this next model uses the monthly change in the index (the "first difference") rather than the index itself as the basic element in the model.

4. Click the red triangle at the very top of the **Time Series** report, and select **ARIMA** again.
5. This time, set both the **Autoregressive Order** and the **Differencing Order** (see Figure 20.11) to 1. The parameter estimates for this model should look like Figure 20.13.

Figure 20.13: Parameter Estimates for AR(1,1) Model

Parameter Estimates

Term	Lag	Estimate	Std Error	t Ratio	Prob>|t|	Constant Estimate
AR1	1	-0.7487255	0.0654773	-11.43	<.0001*	1.48287326
Intercept	0	0.8479737	0.3691592	2.30	0.0235*	

We cannot directly compare these estimates to those for the AR(1) model because prior to estimating the regression, JMP calculated the first differences. At the margin, the incremental change in the monthly *change* in the index is about 75% less than the prior month. The negative coefficient leads to oscillations because if one month has a positive change, the forecast will be for a reduction in the following period and if a month has a negative change, the forecast will be positive. Although the computations are made for the relatively stationary series of first differences, JMP reconstructs the time series for the sake of the forecast graph, showing the general upward trend.

Now, let's compare our autoregressive models to each other and to the smoothing models. Everything we need is in the Model Comparison panel, as shown in Figure 20.14. Because the panel is so wide, we've eliminated some of the reported statistics[2], focusing on the ones that we've discussed earlier.

Figure 20.14: Comparing Five Time Series Methods

Report	Graph	Model	DF	Variance	RSquare	AIC Rank	SBC Rank	MAPE	MAE
✓	☐	Winters Method (Additive)	104	50.177156	0.940	1	2	3.315726	6.109066
✓	☐	ARI(1, 1)	108	46.320348	0.949	2	1	2.877901	5.367753
✓	☐	Linear (Holt) Exponential Smoothing	107	54.421563	0.934	3	3	3.236453	6.043387
✓	☐	Simple Exponential Smoothing(Zero to One)	109	59.966271	0.934	4	4	3.176579	5.966038
✓	☐	AR(1)	109	99.891165	0.878	5	5	4.290917	7.913439

Because every time series is different, we can't say categorically that one method will fit better than another. In this example, the AR(1,1) model outperforms the others on the four criteria that we have studied. It has the smallest variance, MAPE, and MAE of the four as well as the highest R^2. Although we did not discuss the AIC (Akaike's "A" Information Criterion), we might note that it ranks second among the four models using that standard.

Application

Now that you have completed all of the activities in this chapter, use the concepts and techniques that you've learned to respond to these questions.

1. *Scenario:* Let's continue our analysis of the Indian Index of Industrial Production.

    a. We used Winters' method with an assumed season of three months. Re-estimate using this approach with a six-month "season" (that is, two different seasons per year). Compare the performance of this model to the other model.

    b.   Fit a log-linear model using the **Fit Y by X** platform. Does this perform any better than the quadratic model?

    c.   We used simple AR(1) and AR(1,1) models. Try fitting an AR(2,1) model to this data series and report on what you find.

2. *Scenario:* Our **India Industrial Production** data table also contains several other indexes, including one for the production of consumer durable goods. Durables are long-term purchases made by households, including items such as appliances and automobiles.

    a.   Use several of the techniques presented in this chapter to develop a six-month forecast for the index for consumer durables. Explain which technique seems best suited to this time series and justify your choice.

    b.   Save the forecasts using your chosen approach. Then, compare the forecast values from rows 112 through 114 to the observed (but hidden) values in our original **India Industrial Production** table. How well does the model actually forecast the consumer durables index for those three months?

3. *Scenario:* The United Nations and other international organizations monitor many forms of technological development around the world. In Chapter 16, we estimated the annual growth of mobile phone subscriptions in Thailand. Let's apply some of our newer techniques to the data in the **Cell Subscribers** data table.

    a.   Of all the methods we saw in this chapter, we should not use Winters' method for this set of data. Why not?

    b.   Use several of the techniques presented in this chapter to develop a three-year forecast for the number of cell subscriptions in Denmark. Explain which technique seems best suited to this time series and justify your choice.

    c.   Repeat part b for the annual series in Malaysia and report on your findings.

    d.   Repeat part b for the annual series in the United States and report on your findings.

    e.   Did you decide to model all four countries using the same time series technique? Discuss the insights that you've drawn about time series analysis from this exercise.

4. *Scenario:* In prior chapters, we examined cross-sectional data about fertility and birth rates in 2005. In this exercise, we'll look at fertility time series for several countries. The time series are all in the data table **Fertility**; unlike the other

examples in the chapter, these figures are reported every five years rather than annually or more frequently. Within the table, the data through 2005 are actual observed figures, but the 2010 data are U.N. estimates. Accordingly, the rows corresponding to 2010 have been hidden and excluded in the saved data table. As you proceed through this exercise, you'll need to select rows corresponding to specific countries and subset the table to model just that country.

   a. Develop a time series model for the fertility rate in Brazil. Report on your choice of a model.

   b. Develop a time series model for the fertility rate in the Russian Federation and report on your model.

   c. Develop a time series model for the fertility rate in India and report on your model.

   d. Develop a time series model for the fertility rate in China and report on your model.

   e. Develop a time series model for the fertility rate in Saudi Arabia and report on your model.

   f. The U.N. estimate for 2010 in Brazil in 2010 is a fertility rate of 2.1455 children per woman. Based on your analysis, which of our time series techniques might the U.N. have used to develop this estimate?

5. *Scenario*: In the United States, the management of solid waste is a substantial challenge. Landfills have closed in many parts of the nation, and efforts have been made to reduce waste, increase recycling, and otherwise control the burgeoning stream of waste materials. Our data table **Maine SW** contains quarterly measurements of the tonnage of solid waste delivered to an energy recovery facility in Maine. The data come from five municipalities in the state.

   a. Develop a time series model for quarterly solid waste delivered from Bangor. Report on your choice of a model.

   b. Develop a time series model for quarterly solid waste delivered from Bucksport. Report on your choice of a model.

   c. Develop a time series model for quarterly solid waste delivered from Enfield. Report on your choice of a model.

   d. Develop a time series model for quarterly solid waste delivered from Orono. Report on your choice of a model.

   e. Develop a time series model for quarterly solid waste delivered from Winslow. Report on your choice of a model.

    f.   Did the same method work best for each of the five communities? Speculate about the reasons for your finding.

6. *Scenario*: The World Bank and other international organizations monitor many forms of technological development around the world. Elsewhere we've looked at other Millenium Development Indicators. In this exercise, we'll study carbon dioxide emissions between 1990 and 2004. The relevant data table is called **MDGCO2**. The time series column is called **CO2**, which measure the number of metric tons of CO2 emitted per capita annually in the country. As you've done in earlier problems, you'll need to subset the data table to focus on one country at a time.

    a.   Develop a time series model for the **CO2** in Afghanistan. Report on your choice of a model.

    b.   Develop a time series model for the **CO2** in The Bahamas. Report on your choice of a model.

    c.   Develop a time series model for the **CO2** in China. Report on your choice of a model.

    d.   Develop a time series model for the **CO2** in Sudan. Report on your choice of a model.

    e.   Develop a time series model for the **CO2** in the United States. Report on your choice of a model.

    f.   Did the same method work best for each of the five nations? Speculate about the reasons for your finding.

7. *Scenario*: Our data table called **MCD** contains daily share prices for McDonald's common stock for the first six months of 2009. The **Daily Close** column is the price of one share of stock on any given day.

    a.   Create a time series plot of the **Daily Close** column. Looking at this series, which time series modeling method or methods seem particularly inappropriate? Explain.

    b.   Develop a time series model for the daily closing price and forecast the next five days beyond the series. Report on your choice of a model.

8. *Scenario*: Our data table called **Stock Index Weekly** contains weekly broad market indexes and trading volumes for six major stock markets around the world. Specifically, we have observations for every week of 2008, a year in which financial markets across the globe shuddered as a consequence of lending practices in the United States. The six market indexes are for the markets in

Tokyo (NIK225), London (FTSE), New York (SP500), Hong Kong (HangSeng), Instanbul (IGBM), and Tel Aviv (TA100).

a. Compute correlations for the index values of all six stock markets. Which market is most closely correlated with the S&P 500 index?

b. Develop separate time series models for the S&P 500 and the market you identified in part a. Report on your findings.

c. Does the same method work best for the New York market and the market you identified in part a? What do you think explains this result?

d. Develop a time series forecast for weekly trading volume in the Hong Kong market for the first three weeks of 2009. How confident are you in these forecasts?

9. *Scenario*: The data table **Global Temperature and CO2** contains annual measures of two variables: the concentration of carbon dioxide in the earth's atmosphere and the mean global temperature in degrees Fahrenheit.

a. Create a time series plot of the **Concentration** column. Looking at this series, which time series modeling method or methods seem well suited to the data? Explain.

b. Using one of the ARIMA or smoothing methods in the Time Series platform, develop a 5 year forecast for CO2 concentrations.

c. Repeat parts a and b using the temperature data.

---

[1] In 2005, India was reported by the World Bank as ranking 12th in the world in terms of total Gross Domestic Product (GDP).

[2] As with other JMP reports, users can select table columns by hovering over the table, right-clicking once, and choosing **Columns**. This opens a list of column names, and the user may specify the contents of the reported table.

Elements of Experimental Design

Overview

In Chapter 2, we discussed the difference between *experimental* and *observational* studies. In an experiment, the investigator creates a controlled setting in which she or he can deliberately manipulate the values of some variables (*factors* or *independent* variables) and measure the values of one or more *response* (or *dependent*) variables. We can confidently draw inferences about causation based on well-designed experiments.

In contrast, researchers exercise little or no control in observational studies. They gather data, sometimes after the fact, about events or subjects and then use methods that we've studied to analyze the evidence of interdependence among variables. Even the best observational study does not permit us to conclude that one variable causes another.

The data tables that accompany this book typify observational and experimental research as well as survey research. Often, we have no choice but to use observational data because a properly designed experiment would be prohibitively expensive, would raise serious ethical issues, or would be impractical or impossible for one reason or another.

For example, engineers concerned with buildings that can withstand earthquakes have no ability to randomly assign some cities to an earthquake. Even if it were possible to do so, it is unethical for public health officials to randomly expose some citizens to pollution or disease. This chapter is devoted exclusively to the design of experimental studies.

Why Experiment?

When we can feasibly conduct an experiment, we should. If we need to establish the efficacy of a new disease treatment, we intentionally need to control relevant factors so that we can assess and understand their impact on the course of a disease or on patients' symptoms. In this chapter, we'll begin with a simplified version of the concrete experiment first presented in Chapter 2. Concrete is a mixture of ingredients that always include water, cement, some kind of coarse aggregate like gravel, and a fine aggregate like sand. The investigator, Dr. Yeh, was interested in predicting the compressive strength of concrete, particularly when substituting other ingredients for the material called *fly ash,* a by-product of coal combustion. He theorized that the compressive strength would vary predictably as a result of the particular mixture of cement, water, and other component materials, and the goal of the experiment was to develop a reliable model of compressive strength.

In this chapter, after a brief discussion of the purposes of experimental design and some definitions, we'll learn several standard approaches to DOE—the Design of Experiments. As we do so, our emphasis is on the structure of the experiments and the generation of data. We won't actually analyze the data that a design would generate, but instead devote this chapter to some of the fundamental concepts and issues in the design process. It is also important to note that DOE is a large and complex topic. This single chapter cannot deal with all aspects. The goal is to introduce the reader to some foundational concepts.

Goals of Experimental Design

DOE refers to the systematic process of planning the variation among factors so that suitable data can be collected for analysis in such a way that an investigator is able to generalize from their experiment. In other words, one major objective in the statistical design of experiments is to ensure that our experiments yield useful data that can be analyzed to draw valid conclusions and make better decisions.

Box, Hunter, and Hunter (1978) laid out a process of progressive experimentation by moving between generating ideas and then gathering facts, which would in turn generate new insight and ideas. In so doing, they described differing goals of experiments. Hoerl and Snee (2002) summarize the process by identifying three experimental contexts, each of which is characterized by its own fundamental question, although any particular experiment might be a hybrid of the three categories.

- *Screening*: Which variables have the largest effect on the response variable?

- *Characterization*: How large are the effects of the few most important variables? Are there any important interactions among these variables? Are relationships linear or curvilinear?

- *Optimization*: At what values of the independent and control variables does the response variable attain a desired minimum, maximum, or target value?

In this example, the ultimate goal is characterization: to build a model that predicts the compressive strength of a particular mixture of concrete. How do we do this? There are many components that might influence strength: types of coarse and fine aggregates, water, cement, and other additives. A screening experiment would seek to identify a few key factors affecting the strength. A characterization experiment would focus on factors identified by screening, and seek to understand how they collectively affect the strength. Moreover, it might also assess possible interactions whereby the effect of one factor changes depending upon the values of other factors. With the relevant variables and relationships identified, we would finally use an optimization approach to fine-tune the variables to design a batch of cement with the desired compressive strength for a particular structure. We'll begin with some standard approaches to screening experiments and define some important terms.

Factors, Blocks, and Randomization

We'll refer to the components that might affect compressive strength as *factors*. In general, a factor is a variable that can affect the response variable of interest. Factors might be categorical or numerical. The strategy in a screening experiment is to test a number of factors in order to discover which ones have the greatest impact on the response variable. When we design an experiment, we ideally take into account all potential sources of variation in the response variable. We generally do this in several ways within an experiment.

- We deliberately manipulate—that is, set the values of—some factors (sometimes called *treatments*) that we believe affect the response variable.

- We *control* other factors by holding them constant. For instance, because we understand that the strength of concrete depends on the temperature at which it cures, we might test all combinations at the same temperature.

- We often repeat, or *replicate*, measurements of the response variable for a given treatment condition through a series of *experimental runs*. We do this to gauge and control for measurement error.

- We might *block* some runs to account for other "nuisance" factors that cause similarities among runs. We block when we expect that some response measurements will be similar due to the shared factor being blocked.

- We *randomize* experimental runs to account for those factors that we cannot control, block, or manipulate. Thus, if we expect that changes in barometric pressure could affect the outcome, we randomly assign the sequence of testing different formulations.

Multi-Factor Experiments and Factorial Designs

It might be natural to think that one should experiment by holding all factors constant except for one, and systematically alter the values or levels of that single factor. So, for example, we might want to test varying amounts of sand in concrete mixtures with all other components being held at given quantities. However, such an approach is very inefficient and moreover, it can conceal any combined influence of two or more factors. Perhaps the sand quantity is not nearly as important as the ratio of sand to coarser aggregate. When two variables combine to lead to a result, we say that there is an *interaction* between the two, and if, in reality, there is an important interaction to be found, we'll never find it by examining only one factor or the other. We need to vary multiple factors at a time.

Another way to conceive of the issue is in terms of a three-dimensional *response surface*. Imagine that we have a response variable and two factors. If both factors were to contribute to the response in a simple additive and linear fashion (more of component A increases strength, more of component B weakens strength) then it would not matter if we vary A and B individually or jointly. The response variable would vary by strictly increasing one direction and decreasing in the other. The three variables together would form a plane, and by varying A and then B, we'd essentially be sliding up or down the plane in one direction or the other.

But if the situation is more complex, so that the response values vary on a surface that is curved in one or both dimensions, we'd need to explore a great deal of the surface to develop a more complete model. If the surface had multiple peaks and we vary only one factor at a time, it would be analogous to traversing the surface only from north to south or only from east to west. We'd be characterizing one narrow region of the surface, and could potentially miss the peaks completely.

Although single-factor experiments can be inadequate, multi-factor experiments can become very expensive and time-consuming. Let's suppose that we have identified ten possible factors affecting compressive strength. One conventional experimental strategy for screening is to test each factor at two different levels initially—high and low, large and small, or with and without. A *full factorial* experimental design enables us to measure the response variable at all possible combinations of all factor levels. In this case, a full factorial design would require us to create enough concrete blends to cover all possible combinations of the ten different factors. That would require 2^{10} or 1,024 different designs. If we were to replicate each formulation ten times to estimate mean strength, we'd need to have 10,240 experimental runs—at whatever cost of materials and time that would involve.

There are several DOE strategies to accommodate many factors far more efficiently than described above. In practice, researchers can usually rely on their experiences with similar situations to select factors that are most likely to affect the outcomes of interest. Additionally, we typically design experiments to seek answers to specific questions; therefore, a designer allows the research questions to guide the selection of factors and interactions to investigate.

To illustrate the fundamental logic and approach of factorial designs, we'll begin with a modest number of factors. In general, a given factor might have several possible levels or values. For our first example, let's consider just three factors with two levels each. Specifically, we'll design a full-factorial experiment as follows:

- **Coarse Aggregate**: the amount (kg) of crushed stone per one cubic meter of concrete. The levels tested are 850 kg and 1100 kg.

- **Water**: the amount of water (kg) per one cubic meter of concrete. The two levels will be 150 kg and 200 kg.

- **Slag**: Slag is a by-product in the production of iron from iron ore. Pulverized slag is a binding agent often used in combination with or as a substitute for Portland cement, which is the most common cementitious ingredient in concrete. Slag is also measured in kg/cubic meter, and the amounts to test are 0 and 150 kg.

We'll develop an experimental design that ensures that we test all of the combinations of factors and levels. Since we have three factors with two levels each, there are $2^3 = 8$ combinations in all. We can tabulate the eight possible experimental conditions as shown in Table 18.1.

We could mix up eight batches of concrete (holding other ingredients at fixed levels) and measure the compressive strength of each batch. Typically, though, it is wiser to test several batches of each formulation and to compute the mean compressive strength of each. We refer to such repeated trials as *replication*. Replication is a way to accommodate measurement error, to control its effects as well as to estimate its magnitude. So suppose we decide to replicate each batch four times and compute the average for a total of five batches of each mixture. In all, we would measure $5 \times 8 = 40$ experimental runs.

Table 20.1: Eight Experimental Conditions

Formulation	CoarseAgg	Water	Slag
A	850	150	0
B	850	150	150
C	850	200	0
D	850	200	150
E	1100	150	0
F	1100	150	150
G	1100	200	0
H	1100	200	150

We might mix five batches of formulation A, then five batches of formulation B, and so on. However, this might compromise the validity of any conclusions we draw about the formulations because of the possibilities of such things as experimenter fatigue, changing air pressure, unexpected circumstances, or the like. A better approach would be to *randomize* the sequence of runs.

We'll use JMP to create our sequence and build a suitable design[1], called a *full factorial design*. In a full factorial design, we test all factors at the specified levels and collect enough data to examine all possible interactions among the factors.

1. Select **DOE ▶ Full Factorial Design**. In the dialog box (see Figure 21.1), we need to identify a response variable and all of the factors we'll be testing. In addition, we'll indicate that we want to replicate the experimental runs four times.

Figure 21.1: Full Factorial Design Dialog Box

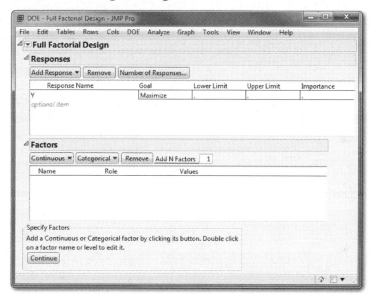

Because JMP eventually randomizes the sequence of trials, each reader of this book could generate a different sequence. For the sake of instructional clarity in this first example, we'll standardize the randomization process by setting the *random seed* value to the same arbitrary value. I have chosen 222.

2. Click the red triangle next to **Full Factorial Design** and select **Set Random Seed**. Enter the value 222 when prompted.

We'll start by naming the response variable and setting the goal in this phase of the experiment.

3. Double-click the **Y** under the heading **Response Name**, and type Compressive Strength.

4. Our eventual goal is to characterize the relationship between strength and factors so that we can achieve a target strength. Click **Maximize** under **Goal** and select **Match Target**. For now leave the **Lower Limit, Upper Limit,** and **Importance** cells blank.

Now let's move on to entering the factors for this experiment. In our initial list of factors, we've identified three continuous variables. In the empty **Factors** panel, we need to enter our three factors.

5. Click the **Continuous** button and select **2 Level.** As soon as you do this, JMP creates an entry for a factor that it provisionally labels as **X1**, with experimental values of –1 and 1 (see Figure 21.2).

Figure 21.2: Specifying a Continuous Factor

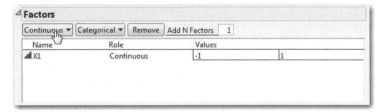

6. Double-click **X1** and change the factor name to CoarseAgg.

7. Click the **–1** under **Values** and change the low value to 850. Similarly, change the high value to 1100.

8. Click **Continuous** and select **2 Level** again to set up our second factor, **Water**, with two levels of 150 and 200.

9. Click **Continuous** and select **2 Level** to set up the final factor **Slag** with two levels of 0 and 150.

When you are done, the **Factors** panel should look like Figure 21.3.

Figure 21.3: The Completed Factors Panel for the Concrete Experiment

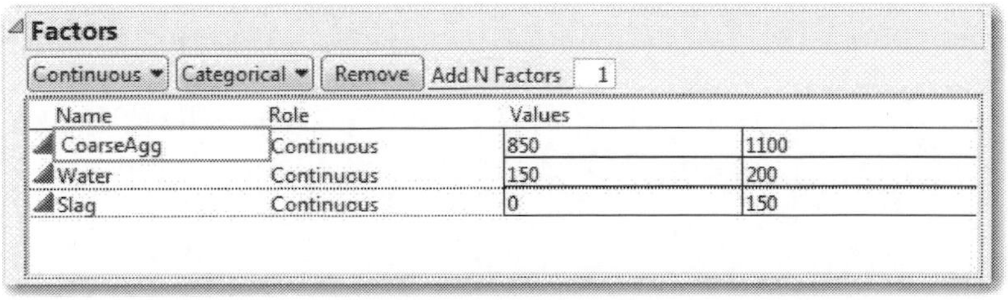

In this chapter, we'll use this list of three factors a few more times. Rather than re-enter the information repeatedly, we can save the three factors and their attributes as follows.

10. Click the red triangle next to **Full Factorial Design**, and choose **Save Factors**. This opens a new data table (not shown) containing the factor names and levels. Save it to your desktop as Chap 21 Factors.

11. Click the **Continue** button in the lower left of the dialog box. This expands an output options panel. (See Figure 21.4.) By default, JMP randomizes the sequence of experimental runs and does not replicate the runs at all.

Figure 21.4: Output Options Panel for a Three Factor, 2-Level Experiment

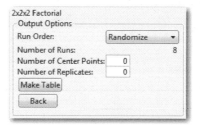

12. Type a 4 in the box next to **Number of Replicates:** to replace the 0, reflecting our intention to test each formulation five times (one original pattern and four additional replicates).

13. Click **Make Table**. A new data table opens containing our randomized sequence of 40 experimental runs. Figure 21.5 displays the first 19 rows of the table.

Figure 21.5: The Full Factorial Design

	Pattern	CoarseAgg	Water	Slag	Compressive Strength
1	++-	1100	200	0	•
2	---	850	150	0	•
3	++-	1100	200	0	•
4	+-+	1100	150	150	•
5	+--	1100	150	0	•
6	-+-	850	200	0	•
7	+++	1100	200	150	•
8	-+-	850	200	0	•
9	---	850	150	0	•
10	+-+	1100	150	150	•
11	+--	1100	150	0	•
12	+-+	1100	150	150	•
13	+--	1100	150	0	•
14	++-	1100	200	0	•
15	+++	1100	200	150	•
16	++-	1100	200	0	•
17	+-+	1100	150	150	•
18	---	850	150	0	•
19	---	850	150	0	•

Left panel of the data table:

- 2x2x2 Factorial
- Design — 2x2x2 Factorial
  - Screening
  - Model
  - DOE Dialog
- Columns (5/0)
  - Pattern
  - CoarseAgg ✱
  - Water ✱
  - Slag ✱
  - Compressive Strength ✱
- Rows
  - All rows — 40
  - Selected — 0
  - Excluded — 0
  - Hidden — 0
  - Labelled — 0

The table contains five columns. The first column (**Pattern**) represents the levels of the three factors. So, for example, the first run is marked **++–**, meaning the "high" setting of coarse aggregate and water and the "low" setting of slag. In our Table 20.1, this corresponds to Formulation G.

The next three columns of the data table show the settings of the factors, and the fifth column is ready for our measurements. Working from the sequence specified in this data table, we would then combine ingredients in each formulation, allow the concrete to cure, and measure compressive strength. We would then return to this data table, type in the compressive strength measurements, and estimate the model.

We can visualize this full factorial design as a cube, and the pattern variable refers to the different corners of the cube, as shown in Figure 21.6. Each factor has two levels, and each corner of the cube corresponds to a different combination of the levels of each of the three factors.

Figure 21.6: Visualizing the Full Factorial Design

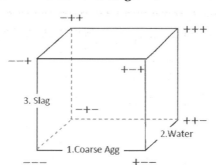

Blocking

In this example, we assumed that a single experimenter performed all of the experimental runs. It is more likely that this work would be done in a team of people, each of whom might execute a share of the work. Suppose this experiment were being conducted by a four-member team, and we want to control for the possibility that different individuals could introduce an unintended factor into the mix.

We do this by *blocking* the experimental runs. A *block* is a group of runs that we expect to be relatively homogenous, but possibly different from other blocks. When we block, we eventually compare the results of each experimental condition within each block. In this case, it makes sense to block by person, enabling each experimenter to run several different formulations. We will randomly assign ten runs to each experimenter. Each team member will run different formulations, and again the experiment will have 40 runs.

1. Select **DOE ▸ Custom Design**. This dialog box works much like the **Full Factorial** dialog box that we worked with earlier. As before, we'll specify that our response variable is **Compressive Strength**.

2. Instead of re-typing all of the factors, click the red triangle next to **Custom Design** and choose load **Factors**. When the **Open Data File** dialog appears, locate the Chap 21 Factors data table file and open it. This reads in all of the factor specifications.

3. In the **Factors** panel, click **Add Factor** again and select **Blocking ▸ Other...** and specify 10 runs per each of the four blocks.

4. This adds a new factor, **X4**, to the model. Change the name of **X4** to **Team Member** and move to the **Model** panel.

5. This is a full factorial experiment, so we want to be able to examine all interactions between the factors. Therefore, in the **Model** panel click the **Interactions** button and select **2nd** as shown in Figure 21.7.

> You will see a warning that JMP does not support interactions involving blocks. Click **Continue**, and the model will include interactions for the three continuous factors.

Figure 21.7: Specifying Second-Order Interactions

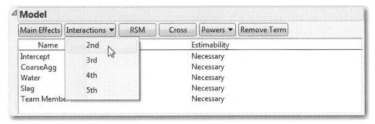

6. Finally, in the **Design Generation** panel, under **Number of runs**, select the **User Specified** option and enter 40.

7. Scroll back up to the list of factors. The **Team Member** blocking factor now shows four cells. Enter these names: Jiao, Pranav, Daniel, and Li.

8. Click **Make Design**.

This time, we still have a full factorial design with four replicates and a total of 40 runs, but now see that the runs are grouped into four blocks. To compare your work to the author's, we will first set the random number seed once again.

9. Click the red triangle next to **Custom Design** and select **Set Random Seed**. Set the seed value to 5911.

10. Scroll to the bottom of the **Custom Design** window and select **Make Table**. In the new data table (Figure 21.8) the random assignment of runs to team members is clear.

Figure 21.8: Portion of a Data Table with Blocked Design

	CoarseAgg	Water	Slag	Team Member	Compressive Strength
1	1100	150	0	Daniel	•
2	1100	150	150	Daniel	•
3	850	150	150	Daniel	•
4	1100	200	150	Daniel	•
5	1100	150	150	Daniel	•
6	850	150	0	Daniel	•
7	850	200	0	Daniel	•
8	1100	200	0	Daniel	•
9	850	200	150	Daniel	•
10	1100	200	0	Daniel	•
11	850	150	0	Jiao	•
12	850	150	0	Jiao	•
13	850	150	150	Jiao	•
14	1100	150	150	Jiao	•
15	1100	150	0	Jiao	•
16	1100	200	0	Jiao	•
17	1100	200	150	Jiao	•
18	850	200	150	Jiao	•
19	850	200	0	Jiao	•

Left panel:
- Custom Design
- Design — Custom Design
- Criterion — D Optimal
- Model
- DOE Dialog
- Columns (5/0)
  - CoarseAgg *
  - Water *
  - Slag *
  - Team Member *
  - Compressive Strength *
- Rows
  - All rows — 40
  - Selected — 0
  - Excluded — 0
  - Hidden — 0
  - Labelled — 0

In this blocked design, we still have 40 runs in all, with four replicates of the eight possible formulations, and each block (team member) has an assignment of 10 formulations to test in randomized sequence. The analysis of the strength data would proceed as before, with the blocks accounted for.

Fractional Designs

In this example, we have not considered any limits on the number of experimental runs to be performed. This is not the case for all experiments, particularly those with more factors or levels, or those with time constraints or with scarce or very costly materials.

We might run fewer replicates to save time or materials, but then we'd lose the advantages of replication. Instead, if we can confidently forego the testing of all possible interactions, we can reduce the total number of runs by using a *fractional* design. In a fractional design, we deliberately omit some interaction combinations in exchange for the savings in experimental runs. We plainly collect less information than with a full factorial design, but we can still obtain useful data that we can analyze fruitfully.

We'll illustrate with an example in which we drop the number of experimental runs in half, generating only 20 of the possible 40 runs. We'll retain our replication scheme as well as our four-member team from the prior example.

1. Select **DOE ▸ Screening Design**. This dialog box (see Figure 21.9) works similarly to the earlier ones. Name the response variable, and reload the three continuous factors as you did in the previous example.

Figure 21.9: Screening Design Dialog Box

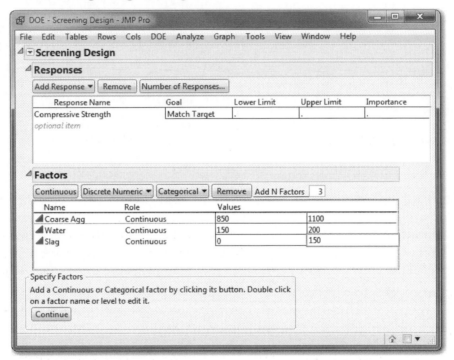

2. Click **Continue** twice. This expands the **Factors** panel as shown in Figure 21.10, enabling us to choose a design. Select **Fractional Factorial** and **Continue**.

Figure 21.10: Available Factorial Designs

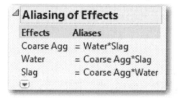

Once again, the panel expands revealing several new items, including a minimized panel labeled **Aliasing of Effects**. By clicking on the disclosure arrow, you see the alias structure illustrated in Figure 21.11.

Figure 21.11: Alias Structure

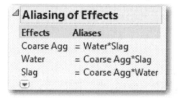

We read this table to mean that we will not be able to distinguish between main effects associated with coarse aggregate and the interaction (if any) of water and slag. In other words, if high volume of coarse aggregate is associated with greater compressive strength, we won't be able to decide whether the aggregate causes greater strength or whether the interaction of water and slag are the cause. To see why, let's generate the entire table of runs.

3.   Now specify that we want 4 replicates, and click **Make Table**.

Look closely at the data table that you have generated. Notice that there are now 20 runs (half the original number in the full factorial model), still with three factors and four replicates. A close look reveals that there are only four different patterns in use. These happen to be just four corner points of the full factorial cube that we looked at earlier in Figure 21.6, as shown now in Figure 21.12. In particular, notice that we will test *opposite* corners of the cube. Notice that on each face of the cube, two opposite corners are included in the design. Within the four experimental combinations, we have two with high values and two with low values for each of the three factors. This enables us to estimate the main effects of the factors.

Figure 21.12: The Half-Fractional Factorial Design

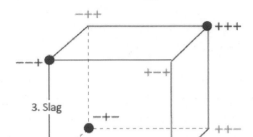

As with any compromise, we gain something and lose something in this half-factorial design. We gain the efficiency by using only half of the possible runs. This comes at the loss of valuable information; because we run only some of the possible combinations, we will not test all possible interactions, and thus some effects will be confounded with others. We might not be able to determine whether a particular response value is due to the main effects of the factors or to their interactions. We refer to the confounded main effects and interactions as *aliases*.

Now we might be able to understand the alias issue better. Look at the formulations that we are testing. We test only high coarse aggregate with (a) low water and low slag and (b) high water and high slag. We never test high aggregate with, say, low water and high slag.

Because there will be effects that cannot be clearly identified, we say that this half-fraction design has a weaker *resolution* than the comparable full-factorial design. We say that this is a *Resolution III* design. Fractional experimental designs commonly are classified as having the following resolutions:

- **Resolution III**: Main effects can be confounded with two-factor interactions, but not with other main effects.

- **Resolution IV**: Main effects are not aliased with each other or with any two-factor interactions, but two-factor interactions are aliased with one another.

- **Resolution V**: Neither main effects nor two-factor interactions are aliased with one another, but two-factor interactions are aliased with three-factor interactions.

Selection of appropriate design resolution goes well beyond the scope of this chapter, but it is important to understand that some experimental designs provide more information to an investigator than others. Designs of different resolution all have their places in practice.

Response Surface Designs

Earlier in the chapter, we introduced the idea of a *response surface* that might be non-linear. In our examples up to this point, we've tested high and low values of factors with the implicit assumption that the response variable would change in linear fashion. For example, if the response increases when a factor level shifts from low to high, we might infer that at an intermediate factor level, the response variable would also take on an intermediate value. But what if the response actually curves, so that between the two tested factor levels it actually declines?

It is possible to design an experiment explicitly for the purpose of exploring curvature in the response surface. To illustrate the concept and basic structure of a response surface experimental design, we'll introduce another example. Many retailers operate customer loyalty programs whereby they offer rewards and incentives to regular customers. Most often customers accept a complementary bar-coded card or transponder, which they present at the register when making a purchase. The retailer adds the specific details of each transaction to their data warehouse, tracking consumer purchasing patterns.

Among the common rewards programs are special offers generated at the register. A retailer might present the member-customers with a coupon worth a discount on their next purchase of a specific product, or a free item if they return to the store within so many days.

For this example, suppose that a retailer wants to conduct a two-factor experiment with their customer loyalty program. The response variable is *sales lift*, a common term for the increase in the amount customers spend in a given period above and beyond their normal purchases. For example, a grocery store could have a great deal of data about prior buying behavior. If customers receiving special offers end up spending an average of $20 per month more than customers not receiving the offers, the sales lift would be $20.

Suppose further that the two factors are the dollar amount of the discount offered and the length of time allowed before the offer expires. With each of these factors, it might be reasonable to anticipate curvature in the response. A very small discount might not provide sufficient incentive, but too large a discount might offset any sales lift. Similarly, if the coupon expires too soon, customers might not schedule a return visit in sufficient time. If the coupon expires too far into the future, customers might forget about it and not use it. The retailer might want to dial in just the right combination of dollars and days to maximize the sales lift and will randomly distribute coupons with different offers over the next month. As a first approximation we'll assume that the response curves are quadratic (in other words, parabolic in shape).

Because the factors are continuous and could conceivably be set at any number of levels, we need a strategy to select starting levels and to evaluate the quality of those choices. One measure of the design quality is the variance or standard errors of the coefficients that our design will ultimately enable us to estimate. JMP features an interactive graph called a *Prediction Variance Profile Plot* that enables us to visualize the effects of different choices and to compare competing designs.

1. Select **DOE ▸ Custom Design**. Name the response variable **Sales Lift**, and note that we want to maximize it.

2. Add two continuous factors, much as we did earlier. The first factor is **Discount**, which ranges in value from –1 to –10 dollars, and the second factor is **Days** ranging from 2 to 30 days before the coupon expires. Click **Continue**.

3. In the **Model** panel (see Figure 21.13), click **Powers** and select **2nd**. This introduces two additional terms into the model, each being the square of the factors (*i.e.,* Discount*Discount and Days*Days)

Figure 21.13: Adding Quadratic Terms for Factors

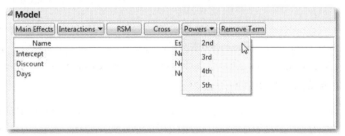

4. Under **Design Generation,** click the radio button labeled **User Specified** beneath **Number of Runs** and enter 100. We'll test these coupons on the next 100 members of the customer loyalty program.

5. Click **Make Design.**

6. You'll now see the list of 100 runs that show three different values of each factor. At the end of the list, find **Design Evaluation**

7. Now click the disclosure icon next to **Prediction Variance Profile**.

   Prior to running the experiment, we cannot estimate the magnitude of the sampling error. However, the profiler can show us the relative size of the variance associated with specific values of the factors. Initially, we see that if both factors are set at the central points of the ranges we specified (indicated by the red dashed cross-hairs), then the resulting relative variance of prediction is 0.050117.

Figure 21.14: Prediction Variance Profiler for the Coupon Experiment

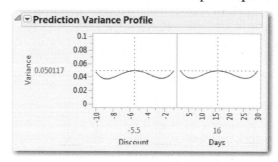

Notice in Figure 21.14 that the prediction variance reaches this value at each of the three values pre-set for the factors: for discount amounts of –1, –5.5, and –10 and expiration days of 2, 16, and 30. Typically, we'd like the prediction variance to be relatively small over the entire range of factor values, and we can examine the profiler to discover the worst case for the prediction variance.

8. Move your cursor over the vertical red dashed line in the left-hand profiler plot, and click and drag the line left and right. As you do so, notice that two things happen: the horizontal dashed line moves up and down, as does the entire profiler graph to the right.

   At most, the relative variance barely rises to about 0.05. We could evaluate alternative designs against this standard by trying, for example, different initial factor levels or a different number of runs.

9. Click the **Back** button at the bottom of the **Custom Design** window, select the default number of just 12 runs (rather than 100), and then click **Make Design.**

10. As you did above, disclose the **Prediction Variance Profiler.**

    In this profile (see Figure 21.15), the variance profile is quite different and much higher than in the 100-run design. The initial relative variance value is approximately 0.48, more than nine times larger than before.

11. Once again, slide the red vertical line in the **Discount** graph within the **Prediction Variance** profiler. Notice that as you move to the right, the level and the shape of the other curve change.

With just twelve runs, the prediction variance ranges from approximately .21 to .48, which indicates that an eight-run design is inferior to the 100-run design— the variability of predictions is much smaller in the 100-run design.

Figure 21.15: A Second Prediction Variance Profile

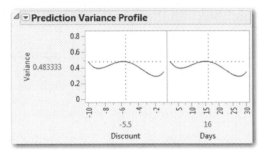

Once the investigators are satisfied with the level of risk as represented by the prediction variance, they could proceed to make the table, carry out the experiment, and collect response data.

Application

In this session, we have not worked with pre-saved data tables, but instead have created data tables based on different DOE principles. Use what you've learned in this chapter to design some experiments suitable for the scenarios presented here. Each exercise specifies a random seed so that your design can be compared to a solution prepared by the author.

1. *Scenario:* We'll continue our work on concrete compressive strength.

    a. Return to our initial full factorial design and add a fourth categorical factor representing the type of stone used as a coarse aggregate. Set the initial values as **Granite** or **Marble**. Leave all other settings as they were initially, but set the random number seed to 27513. Make the data table and report on the **Patterns** in the first five rows of the table.

b. How did the addition of a fourth factor change the design of this experiment?

c. Now modify the design to accommodate five blocks (for a five-person team). Report on your design.

d. Modify the design to be a fractional factorial design with only half as many runs. What information do we give up with this design?

2. *Scenario:* We'll continue our work on a customer loyalty program experiment. We typically apply the response surface methodology after we have identified relevant factors. In this problem, we'll assume we're still working at the screening stage of the experiment. Use the **DOE ▸ Screening Design** command and set the random number seed initially at 1918.

a. In addition to the two factors of discount amount and days, we'll add three more two-level factors to be screened in a full-factorial design:

   ○ A categorical factor indicating whether the discount applies to a specific product or to the entire purchase.

   ○ A continuous factor indicating the minimum purchase dollar value required to qualify for the discount (for example, "$5 off your next purchase over $20"). This factor should range from $2 to $20.

   ○ A categorical factor indicating whether the discount is valid only at the same store or at any store in the chain.

   In a full factorial, full resolution design with four replicates, how many experimental runs would there be total? Why might an experimenter opt for a weaker resolution design?

b. What are the **Patterns** for the first two runs?

c. Briefly explain what happens when we move from a three-factor screening design to a five-factor design.

d. Now revise the design slightly by changing the first categorical factor (the one indicating whether the discount is for a product) and make it a three-level categorical factor. The choices are manufacturer, product type (for example, any soft drink), or entire purchase. How does this revision change the resulting experimental design?

3. *Scenario:* There is a growing body of evidence to suggest that sleep plays an important role in the learning process. Some neuroscientists think that our brains consolidate newly acquired learning during the sleep process, so that interrupted or insufficient sleep can actually prevent us from completing the full cycle of

learning a new skill or set of concepts. This exercise is inspired by two articles reporting on studies of the effects of sleep on learning. (Stickgold & Walker, 2005; Walker, Brakefield, Morgan, Hobson, & Stickgold, 2002). The original experimental conditions have been modified slightly.

This experiment is conducted with 72 right-handed adults (36 men and 36 women). Researchers teach subjects to perform a complex motor skills task, consisting of repeating a five-number sequence at a computer keyboard. Subjects learn the sequence and then repeat it as many times as they can in a 30-second interval. Training consists of 12 repeated trials, with 30-second breaks in between. Subjects are then retested at a later time. Retesting consists of two trials within two minutes, with the performance score being the average of the two trials; the dependent variable is the improvement in performance scores.

Half of the subjects received training at 10 a.m. with retesting at 10 p.m. the same day and again at 10 a.m. after a night's sleep (the Awake First condition). The other half of the subjects were trained at 10 p.m., allowed a night's sleep, and retested at 10 a.m. and 10 p.m. the next day (Sleep First). The groups were further differentiated in that 50% of the individuals in each group were awakened briefly during the night, interrupting their sleep ("Full Night" versus "Interrupted"). The goal was to identify what role sleep might play in learning.

Set the random seed set to 4495.

a. Create a full factorial design for 72 subjects using the three categorical factors listed below. Make the data table, and then report the **Patterns** appearing in the first five rows of the data table.

    o Gender

    o "Awake First" versus "Sleep First"

    o "Full Night" versus "Interrupted"

b. Now let's consider altering the third factor by allowing the possibility that they are interrupted a varying number of times during the night. Specifically, now the third factor refers to the number of interruptions, which can be set at 0, 1, or 2. Make the data table, and report on the **Patterns** in the first five rows of the table.

c. Use the prediction profiler to compare the performance of this design to one in which researchers work with 144 subjects rather than 72. Comment on the benefit of doubling the sample size.

4. *Scenario:* Social scientists, opinion pollsters, and others engaging in survey research are often concerned with low response rates in household surveys. Some survey researchers offer incentives to encourage responses, and some have investigated the effectiveness of different types of incentives. A survey article (Singer, 2002) suggests that potential factors that affect response rates include the following:

    o   Type of incentive (monetary versus non-monetary gifts)

    o   Burden of the survey (length, complexity)

    o   Whether incentive is offered at the time of contact or contingent upon the completion of the survey

    o   Number of contacts made (that is, how many times the respondent is contacted about the survey)

    o   Survey mode: mail, telephone, face-to-face, e-mail[2]

    o   Amount of money offered (for monetary incentives)

    o   Guaranteed gift versus entry in lottery.

    For this exercise, set the random seed to 10501.

    a.   Which of the factors are continuous and which are categorical?

    b.   For each of the categorical factors, identify (or define) plausible factor levels that might be suitable in an experimental design.

    c.   Consider an experiment in which the response variable is dichotomous[3]: respondents either complete a survey or they do not. If we were to create a full factorial design using all of the factors listed above, how many runs would be required? (Assume no replications.)

    d.   Create a fractional design that measures the main effects of each factor. Use 200 replicates, and report descriptively on the design.

5. *Scenario:* Manufacturers frequently test materials and products to determine how long they will stand up to repeated use or to field conditions. This exercise is inspired by an article (Fernandes-Hachic & Agopyan, 2008) describing the use of a material called unplasticized polyvinyl chloride (uPVC), a plastic that is sometimes used in the manufacture of windows and home siding. Specifically, building products manufacturers in Brazil want to assess the durability of the product when exposed to natural weather in Sao Paulo over a period of five years. We will assume the role of designers for this experiment.

The response variable is a standard measurement known as the Charpy Impact Resistance Index. The index is a continuous measurement of the amount of energy lost when a pendulum strikes the material from a fixed distance.

The experimenters will formulate batches of uPVC while varying three components of the material (described below), measure the sample specimens using the Charpy Index at the time of manufacture, and then measure annually for five years. All sample specimens are white in color. During the five-year experiment, all specimens will be left outdoors at the natural weathering station in Piracicaba, Brazil and fixed at a 45° angle facing north.

The experimental factors are these:

- o **Impact modifier.** These are chemicals added to the compound formulation. Four different modifiers will be used: acrylic (ACR), chlorinated polyethylene (CPE), acrylonitrile-butadiene-styrene (ABS), and metacrylate-butadiene-styrene (MBS).

- o **Thermal stabilizer.** These compounds influence the extent to which the uPVC expands and contracts at varying temperatures. The formulations tested were: (1) a complex of lead, barium, and cadmium (Pd/Ba/Cd); (2) lead (Pb); (3) a complex of barium and cadmium (Ba/Cd); and (4) tin (Sn).

- o **Anti-UV additive**. Titanium dioxide protects against degradation caused by ultraviolet radiation. Because this additive is quite costly, three small quantities will be tested: 3, 5, and 10 parts per hundred resin (phr).

For this exercise, set the random seed value to 225196.

a. Design a full-factorial experiment with 10 replications. Make the data table, and report on the **Patterns** in the first five rows of the table.

b. In the actual situation, practical considerations limited the scale of this experiment, so that a full-factorial model was not feasible. Use a half-fractional design. Make the data table, and report on the **Patterns** in the first five rows of the table.

c. Use the prediction profiler to compare the performance of the half- and full factorial designs. Comment on the effects of using the reduced design.

6. *Scenario:* A *Harvard Business Review* article described an e-mail survey conducted by Crayola.com that illustrates DOE in marketing. (Almquist & Wyner, 2001). The response variable was simply whether the e-mail recipient completed the survey questionnaire[4]. During a two-week period, surveys were e-mailed to

customer addresses within the company database according to a design that took into account the following factors:

- o **Subject Line:** the subject line of the e-mail was either "Crayola.com Survey" or "Help Us Help You".

- o **Salutation:** the first line of the message was either "Hi [user name] :)", "Greetings! [user name]", or just "[user name]".

- o **Call to Action:** "As Crayola grows…" or "Because you are…"

- o **Promotion:** Three different conditions—(1) no offer; (2) $100 product drawing; (3) $25 Amazon.com gift certificate drawing.

- o **Closing:** the e-mail was signed either "Crayola.com" or "EducationEditor@Crayola.com".

For this exercise, set the random seed to 688098.

a. First, create a full factorial design with 1000 replications. Make the data table, and report on the **Patterns** in the first five rows of the table.

b. When Crayola conducted the experiment, they opted for a smaller number of combinations. Use the **Custom Design** to specify an experimental design with 27 runs and 1000 replicates. Make the data table and report on the **Patterns** in the first five rows of the table.

c. Comment on the comparison of these two designs.

[1] The DOE menu in JMP includes a **Screening Design** command, which would save a few steps in this example. Because the goal here is to understand how DOE works, we'll use the **Full Factorial** command, which calls upon us to make a few more choices. We'll see the **Screening Design** dialog box shortly.

[2] E-mail was not a major modality of surveying in 2002, but it is included in this exercise to illustrate the concepts of experimental design.

[3] Please note that JMP's DOE platform is designed for continuous response variables; this exercise is intended to illustrate how the number and complexity of factors influences the scale of a design.

[4] Please note that JMP's DOE platform is designed for continuous response variables; this exercise is intended to illustrate how the number and complexity of factors influences the scale of a design.

Quality Improvement

Overview

We expect to observe variation in the continuous functioning of natural and artificial processes. In this chapter, we'll study several techniques commonly used in the world of business, often in the context of manufacturing and service operations. Each of these techniques plays a role in efforts to monitor, assure, and improve the quality of processes that provide services or manufacture goods. As you encounter these techniques, you will recognize concepts and approaches that we have seen earlier, but now we'll combine them in different ways. Collectively, these techniques help managers control their operations in order to maintain consistent levels of quality and to discover opportunities for improvement.

Processes and Variation

One of the core ideas in the realm of quality improvement is the distinction between **common cause** variation and **special** (or **assignable**) cause variation. Take, for example, the case of a company that manufactures plastic tubing for hospitals and medical uses. Tubing is made by an extrusion process where liquefied material is forced through an opening to form a continuous tube with pre-specified inside and outside diameters. By controlling the temperature, composition of the material, the rate of flow, and the cleanliness of the equipment, the manufacturer can assure a high degree of consistency in the diameters. Nevertheless, variation occurs after many hours of continuous high-speed operation.

One challenge in managing any process is to distinguish between variation that indicates a problem and variation that is simply part of the fluctuations of any human system. We'll conceive of the routine chance variation that characterizes an imperfect process as being attributable to common causes. Special cause variation is typically attributable either to a problem that requires attention or a temporary, passing event. If special causes are rare and most of the variation is attributable to common causes, we say that the process is "in control". If special cause variation is unacceptably frequent or pronounced, the process is "out of control" and the manager needs to intervene and restore the process to its desired state. Sometimes management will want to improve a process either by shifting its overall level or by understanding and reducing common cause variation.

Control Charts

A control chart[1] is a time series graph of measurements taken from an ongoing process. Control charts are visual tools that were created in the 1930's so that supervisors on a factory floor could make real-time judgments about whether to intervene in a production process to adjust equipment or otherwise respond to a process that was slipping out of control. There are several types of control charts suitable for different data types and different analytical goals. The design and computations required for many control charts are quite simple, reflecting their historical origins as "quick and dirty" methods to track and improve production. In this section, we'll use a single example to introduce several techniques.

Depending on the product or process being observed, we'll first distinguish between charts for measurements and charts for counted events. If we are observing a process and measuring a key outcome (for example, the size or weight of a product), we'll be charting continuous data, principally by charting means. On the other hand, if we are observing the frequency with which particular attributes occur (for example, defects of a certain type), then we'll chart nominal data, typically by charting frequencies or relative frequencies.

A company manufactures catheters, blood transfusion tubes, and other tubing for medical applications. In this particular instance, we have diameter measurements from a newly designed process to produce tubing that is designed to be 4.4 millimeters in diameter. Open the data table **Diameter** and look at the data.

Each row of the table is a single measurement of tubing diameter. There are six measurements per day, and each day one of four different operators was responsible for production using one of three different machines. The first 20 days of production were considered Phase 1 of this study; following Phase 1, some adjustments were made to the machinery, which then ran an additional 20 days in Phase 2. In total, there are 240 individual observations (40 days x 6 measurements daily).

In analyzing the data, we'll want to place ourselves in the mindset of the operations manager who wants to evaluate the effectiveness of the adjustments. Initially, we'll do so by examining a *Run Chart* of individual diameter measurements during the two phases.

Charts for Individual Observations

The simplest control chart plots each measurement over time. It is a time series graph of individual measurements. For this first example, we'll use JMP's **Control Chart Builder**, which operates much like **Graph Builder,** which we have used throughout the book.

Unlike **Graph Builder**, **Control Chart Builder** has just three main graph zones corresponding to the phase of a study, the response variable, and one or more subgrouping variables.

1. Select **Analyze ▸ Quality and Process ▸ Control Chart Builder**. Drag and drop **DIAMETER** into the **Y** graph zone, and **Phase** into the **Phase** zone.

2. Click **Done**.

The result will look like Figure 22.1. The report window contains two graphs and a table of statistics. On the left side of the report, the upper graph is a run chart of individual measurements, with the Phase 1 results on the left and the Phase 2 results on the right. Below the graph of individual diameters is a Moving Range chart, which shows the differences between successive observations starting with the second row of the data table. Because each panel of the report provides different perspectives, we'll discuss each of them in turn.

Figure 22.1: Control Chart Builder for the Tubing Diameters

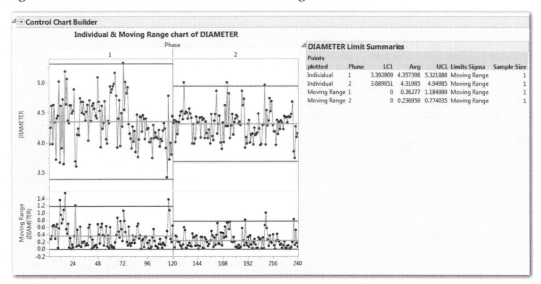

A run chart is a time-series graph of individual values with a line connecting each observation. Additionally the charts in Figure 22.1 display a green horizontal line at the mean diameter value and two red lines representing *upper* and *lower control limits.* Control limits are set at approximately two standard deviations above and below the mean. These limits are not exactly equivalent to two standard deviations, but are based on easily-computed approximations, reflecting the origins of control charts. When a process is in statistical control, it oscillates around the mean line and approximately 95% of the measurements consistently lie within the control limits.

We can see in the Phase 1 Individual run chart that the average diameter is slightly less than 4.4 mm (4.357 mm according to the **Diameter Limit Summaries** in the right panel), but that there is quite a lot of variation and instability in the process. In Phase 2, after adjustments to the process were made, the mean remained very nearly unchanged but the control limits were drawn in closer to the mean. In other words, the common cause variation was reduced in magnitude. Observations lying within the control limits are unremarkable, but quality analysts take special note of observations beyond the control limits because there may be *special* (or *assignable*) causes of those outlying observations.

The moving range chart also indicates that the Phase 2 modifications reduced process variability as measured by the range between observations. It is clear that the measurement differences between consecutive observations tended to be considerably smaller in Phase 2 than in Phase 1.

Before searching for possible special causes, let's take a closer look at the question of *stationarity*, or stability of the process. We'll examine the degree of stationarity by looking at averages of several measurements at a time. Specifically, we'll use an *X-bar Chart*, which graphs the means of successive samples through time.

Charts for Means

Inasmuch as this tubing example involves a continuous product, the individual diameter measurements could be subject to considerable sampling variation. In cases like this, it often makes sense to compute the mean of a sample of measurements and use the sample mean as a gauge of process stability. In our data table, we have six daily measurements all recorded by a different operator each day, so it seems logical to define subsamples by day (*i.e.* our sample size will be $n = 6$).

1. Modify the current Control Chart Builder graph by dragging **Day** to the **Subgroup** drop zone at the bottom; just drop it directly over the horizontal axis. The title, the graph, and the **Limit Summaries** change, as shown in Figure 22.2.

Figure 22.2: XBar-R Chart

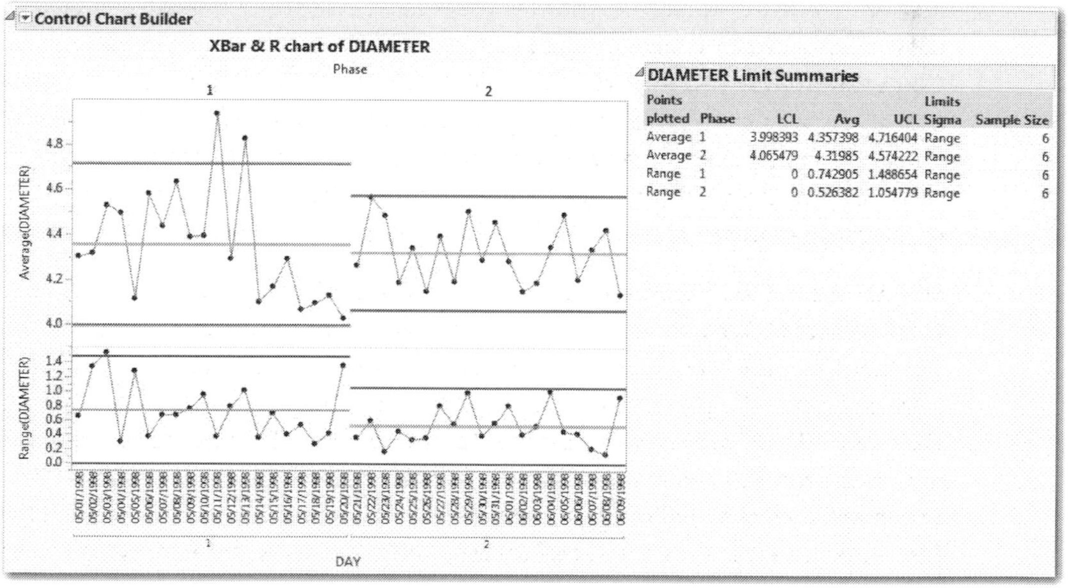

Points plotted	Phase	LCL	Avg	UCL	Limits Sigma	Sample Size
Average	1	3.998393	4.357398	4.716404	Range	6
Average	2	4.065479	4.31985	4.574222	Range	6
Range	1	0	0.742905	1.488654	Range	6
Range	2	0	0.526382	1.054779	Range	6

To interpret the resulting control chart, it is valuable to understand a few essential ideas about the structure and use of these charts. An XBar chart, as the name implies, is a graph of sample means. In this example, we have six daily measurements of diameter over a period of 40 days. We're specifying that we want to chart the daily sample means during the two phases, so the chart will display 20 points within each phase.

In the parlance of statistical process control, a process is *in control* if the variation exhibited by the process (as measured by the sample means) remains within the bounds of an *upper* and *lower control limits.* There are different methods for establishing the control limits, and the default method in in a JMP XBar chart is to base the limits on the observed range of values. On the left side of **Control Chart Builder**, note the **Limits** in the upper and lower graphs (*i.e.,* graphs [1] and [2]) refer to Range as the estimate of sigma. Because we specified phases in this study, the center lines and control limits are computed separately for each phase of the investigation.

A process cannot be in control if the variability is not constant throughout the observation period. In other words, there is a requirement of homoskedasticity in order to draw conclusions about whether a process is in control. Traditionally, there are two methods to judge the stability of the process variation: by charting the *sample ranges (R chart)* or by charting the sample *standard deviations (S chart).*

The chief virtue of an R chart over an S chart in a factory setting is the computational simplicity: if an operator makes six sample measurements, then plotting the range of the sample requires only the subtraction of two measurements. Computing a sample standard deviation might have been an intrusive burden on a production worker when control charting was first introduced into factory work. With software, the S chart is readily available and probably the better choice.

The default setting, in a nod to history, is the R chart. Now, look again at your screen or Figure 22.2. You should develop the habit of interpreting these charts by looking first at the lower R chart to decide whether this manufacturing process exhibits stable variability. In the Range Chart, we see a center line that represents the mean of the ranges of the twenty samples. Above and below the center line are control limits.

When we scrutinize a control chart, we're looking for specific indications of non-random variation[2] such as:

- Do an excessive number of points fall outside of control limits?
- Is there systematic clustering of consecutive points either above or below the center line?
- Does the series line consistently drift either upward or downward? An increase in sample ranges indicates greater inconsistency in the process.
- Is there a sudden shift in the level of the curve, either upward or downward? Such a shift might suggest a one-time shock to the process.

As we look at this range chart in Phase 1, we note that the range of just one sample of the 20 lies above the UCL—exactly 5% of the observed samples. Of course, the LCL in the R-chart falls at 0, and it is impossible that any sample range would fall below the LCL. On the first several days, the variability appears more extreme than on later days, although the final day shows a range comparable to those earlier ones. Toward the right, there are six consecutive days with below-average ranges, which might suggest a special cause. Overall, though, there are no large red flags in the range chart.

With a stable range, we can meaningfully interpret the XBar chart. Here again, there is a center line at the grand mean of the first 20 samples. The upper and lower control limits lie at roughly three sigmas from that mean, where sigma is estimated based on the sample ranges.

In the first ten samples, the means fluctuate without any noteworthy pattern. Then we find two out of three in a row with means above the UCL, followed by seven consecutive means below the center line, and a gradual descent toward the LCL. A non-random pattern such as this one suggests that the process needs some adjustment.

Figure 22.3: The Control Panel in Control Chart Builder

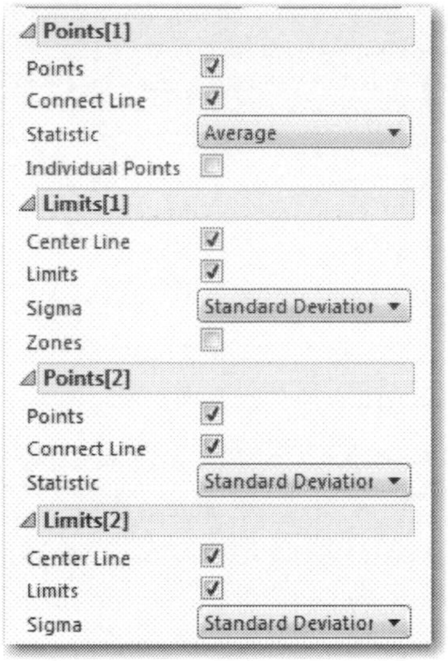

Let's take a second look at the same data, this time using an S chart to evaluate the variability within the process. This is easily accomplished just by changing some settings in **Control Panel** on the left side of **Control Chart Builder**, shown above in Figure 22.3. The upper and lower graphs each display points and limits. These are set by default, but we can alter them. The limits in both graphs are based on observed ranges in each phase and the plotted points in the lower graph [2] are also based on ranges. We want to change the basis of computations from ranges to standard deviations.

2.   In the **Control Panel** (see Figure 22.3 above), click the three instances where **Statistic** is set to **Range** and change them to **Standard Deviation**.

The new report is shown below in Figure 22.4. The general visual impression is very much like the Xbar and range chart. Upon closer inspection, you should notice three key differences. First, the vertical scale in the S chart is different from that shown in the R chart because we are plotting sample standard deviations rather than ranges. Second, because the standard deviation is less sensitive to extreme values than the range, we no longer observe a run of consecutive six points below the center line in the lower graph. Finally, the plotted means in the XBar chart are identical to those in Figure 22.2, but the UCL and LCL are located slightly farther from the center line, reflecting a different (and more accurate) estimate of the process standard deviation.

Figure 22.4: Xbar & S Chart for the Diameter Data

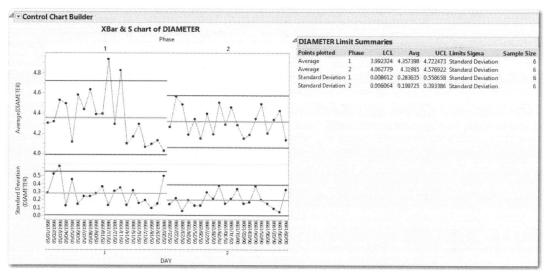

The conclusions that we draw from these control charts are the same as those noted above. The Phase 1 testing indicates a process that is not in statistical control. At this point, management intervened to fine-tune the process.

The S chart shows a smaller mean and standard deviation in Phase 2, indicating less process variability overall. Moreover, the standard deviation is more stable, consistently remaining within the newer and tighter control limits. This is a process with stable variability.

The XBar chart also shows improvement with respect to variability in the mean diameters. The charted line remains within the narrower control limits and the oscillations are random. However, recall that the nominal diameter of this tubing is 4.4 mm and the mean of these samples is smaller than in Phase 1, when the grand mean was 4.36 mm. Whether this is a problem for the firm depends on the specifications that they have established with and for their customers. At any rate, the revised process is more consistent than the Phase 1 tests, but appears to be producing a product that does not conform to the intended specification of diameter.

Charts for Proportions

For some processes, we might be concerned with a categorical attribute of a product or service rather than a measurable variable. For example, manufactured products might be unacceptable for distribution to customers because of one or more defects. The two major types of attribute charts are known as *NP charts*, which illustrate simple counts of events, and *P charts*, which represent proportions or relative frequency of occurrence. To illustrate how control charting can help monitor attribute data, let us turn to some additional data about airline delays.

Air travel inevitably involves some chance that a flight will be delayed or cancelled. Our data table **Airline delays 2** has daily summary information about flights operated by United Airlines that are scheduled to arrive at Chicago's O'Hare International Airport during one month. United is the largest carrier using O'Hare and accounted for just over 25% of all arrivals at the airport that month. Conversely, O'Hare is United's largest hub, with nearly 20% of all United flights touching town there. United was chosen to illustrate these control charts rather than to suggest that their performance is problematic. In fact, during the month observed, United's on-time record was superior or equivalent to other major carriers.

The data table contains information about delayed, cancelled, and diverted flights. Each row represents a single day, and we have both counts and relative frequencies for the occurrence of these flight disruptions. Let's first look at the number of delayed arrivals among all of the flights, taking into account that the total number of flights varied from day to day. The two columns of interest in this analysis will be the number of late flights on a given day (**NumLate**) and the total number of flights that day (**TotFlight**).

1. Open the data table **Airline delays 2.**

2. Select **Analyze ▸ Quality and Process ▸ Control Chart Builder**.

3. Click the **Shewhart Variables** button and select **Shewhart Attribute** to indicate we are dealing with a categorical (attribute) variable.

4. Drag **NumLate** to the **Y** drop zone and drag **TotFlight** to the **n Trials** drop zone in the lower right of the graph. Your graph will now look like Figure 22.5.

Figure 22.5: NP Chart for Flight Delays

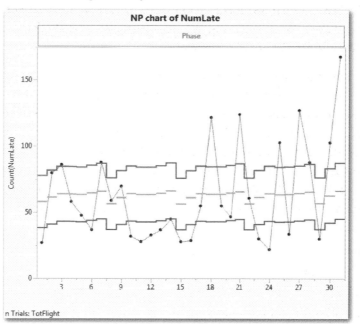

With attribute charts, there is no equivalent for the S or R charts. The P chart is structured identically to other control charts shown earlier. In this chart, though, the center line and control limits are jagged due to the different sample sizes (number of flights) each day. We see a green center line indicating a varying average. We also see varying upper and lower control limits.

In the first part of the month, the process appears to have been relatively stable, but then we see a run of consecutive days of very few delays (a good thing) followed by two weeks of the month with much larger oscillations. With flight delays, values near the UCL are the troubling ones, and this process is an example of one that is out of statistical control. The control chart cannot reveal the reasons for the delays, but does highlight that the extent of ordinary variation grew late in the month.

An alternative to the NP chart is the P chart, which analyzes the *proportion* rather than absolute *counts* of late flights. To convert the NP chart into a P chart, one just needs to make one small change.

5. In the **Points Panel** in the lower left of the **Control Panel**, change the **Statistic** from **Proportion** to **Count**. The revised graph appears in Figure 22.6.

In this chart, we see that both the center line is continuous but the control limit lines are jagged. The overall proportion of flights is based on all of the data, and we can see that 25.6% of all flights experienced a delay. Because the daily sample sizes are different, the limits are recomputed from left to right, reflecting changing estimates of the underlying sampling distribution. As before, it is clear that flight delays were more common and more variable in the latter part of the month.

Figure 22.6: P Chart of Flight Delays

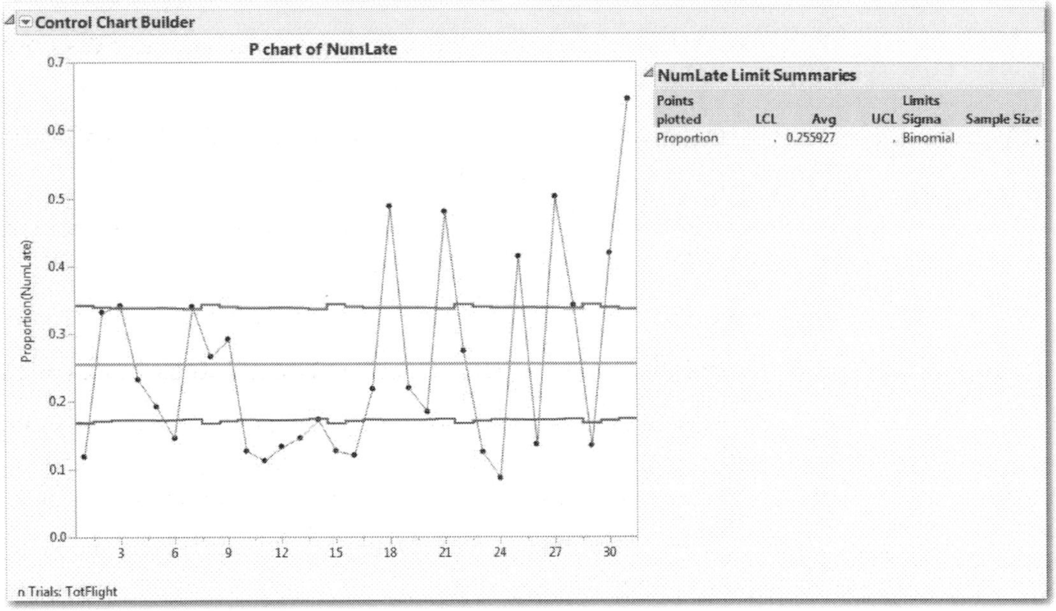

Capability Analysis

In most settings, it is not sufficient that a process be in statistical control. It is also important that the results of the process meet the requirements of clients, customers, or beneficiaries of the process. To return to our earlier example, if the extrusion process consistently produces tubing that is either the wrong diameter or highly inconsistent, the tubing might not be suitable for the intended medical purposes.

For a process that is in statistical control, *process capability* refers to the likelihood that the process as designed can actually meet the engineering or design criteria. We analyze capability only for numerical measurement data, not for attributes. Thus, for example, if our tubing manufacturer needs the tubing to have an average diameter of 4.4 mm with a tolerance of ± 0.1 mm, then once the process is in control (that is, Phase 2), they want the individual observations to remain in the range between 4.3 and 4.5 mm.

In a capability analysis, the fundamental question is whether the process is capable of delivering tubing that meets the standard. There are several methods for generating a capability analysis within JMP, and we'll illustrate one that is an extension of the Phased XBar-S control chart that we made earlier.

Because we want to use only the Phase 2 data, we'll hide and exclude the Phase 1 data.

1. We need to return to the data table of tubing diameters; select **Window ▸ Diameter** to make this data table active.

2. Move your cursor over any row where **Phase** is equal to 1. Right click in the **Phase** cell and click **Select Matching Cells**.

3. Then choose **Rows ▸ Hide and Exclude**. Now only the Phase 2 measurements will be used in the calculations and graphs.

4. Now select **Analyze ▸ Quality and Process ▸ Capability.**

5. In the Capability dialog (not shown), select **DIAMETER** as **Y** and click **OK**.

6. This opens a second dialog (shown below in Figure 22.7). Recall that we want to investigate whether this system is capable of producing tubing that is consistently between 4.3 and 4.5 mm in diameter, with a mean of 4.4 mm. In this dialog, enter the limits as show in Figure 22.7 and click **Next**.

Figure 22.7: Setting Capability Specifications

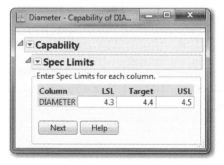

The Capability report window (in Figure 22.8) shows that slightly more than 43% of the Phase 2 diameters were below the lower specification of 4.3 and almost 16% were above the upper specification limit, for a total of 59% outside of the specifications. So despite the fact that the process was under control in Phase 2, it does not appear to be meeting the producer's needs.

Figure 22.8: Capability Analysis Results

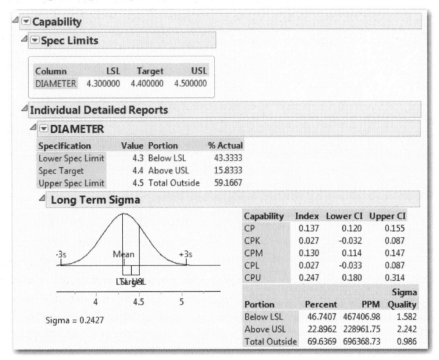

Below the numerical summary of the distribution, we find a graph in the **Long Term Sigma** panel. Use the grabber tool to adjust the horizontal axis and improve the readability of the graph. The graph shows the target marked at 4.4 mm with two vertical lines at the upper and lower specification limits. We can see the comparatively small proportion of values falling within the desired specification limits, and this casts doubt on whether the system in phase 2 is capable of regularly producing the desired tubing diameters.

In addition to considering the proportion of observations within the specified limits, there are several common indexes of process capability, and we'll look at just three of them. We should interpret these statistics only if we can assume normally distributed data and have a moderately large sample of independent observations. We might fall short on the issue of independent observations, because we might have operator effects here, but the normal quantile plot indicates that normality is a reasonable assumption.

The three indexes of process capability are CP, CPK, and CPM. For this discussion it is less important to know the specific formulas[3] for these indexes than it is to understand what they measure. In a capable process, the large majority of observed measurements fall within the specification limits. In a capable process, all three indexes would be close to or greater than 1, indicating that all observations lie within the specification limits. CPK and CPM adjust for the possibility that the process is not centered on the target.

In this case, all three measures are considerably less than 1. This echoes the visual impression we form from the graph.

What do we conclude about this process? It is still a work in progress: the good news about Phase 2 is that the firm has now brought the process into statistical control, in the sense that it is stable and predictable. On the other hand, there is further work to do if the firm needs tubing that is uniformly between 4.3 and 4.5 mm in diameter.

Pareto Charts

Yet another simple charting tool for attribute data is known as a *Pareto chart*. A Pareto chart is a bar chart showing the frequency or relative frequency of specific attributes, often those representing some type of process or product defect. The bars in the chart are always sorted from most to least frequent to draw attention to the most commonly occurring problems. Additionally, Pareto charts display the cumulative relative frequency of the various problem categories to enable managers to identify the critical few issues that account for the greatest proportion of defects.

To illustrate, let's look back at the pipeline disruption data. We've analyzed the data previously, focusing on several different variables. This time, we'll examine the reported types of pipeline disruptions.

1. Open the **Pipeline Safety** data table.

2. Select **Analyze ▸ Quality and Process ▸ Pareto Plot**. Select **LRTYPE_TEXT** as the **Y, Cause** column and click **OK**.

You'll now see the graph that is shown in Figure 22.9. The data table contains three categories of disruptions and for some of the incidents, no classification was reported (N/A). Looking at the left vertical axis, we can see that approximately 220 of the incidents were labeled "other," and ruptures and leaks occur with similar frequency.

Figure 22.9: Pareto Plot of Pipeline Disruption Types

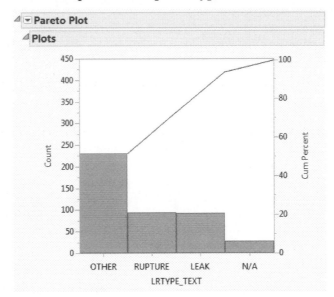

The right vertical axis displays cumulative percent, or relative frequency. The kinked diagonal line indicates that approximately 50% of the incidents are in the "other" category, and that about 95% of all incidents were either other, ruptures, or leaks.

What does this plot imply? First, it might be useful for those managing the data collection process to consider refining the category now called "other" because so many incidents fall into this catchall grouping.

Second, from the standpoint of emergency preparedness, we might wonder whether these disruption types occur with equal frequency in different regions of the county. In an area where ruptures are more common than leaks, emergency responders might need different resources or training than those in a different area.

3. Click the red triangle and select **Script ▸ Relaunch Analysis**. Select **IREGION** as the **X, Grouping** column and click **OK**.

In the new results, we see four separate panels corresponding to the four regions. We can quickly see that leaks are more prevalent than ruptures everywhere except in the East. More disruptions occur in the Central and Eastern regions, and the South and Southwest are relatively unlikely to classify disruptions as "other." You should begin to see the potential uses of this type of chart in suggesting areas for inquiry about possible operational problems.

Application

Now that you have completed all of the activities in this chapter, use the concepts and techniques that you've learned to respond to these questions.

1. *Scenario:* Let's continue our analysis of the tubing diameter data (**Diameter** data table). As before, we'll use sample sizes of 6, corresponding to the six daily measurements.

   a. In Phase 1, we concluded that the process was not in control. Restrict your attention to Phase 1 data once more and create XBar-S charts **BY Machine**. Is there any indication that the problems are caused by one or more specific machines? Describe what you find and what you conclude.

   b. Now, create a set of similar control charts breaking out the data by **Operator**. Do any of the operators seem to need attention? Explain.

   c. Now let's refocus on just the Phase 2 data. Earlier, we concluded that the process was not capable of producing 4.4 mm tubing within tolerances of 0.1 mm. Perform a capability analysis to determine whether the process is capable within tolerances of 0.5 mm. Report your findings.

2. *Scenario:* Let's return to the airline data in **Airline delays 2**. In this exercise, we'll focus on flight cancellations.

   a. Create a P chart for the proportion of cancelled flights and report on what you see.

   b. Create an NP chart using the columns **NumCancelled** and **TotFlight** and report on what you see.

3. *Scenario:* Let's return to our analysis of the Indian Index of Industrial Production and look more closely at some of the sub-indexes for specific types of products. For this exercise, we'll treat each three-month quarter as a sample and therefore specify a sample size of 3 observations.

   a. Create an XBar-R chart for **Basic Goods**. Before going further, click the red triangle in the upper left and choose **Set Sample Size**; specify n = 3. This chart looks very different from the ones that we've seen earlier; does this indicate a problem? Why is this chart so different from the ones in the chapter?

b. Look at the R chart more closely. What pattern or patterns do you see, and what might account for them?

c. Create an XBar-R chart for **NonDur** (non-durable goods). How does this chart compare to the one for basic goods? What might explain the similarities and differences?

d. Look at the R chart more closely. What pattern or patterns do you see, and what might account for them?

4. *Scenario:* Let's continue our analysis of the **Pipeline Safety** data.

a. Create one or more control charts to examine variation in the **STMin** (time required to restore the site to safety) by **Year**. Then click the red triangle in the upper left and choose **Script ▶ Local Data Filter**, and filter by **IREGION**. Click through the different regions and report on what you see in this group of control charts.

b. Create one or more control charts to examine variation in the **Prpty** (dollar estimate of property damage) by year, again filtering by **IREGION**. Report on what you find.

5. *Scenario*: In the United States, the management of solid waste is a substantial challenge. Landfills have closed in many parts of the nation, and efforts have been made to reduce waste, increase recycling, and otherwise control the burgeoning stream of waste materials. Our data table **Maine SW** contains quarterly measurements of the tonnage of solid waste delivered to an energy recovery facility in Maine. The data come from five municipalities in the state. Each part of this scenario asks you to use a sample size of 4. To do that, choose your Y variable, click the red triangle in the upper left, and choose **Set Sample Size**; specify n = 4.

a. Create an appropriate control chart for annual solid waste delivered from Bangor (that is, use a sample size of 4 quarterly observations). Comment on important or noteworthy features of the process.

b. Create an appropriate control chart for annual solid waste delivered from Bucksport (that is, use a sample size of 4 quarterly observations). Comment on important or noteworthy features of the process.

c. Create an appropriate control chart for annual solid waste delivered from Enfield (that is, use a sample size of 4 quarterly observations). Comment on important or noteworthy features of the process.

    d. Create an appropriate control chart for annual solid waste delivered from Orono (that is, use a sample size of 4 quarterly observations). Comment on important or noteworthy features of the process.

    e. Create an appropriate control chart for annual solid waste delivered from Winslow (that is, use a sample size of 4 quarterly observations). Comment on important or noteworthy features of the process.

6. *Scenario*: The World Bank and other international organizations monitor many forms of technological development around the world. Elsewhere, we've looked at other Millenium Development Indicators. In this exercise, we'll study carbon dioxide emissions between 1990 and 2004. The relevant data table is called **MDGCO2**. The time series column is called **CO2**, which measures the number of metric tons of CO2 emitted per capita annually in the country. Let's examine world-wide CO2 emissions.

    a. Create an XBar-R chart for **CO2**, using **Year** as the **Subgroup**. Then click the red triangle in the upper left, choose **Script ▶ Local Data Filter**, and filter by **Region**. Click through the different regions and report on what you see in this group of control charts.

    b. Repeat this analysis with XBar-S charts. How does this analysis differ from the previous one?

7. *Scenario*: Let's look again at some airline delays data from Chicago. This time, we'll consider a single week during March 2009 and focus on *departure* delays for United Airlines flights. There were more than 1,500 scheduled departures during the week. 58 were cancelled and have been omitted from the table called **Airline delays 3**.

    a. The departure time of a flight refers to the difference, if any, between the scheduled departure and the time at which the plane rolls away from gate. Create an XBar-S chart for **DepDelay** and report on the extent to which this process seems to be under control. Use a sample size of 20 flights. To specify the sample size, choose your Y variable and then click the red triangle in the upper left and choose **Set Sample Size**; specify n = 20.

    b. The data table also contains a column labeled **SchedDeptoWheelsOff**, which is the elapsed time between scheduled departure and the moment when the wheels leave the runway. Create an XBar-S chart for this variable ($n = 20$) and report on the extent to which this process seems to be under control. How does it compare to the previous chart?

    c. (Challenge) Imagine that the airline would like to control departure delays with a goal of having 90% of all non-cancelled *weekday* flights leave the gate 20 minutes or less from the scheduled time. Run a capability analysis, and report on your findings. Hint: the column called **Weekend** is a dummy variable equal to 1 for Saturday and Sunday flights.

8. *Scenario*: The data table called **Earthquakes** contains earthquake measurements across the entire globe for one month of 2009. Although this refers to a natural process of geologic activity, we can use control charts to look for changes in earthquake activity. The key variable is **Magnitude**, which measures the size of the earthquake on the logarithmic scale developed by Charles Richter.

    a. Use a means control chart and sample size of 10 to monitor the magnitude of earthquakes during this period. To set the sample size, choose your Y variable and then click the red triangle in the upper left and choose **Set Sample Size**; specify n = 10. Comment on noteworthy features of the chart.

    b. Repeat the analysis with a sample size of 20. How does this change alter the analysis, if at all?

---

[1] Control charts are sometimes referred to as Shewhart charts after Walter Shewhart, an engineer at Bell Laboratories who introduced them in the 1930s.

[2] There are some standard criteria against which JMP can report important deviations. This list is intended to introduce the concept of specific tests by which to judge process stability.

[3] Here are the formulas, in which USL and LSL are the upper and lower specification limits.

$$CP = \frac{USL - LSL}{6s}$$

$$CPK = \min\left(\frac{USL - \bar{X}}{3s}, \frac{\bar{X} - LSL}{3s}\right)$$

$$CPM = \frac{\min(T - LSL, USL - T)}{3\sqrt{s^2 + (\bar{X} - T)^2}}, \quad \text{where T is the target mean.}$$

Data Sources

Overview

Almost all of the data tables used for chapter examples and for Application Scenarios contain real data. This appendix provides the citations and initial source of each table. Most are from public-domain Web sites, and those that are not also cite the original author or compiler who granted permission for the data table to be used in this book. Those generous individuals are also listed in About this Book. The bibliography includes citations of publications based on datasets as relevant. Data new to this edition appears in boldface.

The University of California at Irvine Knowledge Discovery in Databases (KDD) Archive has proved an especially valuable source of data (http://kdd.ics.uci.edu), as noted in the table beginning on the next page. Readers are encouraged to visit the KDD site for additional data tables. Similarly, the Data and Story Library (lib.stat.cmu.edu/DASL), maintained at Carnegie Mellon University, is a very rich source of data sets.

Data Tables and Sources

Table Name	Source	URL
Airline delays	US Department of Transportation, Bureau of Transportation Statistics.	http://www.bts.gov/xml/ontimesummarystatistics/src/index.xml
Airline delays 2	US Department of Transportation, Bureau of Transportation Statistics.	http://www.bts.gov/xml/ontimesummarystatistics/src/index.xml
Airline delays 3	US Department of Transportation, Bureau of Transportation Statistics.	http://www.bts.gov/xml/ontimesummarystatistics/src/index.xml
Anscombe	JMP sample data tables.	

Table Name	Source	URL
Binge	Based on tabulations reported by R.W Hingson, E.M. Edwards, T. Heeren, and D. Rosenblum (2009).	
Birthrate 2005	United Nations.	http://unstats.un.org/unsd/default.htm
Cell Subscribers	United Nations Millennium Development Goals Indicators.	http://unstats.un.org/unsd/mdg/Default.aspx
Concrete	Prof. I-Cheng Yeh (cf. Bibliography)	http://www.sigkdd.org/kddcup/index.php
Concrete 28	Subset of Concrete table.	
Diameter	JMP sample data tables.	
Dolphins	JMP sample data tables.	
Earthquakes	U.S. Geologic Survey.	http://neic.usgs.gov/neis/epic/epic_global.html
FAA Bird Strikes	Federal Aviation Administration. National Wildlife Strike Database.	http://wildlife.faa.gov/
FAA Bird Strikes CA	Subset of FAA Bird Strikes.	
Fertility	United Nations	http://unstats.un.org/unsd/default.htm
FTSE100	Yahoo! Finance.	http://www.finance.yahoo.com
Global Temperature and CO2	Earth Policy Institute, NASA and NOAA.	http://data.giss.nasa.gov/gistemp/tabledata_v3/GLB.Ts+dSST.txt and www.esrl.noaa.gov/gmd/ccgg/trends/co2_data_mlo.html
Gosset's Corn	Based on tables in W.S. Gosset, "The Probable Error of a Mean," Biometrika, 6 (1908).	
Haydn	Subset of Sonatas data.	
Hubway	Hubway and Metropolitan Area Planning Council, Boston, MA.	http://hubwaydatachallenge.org/trip-history-data/
India Industrial Production	Government of India Ministry of Statistics and Programme Implementation.	http://www.mospi.gov.in/mospi_iip.htm
Insulation	JMP Software: ANOVA and Regression course notes, SAS Institute. Used with permission.	
Life Expectancy	United Nations World Population Division.	http://un.data.org
Maine SW	George Aronson. Used with permission.	
MCD	Yahoo! Finance.	http://www.finance.yahoo.com
MDG Technology	United Nations Millennium Development Goals Indicators.	http://unstats.un.org/unsd/mdg/Default.aspx
MDGCO2	United Nations Millennium Development Goals Indicators.	http://unstats.un.org/unsd/mdg/Default.aspx
Michelson 1879	Data and Story Library; based on table in S.M. Stigler (1977).	lib.stat.cmu.edu/DASL
Military	Diez et. al, OpenIntro Stats; U.S. Department of Defense.	http://www.openintro.org/stat/extras.php and data.gov
Mozart	Subset of Sonatas data.	

Table Name	Source	URL
NC Births	Diez et. al, OpenIntro Stats; North Carolina State Center for Health Statistics.	http://www.openintro.org/stat/extras.php
NHANES	National Health and Nutrition Examination Survey (2005).	http://www.cdc.gov/nchs/nhanes.htm
NHANES expanded	National Health and Nutrition Examination Survey (2005).	http://www.cdc.gov/nchs/nhanes.htm
NHANES SRS	Subset of NHANES table.	
Parkinsons	Prof. Max Little (cf. Bibliography)	http://www.sigkdd.org/kddcup/index.php
Pipeline Safety	US Dept of Transportation, Pipeline and Hazardous Materials Safety Program (PHMSA) Pipeline Safety Program.	http://primis.phmsa.dot.gov/comm/reports/safety/SIDA.html
Pipeline Safety SouthSW	Subset of Pipeline Safety table.	
Planets	Smithsonian Institution National Air and Space Museum.	http://www.nasm.si.edu/etp/ss/ss_planetdata.html
Popcorn	Artificial data inspired by experimental data from Box, Hunter, and Hunter (1978). Factors are brand (popcorn), amount of oil (oil amt), and batch size (batch).	
Sleeping Animals	JMP sample data tables.	
Sonatas	Prof. Jesper Rydén (see Bibliography).	
States	Demographic and economic data about the 50 U.S. states and the District of Columbia, compiled from several sources	http://quickfacts.census.gov/qfd/index.html http://www.infochimps.com http://www-fars.nhtsa.dot.gov/ http://www.bls.gov/web/laus/laumstrk.htm
Stock Weekly Index	Yahoo! Finance.	http://www.finance.yahoo.com
TimeUse	Extract from the 2003 and 2007 American Time Use Survey.	http://www.atus.org
Tobacco Use	United Nations.	http://data.un.org/Browse.aspx?d=WHO&f=inID:RF17
USA Counties	US Census Bureau.	www.census.gov
Used Cars	Prices and Mileage of two-year old used cars in three U.S.markets; downloaded from cars.com by author, March 2009.	http://www.cars.com
WDI	World Development Indicators.	http://data.worldbank.org/data-catalog/world-development-indicators
World Nations	United Nations.	http://data.un.org

Data Management

Overview

Before any data analysis can be undertaken in JMP, one must make the data accessible to JMP either in the form of one or more JMP data tables, or at least in a format that JMP can read. Depending on the particular goals of a study, it is often desirable to integrate data from multiple sources into a single data table. This appendix presents some foundational ideas of data management, including techniques for entering data from a keyboard, reading data from an Excel worksheet, and transferring data from the web. Additionally, it introduces some concepts and methods for combining data from two or more tables into a single table.

Entering Data from the Keyboard

Many readers are using this book in conjunction with a college course in statistics for which they have a principal textbook. Suppose that one such reader encounters a textbook problem based on the following small data set[1] from an online music seller that advertises at sporting events.

Name	Age (yr)	Time Since Last Purchase (Days)	Area Code	Nearest Stadium	Internet Purchase?
Katharine H.		130	312	Wrigley	Y
Samuel P.	24	18	305	Orange	N
Chris G.	43	368	610	Veterans	Y
Monique D.		5	413	Fenway	Y

Even with a small data set, we need a plan before we start tossing data into a table. This is a cross-sectional sample of customers containing both character and numeric data. "Name," "Nearest Stadium," and "Internet Purchase" are clearly categorical data, and "Age" and "Time Since Last Purchase" are continuous. What about "Area Code"? The digits are numeric, but the codes have no numerical meaning. Area codes are categorical labels identifying a geographic region.

We also have a choice to make about the "Internet Purchase" column. We can enter the data as Ys and Ns, as shown above. Alternatively, we can follow the NHANES example and use numbers to represent Yes and No. Either method is acceptable; for the sake of learning about column properties, let's use numeric codes.

1. Select **File ▶ New ▶ Data Table**. You'll see an untitled blank table, as shown in Figure B.1. Initially, a JMP data table contains one empty numeric column.

Figure B.1: A New Data Table

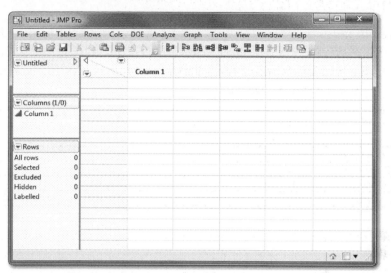

2. Before entering any data, let's name and document this data table. Click on the word **Untitled** in the **Table** panel, and change the name of this table to **Customers**.

3. Click the red triangle in the upper left next to **Customers** and choose **New Table Variable**. A simple dialog box opens (not shown here). Next to **Name**, type Source and next to **Value,** type Textbook example. Then click **OK**.

We'll first set up six columns, name them, and assign appropriate properties to each column in anticipation of typing in our data.

4. Move your cursor to the top of the first column and right-click once to bring up the menu shown in Figure B.2.

Figure B.2: Context-Sensitive Column Menu

Column Info...	
Column Properties	▶
Modeling Type	▶
Preselect Role	▶
Formula...	
Color Cells	▶
Use Value Labels	
Label/Unlabel	
Scroll Lock/Unlock	
Hide/Unhide	
Exclude/Unexclude	
Data Filter	
Sort	▶
Delete Columns	
Copy Column Properties	

5. Choose **Column Info** from this menu to name the column and specify the properties that we want to set.

In this dialog, we specify the name, data type (numeric, character or row state), modeling type (continuous, ordinal or nominal), format, and initial data values. We also have the option to select from a large number of other column properties.

6. Complete the dialog box as shown below in Figure B.3. Click on the **Column Properties** button to add the **Note** field to the column information. When you're done, click **OK**.

Figure B.3: Entering Column Information

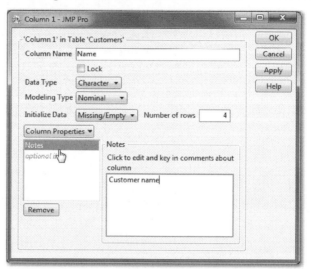

7. Now choose **Cols ▶ New Column** to add a second column for customer ages. Name the column **Age**, and specify that it will contain numeric data. **Modeling Type** will be **Continuous**. Add a note to say that ages are in years, and click **OK**.

    Continue to create new columns until we have six columns in all, with the following names and attributes:

Col	Column Name	Data Type	Modeling Type	Note
1	Name	Character	Nominal	Customer name
2	Age	Numeric	Continuous	Customer age in years
3	LastPurch	Numeric	Continuous	Days since last purchase
4	AreaCode	Numeric	Nominal	Area code
5	Stadium	Character	Nominal	Nearest stadium
6	Internet	Numeric	Nominal	Internet purchase?

8. In the "Internet" column, we'll establish a coding of 0 for No and 1 for Yes by choosing **Value Labels** from the column properties drop-down list.

9. Within the value labels portion of the dialog box (shown in Figure B.4), indicate that values of 0 should be labeled No and then click **Add**. Then associate the Yes label with values of 1, and click **OK**.

Figure B.4: Specifying Value Labels

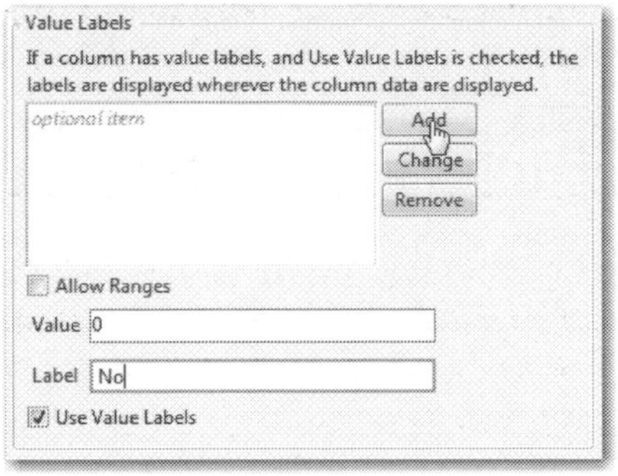

10. With the columns fully defined, type the data into the cells one column or row at a time. For Katherine and Monique, just skip past their "Age" cells. In the "Internet" column, enter a 0 in row 2 and 1 in each of the other rows.

Figure B.5 shows part of the **Data Table** window for this example. When you have finished entering all the data, screen should look much like the one shown in Figure B.5. Having done the work to type in this set of data, you should save it.

Figure B.5: The Completed Table for Customer Data

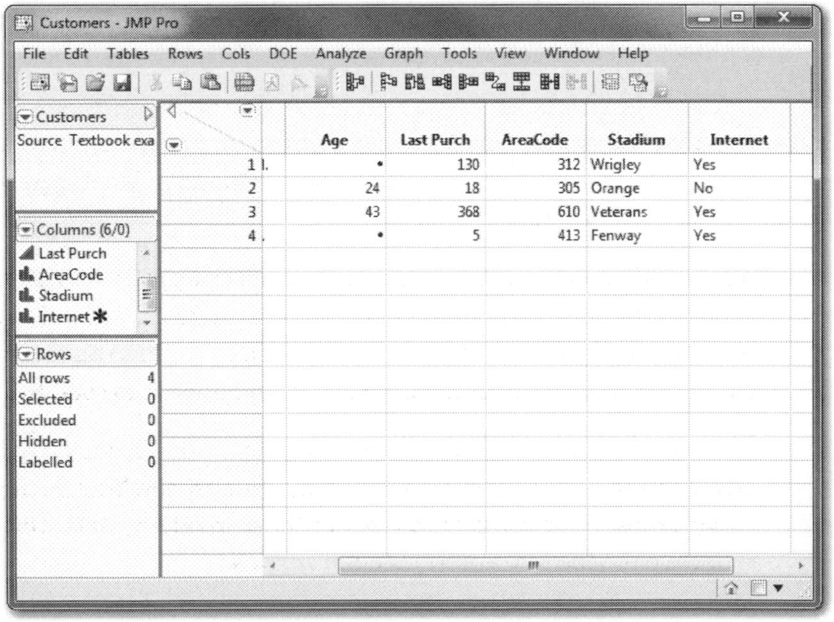

11. Select **File ▶ Save As**. By default, JMP wants to save this file in the JMP data table format. Select a proper location (folder) in which to save it, and this new data table will be preserved under the name **Customers**.

Moving Data from Excel Files into a JMP Data Table

There's a very good chance that you'll want to begin an analysis with some data that you already have within an Excel workbook file. There are several ways to transport data from Excel and other formats into JMP; Chapter 3 of the *Using JMP* (**Help ▶ Books ▶ Using JMP**) covers this topic in depth, but in this section, we'll see two approaches.

Importing an Excel File from JMP

This method works most simply if the Excel spreadsheet is already structured like a JMP data table: each column of the worksheet must represent a single variable, and each row should represent the data from one case or observation. It's best if the top row is reserved for column names and the observations start in row 2. Initially we'll assume there's just a single sheet within the workbook file, and that there are no merged cells.

Finally, it's important to understand that if there are formulas or functions in the Excel file, they will not be imported into JMP. JMP will read the *results* of a formula into a cell. If Column C of the spreadsheet equals the sum of Columns A and B, JMP will read in the sums but not the formula.

To illustrate this method, let's suppose we have the same customer data in a spreadsheet called **Customers.xls**. You may want to experiment in this section with any spreadsheet you already have; for this example, I have constructed an Excel worksheet containing the same data shown earlier.

Be aware that when you import from a spreadsheet file, JMP has no way of knowing whether some numeric columns are intended as nominal data (such as area codes); so, you'll need to proofread the initial settings for the columns. You'll also need to add any column notes or data value coding schemes that you want to establish.

12. Select **File ▶ Open…**. Within this dialog box, first navigate to the folder containing your properly constructed spreadsheet. You'll then see the list of all Excel files in the directory as well as those in other JMP-supported formats. Choose the file that you want and click **OK**.

13. This opens the **Excel Import Wizard**, as shown in Figure B.6. The Wizard allows considerable flexibility in working with multi-sheet workbooks, as well as with worksheets that have somewhat unusual layouts.

Figure B.6: The Excel Import Wizard

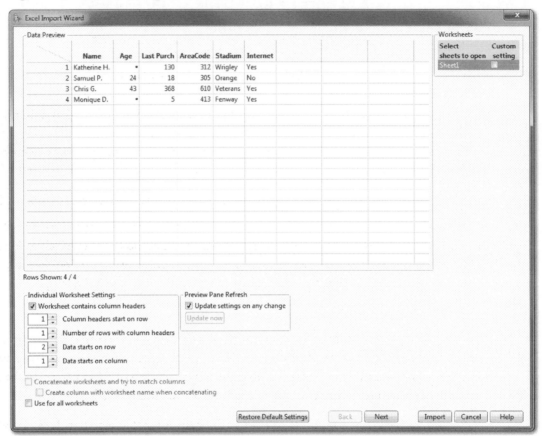

The upper portion of the Wizard displays a preview of the data as JMP will import it. In this example, the Excel file is organized exactly as discussed earlier. In other cases, you may find that an Excel file has column headings that occupy two rows, or that the data of interest begins in, say, column B of the worksheet rather than column A.

Look in the lower left panel of the Wizard (see Figure B.7). This provides a series of options to indicate precisely where the heading rows are and where the data rows and columns begin.

Figure B.7: How to Specify Location of Data Within the Worksheet

Individual Worksheet Settings

☑ Worksheet contains column headers

[1 ⏶⏷] Column headers start on row

[1 ⏶⏷] Number of rows with column headers

[2 ⏶⏷] Data starts on row

[1 ⏶⏷] Data starts on column

Moreover, the **Next** button in the Wizard opens a similar panel (not shown here) allowing further customization. Additionally, if a workbook does contain several worksheets, you can specify either a single layout that applies to all worksheets or specify a different layout for each worksheet.

14. Click the **Import** button in the Wizard, and a new JMP data table will open.

Particularly for a large worksheet, the Import Wizard is quite efficient, but as noted, it has no way to know that some numeric characters are actually categorical, as in the case of area codes. Before starting to work with imported data, be sure to look closely at the **Columns** panel to check the modeling types and make any needed changes.

The JMP Add-in for Excel

Suppose you are working in Excel and want to perform a simple analysis using a JMP platform, then continue your work in Excel. In the simplest case, let's say you have an Excel workbook and want to see the distribution of a numeric column of data. In Excel, several steps are needed to create a histogram and summary statistics.

When JMP 11 is installed on a Windows-based computer that also has Excel installed, JMP automatically installs an add-in within Excel, which creates a JMP tab in the Excel Ribbon. Figure B.8 shows the options available in the JMP add-in; as with the Excel Import Wizard, the add-in is extensively documented in Chapter 3 of *Using JMP*.

Figure B.8: JMP Excel Add-In Commands

These buttons invoke JMP from within the Excel environment, and you can issue JMP commands beginning in Excel. If you select a data column in Excel and click the **Distribution** button, JMP will open and display the standard distribution report for that variable.

Importing Data Directly from a Website

There is a large and growing amount of authoritative data available online. Many websites offer the option of downloading data tables in Excel or other formats; often, a website simply displays tabular data on a single page. One might be tempted to copy and paste such a table either into Excel or directly into a JMP data table, but JMP offers a simpler and more direct approach.

As an illustrative example, consider the World Bank's collection of World Development Indicators (WDI). Several of the data tables accompanying this book come from this database. There are several ways to access the WDI; the following steps show you one way, using one of the 331 different indicators.

1. Open a browser and visit http://data.worldbank.org/indicator. Figure B.9 shows the home page.

2. Under "Agriculture & Rural Development" select the second item, Agricultural land (% of total land area).

Figure B.9: The World Bank Data Home Page

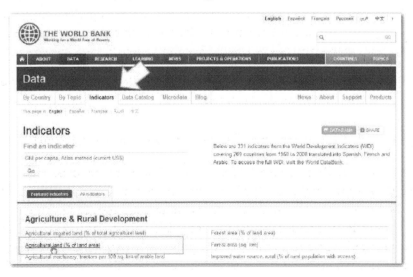

The next screen contains the data table, as well as some descriptive and navigational materials. Although there is a Download Data button, JMP can detect the presence of a table array and directly import the data.

3. Move the cursor to the URL address box of your browser and copy the entire URL.

4. Move back to JMP and choose **File ▶ Internet Open…** This will open the dialog shown in upper panel of Figure B.10; paste the URL as shown.

Figure B.10: Specifying In Internet Data Import

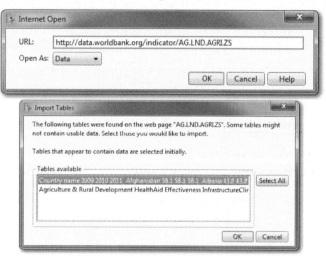

JMP finds two "candidate" data tables on this web page. The first is the one we want, listing the percentage of agricultural land in each country over a period of years; the second is at the very bottom of the page, listing data topic headings.

5. Just highlight the top choice and click OK. You will now have a new data table containing the data from the website.

Combining Data from Two or More Sources

When conducting an observational study, it is common to assemble data from multiple sources into a single data table for analysis. For example, among the data tables provided with this book, there are several that contain data about the nations of the world. For this example, let's consider the data tables **MDG Technology 2005** and **World Nations**. The former table contains data about the prevalence of various technologies (cellular subscribers, internet users, etc.) within countries; the latter contains the U.N. regional classifications and populations for the same countries. Suppose we wanted to describe, say, cellular telephone usage by the regions of the world. The region column is in one data table, but the cell usage is in the other. One might be tempted to copy and paste the relevant columns from one table into the other, but it is not such a simple matter.

1. To see why, open the two data tables.

Look at the rows panels of the two tables. There are 209 countries listed in the **World Nations** table and 231 countries in the **MDG Technology 2005** table. For this reason, the copy and paste approach would require considerable care to ensure accurately aligning of countries and data values. With more than 200 rows, the task would be fussy; with 20,000 rows, it would be entirely inefficient.

A better approach relies on database management, in which it is quite common to merge data residing in several different tables. The approach hinges on the idea of *key fields*, which are simply table columns that uniquely identify an observation and that are common to several tables. In database terminology, we want to *join* data from two different tables, and we can accomplish this if and only if the two tables share a matching key column.

We can summarize the content of our two data tables as follows:

World Nations (n = 209)	MDC Technology 2005 (n=231)
CCODE	CountryCode
Country	Country
Region	Cellular subscribers
Subregion	Cellular subscribers per 100 population
POP2005	Internet users
GDPperCap	Internet users per 100 population
Quartile	Personal computers
	Personal computers per 100 population
	Telephone lines
	Telephone lines per 100 population

In order to join data from the two tables, we need a key field that appears in both and that contains matching information. Looking at the tables, there are two columns that are candidates for use as key fields: Country Code (CCODE in **World Nations**) and Country name. Let's examine the data to decide if one field is more suitable than the other for present purposes.

2. Scroll down to row 81 of the **World Nations** table and row 44 of the **MDG Technology 2005** table.

    Both rows contain an entry for country #344, Hong Kong. In both tables, the country is identified by the very same code number, but the name is different. A human reader can conclude that "Hong Kong SAR of China" is the same country as "China, Hong Kong Special Administrative Region" but that conclusion is not necessarily obvious to software. However, software will reliably recognize the identity of two matching country codes.

    We'll plan to join the tables using country codes.

3. Move your cursor to the **World Nations** table, and choose **Tables ▶ Join**. This opens the dialog shown in Figure B.11. To complete it as shown in the figure, follow these steps:

Figure B.11: Specifying Parameters to Join Two Tables

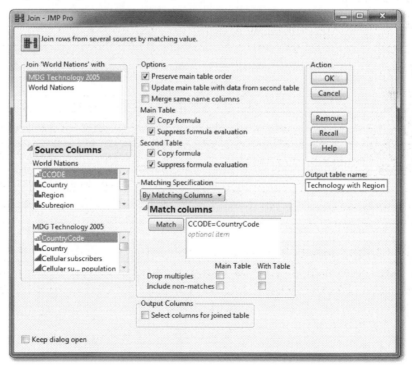

a. In the upper left panel, below Join **World Nations** with, select **MDG Technology 2005**.

b. Next we need to specify the criteria for JMP to use when matching data rows. On the left of the dialog, click on **CCODE** under **World Nations** and **Country Code** under **MDG Technology 2005**. Then click the **Match** button under the **Match columns** heading.

c. Finally, on the right side of the dialog type the words Technology with Region in the **Output table name** box, and click **OK**.

You will now see a new data table with 17 columns and 206 rows. Why 206? Three of the 209 nations in the first table did not appear in the second. By inspecting the data, you should be able to confirm that the rows from the two source tables properly align, so that data values appropriately correspond to the correct country. Now we can perform whatever analyses we deem fit.

Note two additional things. First, the **Join Tables** dialog has several options that you might wish to explore on your own. One particularly useful option appears just below the **Matching Specifications** panel. The option is a checkbox labeled **Select columns for joined table**. By default, the join operation combines all of the columns from both source tables into the output table. For a specific analysis, you may be interested in just a few of the columns. This option allows you to choose just the columns you actually need.

Secondly, the **Tables** menu provides several other data management operations. We have seen some of them earlier in the book, such as the **Subset** and **Stack** functions. Though the objective in this section is to introduce the powerful **Join** function, be aware that practical data analysis often calls for reorganization, consolidation, or other data management activities. JMP anticipates these needs and provides several tools to facilitate them.

---

[1] (De Veaux *et al.*, 2004), p. 7.

Bibliography

Aliaga, Martha, George Cobb, Carolyn Cuff, Joan Garfield, Rob Gould, Robin Lock, Tom Moore, Allan Rossman, Bob Stephenson, Jessica Utts, Paul Velleman, and Jeff Witmer. 2005. "Guidelines for Assessment and Instruction in Statistics Education: College Report." Alexandria VA: American Statistical Association. Available at http://www.amstat.org/education/gaise/

Almquist, Eric, and Gordon Wyner. 2001. *Boost Your Marketing ROI with Experimental Design*. Cambridge MA: Harvard Business Review.

Anscombe, F. J. 1973. "Graphs in Statistical Analysis." *American Statistician*. 27.1: 17–21.

Box, G. E., W. G. Hunter, and J.S. Hunter. 1978. *Statistics for Experimenters*. New York NY: Wiley.

Box, George E. P., and Gwilym M. Jenkins. 1970. *Time Series Analysis, Forecasting, and Control*. San Francisco CA: Holden-Day.

Brown, E. N., and R. E. Kass. 2009. "What is Statistics?" *The American Statistician*. 63.2: 105–110.

Budgett, Stephanie, Maxine Pfannkuch, Matt Regan, and Chris J. Wild. 2013. "Dynamic Visualizations and the Randomization Test." *Technology Innovations in Statistics Education*. 7.2. Available at http://www.escholarship.org/uc/item/9dg6h7wb.

Cable News Network (CNN) 2014. "CNN/ORC Poll" June 3, 2014. Available at http://i2.cdn.turner.com/cnn/2014/images/06/03/cnn.poll.obama.va.pdf. Accessed June 4, 2014.

Cobb, George W. 1998. *Introduction to Design and Analysis of Experiments*. New York NY: Springer.

Cobb, George W. 2007. "The Introductory Statistics Course: A Ptolemaic Curriculum?" Available at http://repositories.cdlib.org/uclastat/cts/tise/vol1/iss1/art1.

De Veaux, R. D., P. F. Velleman, and D. E. Bock. 2004. *Intro Stats*. 1st ed. Boston: Pearson Education.

Diez, David.M., Christopher D. Barr, and Mine Cetinkaya-Rundel. 2012. *OpenIntro Statistics*. 2nd Ed. Online at http://www.openintro.org/stat/textbook.php.

Federal Aviation Administration. 2013. "Wildlife Hazard Mitigation Program." Online at http://www.faa.gov/airports/airport_safety/wildlife/. Accessed 12 July, 2013.

Fernandes-Hachic, V., and V. Agopyan. 2008. "Influence of Composition of uPVC on Mechanical Properties of Window Profiles Exposed to Weathering." *Journal of Materials in Civil Engineering*. 20.3: 199–204.

Freund, R., R. Littell, and L. Creighton. 2003. *Regression Using JMP*. Cary NC: SAS Institute Inc.

Friedman, J. H., and W. Stuetzle. 2002. "John W. Tukey's Work on Interactive Graphics." *The Annals of Statistics*. 30.6: 1629–1639.

Gelman, A., M. Stevens, and V. Chan. 2003. "Regression Modeling and Meta-Analysis for Decision Making: A Cost-Benefit Analysis of Incentives in Telephone Surveys." *Journal of Business & Economic Statistics*. 21.2: 213.

Gosset, William S. 1908. "The Probable Error of a Mean." *Biometrika*. 6:1–25.

Gould, Robert. 2010. "Statistics and the Modern Student." International Statistical Review. 78.2: 297-315.

Hideki, K., Y. Hajime, M. Hiroki, and K. Hiroshi. 2008. "A Longitudinal Study on Vocal Aging - Changes in F0, Jitter, Shimmer and Glottal Noise." *The Journal of the Acoustical Society of America*. 123.5: 3428.

Hildebrand, D. K., R. L. Ott, and J.B. Gray. 2005. *Basic Statistical Ideas for Managers*. Belmont CA: Thompson Brooks/Cole.

Hingson, R. W., E. M. Edwards, T. Heeren, and D. Rosenblum. 2009. "Age of Drinking Onset and Injuries, Motor Vehicle Crashes, and Physical Fights After Drinking and When Not Drinking." *Alcoholism: Clinical and Experimental Research*. 33.5: 783–790.

Holt, Charles C. 1957. "Forecasting Seasonals and Trends by Exponentially Weighted Moving Averages." *Journal of Forecasting*. Reprinted in 2004. 20.1: 5–10.

Horton, Nicholas J., Benjamin S. Baumer, Daniel Theodore Kaplan, and Randall Pruim. 2013. "Precursors to the Data Explosion: Teaching How to Compute with Data." Conference Presentation at Joint Statistical Meetings, Montreal, Quebec, 2013. Abstract retrieved January 2, 2014 from http://www.amstat.org/meetings/jsm/2013/onlineprogram/AbstractDetails.cfm?abstractid=307064.

Johnson, Eugene G., and Keith F. Rust, 1992. "Population Inferences and Variance Estimation for NAEP Data." *Journal of Educational and Behavioral Statistics*. 17.2: 175–190.

Junghanns, K., R. Horbach, D. Ehrenthal, S. Blank, and J. Backhaus. 2009. "Chronic and High Alcohol Consumption Has a Negative Impact on Sleep and Sleep-Associated Consolidation of Declarative Memory." *Alcoholism: Clinical and Experimental Research.* 33.5: 893–897.

Kapucu, A., C. M. Rotello, R. F. Ready, and K. N. Seidl. 2008. "Response Bias in 'Remembering' Emotional Stimuli: A New Perspective on Age Differences." *Journal of Experimental Psychology.* 34.3: 703–711.

Kennedy, Peter. 2003. *A Guide to Econometrics.* 5th ed. Cambridge MA: MIT Press.

Kish, Leslie. Frankel, Martin Richard. 1974. "Inference from Complex Samples*." Journal of the Royal Statistical Society.* Series B (Methodological). 36.1: 1–37.

Little, M. A., P. E. McSharry, E. J. Hunter, and L. O. Ramig. 2008. "Suitability of Dysphonia Measurements for Telemonitoring of Parkinson's Disease." Nature Proceedings.

Little, M.A., P. E. McSharry, S. J. Roberts, D. A. E. Costello, and I. M. Moroz. 2007. "Exploiting Nonlinear Recurrence and Fractal Scaling Properties for Voice Disorder Detection." *BioMedical Engineering OnLine.* 6.23 (26 June). Online at http://www.biomedcentral.com/content/pdf/1475-925X-6-23.pdf

Mead, R., and R. D. Stern. 1973. "The Use of a Computer in the Teaching of Statistics." *Journal of the Royal Statistical Society.* Series A (General). 136.2: 191–225.

Mendenhall, W., and T. Sincich. 2003. *A Second Course in Statistics: Regression Analysis.* 6th ed. Upper Saddle River NJ: Pearson Education.

Mocko, Megan 2013. "Selecting Technology to Promote Learning in an Online Introductory Statistics Course." *Technology Innovations in Statistics Education.* 7.2. Available at http://www.escholarship.org/uc/item/596195sg.

Moore, D. S., G. W. Cobb, J. Garfield, and W. Q. Meeker. 1995. "Statistics Education Fin de Siecle." *The American Statistician.* 49.3: 250–260.

Neumann, David L., Michelle Hood, and Michelle M. Neumann. 2013. "Using Real-Life Data When Teaching Statistics: Student Perceptions of this Strategy in an Introductory Statistics Course." *Statistics Education Research Journal.* 12.2: 59–70.

Reliasoft Corporation. 2008. Experiment Design and Analysis Reference. Online at http://www.weibull.com/doewebcontents.htm

Rydén, Jesper. 2007. "Statistical Analysis of Golden-Ratio Forms in Piano Sonatas by Mozart and Haydn." *Math Scientist.* 32: 1–5.

Singer, Eleanor. 2002. "The Use of Incentives to Reduce Nonresponse in Household Surveys." Survey Methodology Program SMP. Ann Arbor MI: The University of Michigan.

Singer, J. D., and J. B. Willett. 1990. "Improving the Teaching of Applied Statistics: Putting the Data Back into Data Analysis." *The American Statistician*. 44.3: 223–230.

Stickgold, R., L. James, and J. A. Hobson. 2000. "Visual Discrimination Learning Requires Sleep after Training." *Nature Neuroscience*. 3: 1237.

Stickgold, R., and M. P. Walker. 2005. "Memory Consolidation and Reconsolidation: What is the Role of Sleep?" *Trends in Neurosciences*. 28.8: 408.

Stigler, Stephen M. 1977. "Do Robust Estimators Work with Real Data?" *The Annals of Statistics*. 5.3: 1055-1098.

Tukey, John W. 1972. "How Computing and Statistics Affect Each Other." Babbage Memorial Meeting: Report of Proceedings. 21–37. London.

University of Reading (UK) Statistical Services Centre. 2001. "Approaches to the Analysis of Survey Data" (monograph).

Walker, M. P., T. Brakefield, A. Morgan, J. A. Hobson, and R. Stickgold. 2002. "Practice with Sleep Makes Perfect: Sleep-Dependent Motor Skill Learning." *Neuron*. 35: 205–211.

Wild, C. J., and M. Pfannkuch. 1999. "Statistical Thinking in Empirical Enquiry." *International Statistical Review*. 67.3: 223–265.

Winters, Peter R. 1960. "Forecasting Sales by Exponentially Weighted Moving Averages." *Management Science*. 6.3: 324–342.

The World Bank. 2013. World Bank Millennium Development Goals, http://data.worldbank.org/about/millennium-development-goals, accessed June 17, 2013.

The World Bank. 2013. World Bank History, http://www.worldbank.org/en/about/history http://go.worldbank.org/65Y36GNQB0, accessed June 21, 2013.

Yeh, I-Cheng. 1998a. "Modeling Concrete Strength with Augment-Neuron Networks." *Journal of Materials in Civil Engineering, ASCE*. 10.4: 263–268.

Yeh, I-Cheng. 1998b. "Modeling of Strength of High-Performance Concrete Using Artificial Neural Networks." *Cement and Concrete Research*. 28.12: 1797–1808.

Yeh, I-Cheng. 1999. "Design of High-Performance Concrete Mixture Using Neural Networks and Nonlinear Programming." *Journal of Computing in Civil Engineering*. 13.1: 36-42.

Yeh, I-Cheng. 2006. "Analysis of Strength of Concrete Using Design of Experiments and Neural Networks." *Journal of Materials in Civil Engineering, ASCE*. 18.4: 597–604.

Yeh, I-Cheng. 2008. "Generalization of Strength Versus Water-Cementations Ratio Relationship to Age." *Cement and Concrete Research*. 36.10: 1865–1873.

Index

Printed in Great Britain
by Amazon.co.uk, Ltd.,
Marston Gate.